Surface Topography Effects on Functional Properties of PVD Coatings

Surface Topography Effects on Functional Properties of PVD Coatings

Editors

Peter Panjan
Aljaž Drnovšek

Basel • Beijing • Wuhan • Barcelona • Belgrade • Novi Sad • Cluj • Manchester

Editors
Peter Panjan
Jožef Stefan Institute
Ljubljana
Slovenia

Aljaž Drnovšek
Jožef Stefan Institute
Ljubljana
Slovenia

Editorial Office
MDPI
St. Alban-Anlage 66
4052 Basel, Switzerland

This is a reprint of articles from the Special Issue published online in the open access journal *Coatings* (ISSN 2079-6412) (available at: https://www.mdpi.com/journal/coatings/special_issues/STE_PVD).

For citation purposes, cite each article independently as indicated on the article page online and as indicated below:

Lastname, A.A.; Lastname, B.B. Article Title. *Journal Name* **Year**, *Volume Number*, Page Range.

ISBN 978-3-0365-8620-5 (Hbk)
ISBN 978-3-0365-8621-2 (PDF)
doi.org/10.3390/books978-3-0365-8621-2

© 2023 by the authors. Articles in this book are Open Access and distributed under the Creative Commons Attribution (CC BY) license. The book as a whole is distributed by MDPI under the terms and conditions of the Creative Commons Attribution-NonCommercial-NoDerivs (CC BY-NC-ND) license.

Contents

About the Editors vii

Preface ix

Peter Panjan, Aljaž Drnovšek, Peter Gselman, Miha Čekada and Matjaž Panjan
Review of Growth Defects in Thin Films Prepared by PVD Techniques
Reprinted from: *Coatings* **2020**, *10*, 447, doi:10.3390/coatings10050447 1

Peter Panjan, Aljaž Drnovšek, Nastja Mahne, Miha Čekada and Matjaž Panjan
Surface Topography of PVD Hard Coatings
Reprinted from: *Coatings* **2021**, *11*, 1387, doi:10.3390/coatings11111387 41

Peter Panjan, Peter Gselman, Matjaž Panjan, Tonica Bončina, Aljaž Drnovšek, Mihaela Albu, et al.
Microstructure and Surface Topography Study of Nanolayered TiAlN/CrN Hard Coating
Reprinted from: *Coatings* **2022**, *12*, 1725, doi:10.3390/coatings12111725 73

Peter Panjan, Aljaž Drnovšek, Miha Čekada and Matjaž Panjan
Contamination of Substrate-Coating Interface Caused by Ion Etching
Reprinted from: *Coatings* **2022**, *12*, 846, doi:10.3390/coatings12060846 91

Peter Panjan, Aljaž Drnovšek, Pal Terek, Aleksandar Miletić, Miha Čekada and Matjaž Panjan
Comparative Study of Tribological Behavior of TiN Hard Coatings Deposited by Various PVD Deposition Techniques
Reprinted from: *Coatings* **2022**, *12*, 294, doi:10.3390/coatings12030294 107

Peter Panjan, Aljaž Drnovšek and Goran Dražić
Influence of Growth Defects on the Oxidation Resistance of Sputter-Deposited TiAlN Hard Coatings
Reprinted from: *Coatings* **2021**, *11*, 123, doi:10.3390/coatings11020123 131

Suzana Petrović, Davor Peruško, Evangelos Skoulas, Janez Kovač, Miodrag Mitrić, Jelena Potočnik, et al.
Laser-Assisted Surface Texturing of Ti/Zr Multilayers for Mesenchymal Stem Cell Response
Reprinted from: *Coatings* **2019**, *9*, 854, doi:10.3390/coatings9120854 153

Pal Terek, Lazar Kovačević, Aleksandar Miletić, Branko Škorić, Janez Kovač and Aljaž Drnovšek
Metallurgical Soldering of Duplex CrN Coating in Contact with Aluminum Alloy
Reprinted from: *Coatings* **2020**, *10*, 303, doi:10.3390/coatings10030303 165

Bojan Podgornik, Marko Sedlaček, Borut Žužek and Agnieszka Guštin
Properties of Tool Steels and Their Importance When Used in a Coated System
Reprinted from: *Coatings* **2020**, *10*, 265, doi:10.3390/coatings10030265 183

Žiga Gosar, Denis Đonlagić, Simon Pevec, Bojan Gergič, Miran Mozetič, Gregor Primc, et al.
Distribution of the Deposition Rates in an Industrial-Size PECVD Reactor Using HMDSO Precursor
Reprinted from: *Coatings* **2021**, *11*, 1218, doi:10.3390/coatings11101218 201

About the Editors

Peter Panjan

Peter Panjan received his PhD in materials sciences at the Faculty of Natural Sciences and Technologies, University of Ljubljana (Slovenia). He joined the Jožef Stefan Institute in 1980, where he headed the Department for Thin Films and Surfaces and Hard Coating Center for 13 years. He has almost 40 years of experience in PVD hard coating research, development, and manufacturing. His research group has successfully implemented a number of new hard protective coatings in industrial production. In addition to his primary research topic, he has been active in several other scientific areas related to PVD thin films (depth profiling and diffusion phenomena in multilayer thin films, low-pressure plasma characterization, laser modification of solid surfaces and thin films). He has published over 240 scientific papers, 1 book, 4 chapters in books, and 3 patents. He is also intensively involved in editorial work, as he has been the editor, co-editor, and editorial board member of many scientific journals, and the editor of several books and conference proceedings.

Aljaž Drnovšek

Aljaž Drnovšek received his PhD in material science from the Jožef Stefan International Postgraduate School (Ljubljana, Slovenia). In his thesis, he studied the impact of the surrounding atmosphere and growth defects on the tribological properties of PVD hard coatings. The emphasis of his research was on the influence of surface topography, due to growth defects, on the initial stages of tribological sliding. After his PhD, he was employed at Montanuniversitaet Leoben (Austria) for two years as a post-doc researcher. The focus of his research was on high temperature tribology and the high temperature mechanical properties of CrAlN and CrAlSiN coatings. In the scope of the project, he collaborated with Tomas Polcar's research group in Southampton, UK (engineering materials and surface technology) and prof. Peter Hosemann at the Nuclear Engineering Department at the University of California, Berkeley. Since 2019, he has been employed in the Department of Thin Films and Surfaces at the Jožef Stefan Institute (Slovenia). His current research interest is the nanomechanical characterization of PVD coatings at high temperatures.

Preface

Friction and wear occur in all mechanical systems where moving components are in contact with each other. The tribological contact causes the degradation of their performance and reliability due to increased energy consumption and the progressive loss of material. Interactions between solid surfaces depend not only on the properties of the materials in contact but also on the topography of their surfaces. In the majority of cases, the failures of moving components in mechanical systems are surface-initiated. Therefore, most engineering components are enhanced by surface treatments, which improve their tribological properties. One of the most common and effective approaches is the application of coating deposited by physical vapor deposition (PVD) techniques. An important factor influencing the tribological properties of PVD coatings is their surface topography. Different topographical imperfections on the coating surface can degrade its quality; in certain cases, they can even cause a catastrophic failure. The knowledge of the surface topography of the PVD coating is therefore crucial. However, in the literature, the relationship between the surface topography and the functional properties of the coating has only been partially investigated.

This book provides a comprehensive overview of the surface topography of PVD coatings and their role in different tribological contacts. We demonstrate that the coating topography is strongly dependent both on the substrate topography and on the topography induced by the coating deposition process. The former depends on the substrate preparation steps before the deposition of the coating (mechanical pretreatment, ion etching). During the mechanical pretreatment of a substrate, it is impossible to achieve a perfectly smooth surface because each manufacturing technique (e.g. turning, milling, grinding, polishing, electro discharge machining) leaves its own »fingerprint« on the surface. The resulting surface topography is determined by surface roughness (nano- and microroughness) and waviness (macroroughness). The surface roughness is characterized by a series of asperities (local maxima) and valleys (local minima) of a characteristic shape, amplitude, and spacing. According to ISO 14460-1, the real surface of a workpiece has been defined as: »*A set of features that physically exist and separate the entire workpiece from the surrounding medium*«.

All topographical irregularities on the substrate surface form during mechanical pretreatment and ion etching just before the deposition of coating. These irregularities are then transferred onto the coating surface. However, after coating deposition, additional topographic changes occur which are related to the intrinsic micro- and nanomorphology of the coating itself, and especially due to the formation of growth defects. Although the role of growth defects in many thin film applications is crucial, a comprehensive review of this area in the literature is still missing. The present book is intended to fill this void and provide information on the role of defects, especially in the tribological application of hard coatings.

This Special Issue contains ten papers covering a broad range of topics representing the state of knowledge on the recent developments in the area of surface topography of PVD hard coatings. A brief summary of the papers in each category is provided below.

The first review paper »*Review of growth defects in thin films prepared by PVD techniques*« summarizes our studies of the growth defects in PVD coatings. A detailed historical overview is followed by a description of the types and the evolution of growth defects. All topographical irregularities and foreign particles generated during different steps of substrate pretreatment and different coating deposition processes are described because they are seeds for the formation of growth defects. An overview is given on the research related to the influence of growth defects on the functional properties of thin films and coatings.

The second paper »*Surface topography of PVD hard coatings*« presents a study concerning the surface topography of various PVD hard coatings prepared in various types of industrial deposition systems, which differ significantly regarding both the ion etching method and deposition. The authors show how the coating topography depends on the topography of the substrate surface, intrinsic coating microtopography, and growth defects, formed during the deposition process.

In the third paper with the title »*Microstructure and Surface Topography Study of Nanolayered TiAlN/CrN Hard Coating*«, we focused on the microstructure, surface topography, layer periodicity, interlayer roughness, and formation of growth defects in the nanolayer TiAlN/CrN hard coating. These properties were analyzed with dependence on the substrate rotation mode, the type of substrate material, and the method of ion etching. The multilayer coating was chosen because the growth defect formation and other coating surface irregularities are easier to observe in such a structure. In addition, the nl-TiAlN/CrN coating possesses enhanced mechanical and tribological properties as compared to TiAlN and CrN monolayer coatings. The stresses formed at the interfaces due to different lattice constants significantly contribute to the higher hardness of the nanolayer structure coating in comparison with the corresponding monolayers.

In the fourth paper »*Contamination of substrate-coating interface caused by ion etching*«, we described the problem of target surface contamination in an industrial magnetron sputtering deposition system with the residual products from the etching process. Such contamination can be prevented by a movable shutter located close to the targets, but in order to achieve reasonable economics of the deposition process, complicated installations (including shielding and shuttering) are usually avoided. In the initial stage of deposition, this material is re-deposited back on the substrate and causes the formation of a contamination layer at the substrate-coating interface. We also found that many seed particles that cause the formation of nodules are covered with a similar contamination layer. We believe that these weakly bonded particles were formed on the target surface outside of the racetrack, and that they were transferred to the substrate surface immediately after starting the deposition process by the self-repulsion effect.

In the fifth paper »*Comparative study of tribological behavior of TiN hard coatings deposited by various PVD deposition techniques*«, the authors correlate the tribological behavior of TiN hard coatings, prepared using different deposition methods, to their surface topography, microstructure, and mechanical properties. The surface topography of PVD hard coatings is an important factor influencing their tribological performance under sliding contact conditions because the real contact area strongly depends on the roughness of the interacting surfaces. In this paper, it is analyzed how tribological properties depend on roughness and the surrounding atmosphere (ambient air, nitrogen, oxygen).

The next paper »*Influence of growth defects on the oxidation resistance of sputter-deposited TiAlN hard coatings*« discusses the influence of growth defects on the oxidation resistance of sputter-deposited TiAlN coatings. The formation of an oxide scale at temperatures of 800 °C and 850 °C and different oxidation periods was investigated. The authors found that intensive local oxidation takes place at sites of pinholes and pores that are formed at the rim of the nodular defects. During oxidation, they provide direct paths between the coating surface and the substrate for the transport of oxygen inwards and substrate elements towards the surface.

The surface topography and surface roughness of the orthopedic and dental implants have a decisive influence on their integration and biological response in soft and hard tissues. The paper »*Laser-assisted surface texturing of Ti/Zr multilayers for mesenchymal stem cell response*« deals with the surface functionalization of the Ti-base alloy in terms of improving the osteoblast cell response. The authors performed the irradiation of the Ti/Zr multilayer structure via femtosecond laser irradiation in order to form the laser-induced periodic surface structure and intermixing between the titanium and zirconium layers. They found that cell adhesion and growth improve on these modified surfaces.

In the paper »*Metallurgical soldering of duplex CrN coating in contact with aluminum alloy*«, the performance of CrN duplex coatings with different roughness was evaluated using an ejection test performed with conventional (CS) and delayed (DS) casting solidification. They observed that the roughness strongly affected the ejection force in the CS experiment, where the ejection force increased with the decreased roughness. On the other hand, an almost equal ejection force was measured in the DS experiments for samples of different roughness. These observations along with the above provided discussion suggest that in DS experiments, the effects of surface chemistry are more dominant than the effects of surface topography, i.e., metallurgical mechanisms are more dominant than mechanical mechanisms. The decrease in the ejection force, observed in DS tests, is attributed to the formation of a thick chromium oxide layer on the CrN coating which reduced soldering and sliding friction against a thick aluminium oxide casting scale.

The paper »*Properties of tool steels and their importance when used in a coated system*« deals with correlations between different tool steel properties, including fracture toughness, hardness, compressive and bending strength, wear resistance and surface quality and how these substrate properties influence the coating performance. In the case of coated applications, steel substrates must provide sufficient load-carrying capacity and support for the coating. Surface roughness and topography have a major influence on galling resistance during the forming operation, with smoother surfaces and a plateau-like topography providing better results. In the case of typical hard ceramic coatings, the post-polishing of the coated surface and use of a smoothened substrate give about two times better galling resistance.

The last paper »*Distribution of the deposition rates in an industrial-size PECVD reactor using HMDSO precursor*« deals with the problem of non-uniform deposition rates in commercial plasma reactors for the preparation of thin films from organic precursors using the PECVD technique. The plasma was maintained via asymmetric capacitively coupled radiofrequency (RF) discharge using a generator with a frequency of 40 kHz and an adjustable power of up to 8 kW. They found that the deposition rates of hexamethyldisiloxane far from the powered electrodes dropped by more than an order of magnitude for a fully loaded chamber.

The editors are grateful to all authors that contributed to this book. We also acknowledge the Multidisciplinary Digital Publishing Institute (MDPI) in Basel, and especially Ms. Flora Ao for her assistance with the publication of this Special Issue of the journal *Coatings* in book form.

Peter Panjan and Aljaž Drnovšek
Editors

Review

Review of Growth Defects in Thin Films Prepared by PVD Techniques

Peter Panjan [1,*], Aljaž Drnovšek [1], Peter Gselman [1,2], Miha Čekada [1] and Matjaž Panjan [1]

1. Jožef Stefan Institute, Jamova 39, 1000 Ljubljana, Slovenia; aljaz.drnovsek@ijs.si (A.D.); info@interkorn.si (P.G.); miha.cekada@ijs.si (M.Č.); matjaz.panjan@ijs.si (M.P.)
2. Interkorn d.o.o, Gančani 94, 9231 Beltinci, Slovenia
* Correspondence: peter.panjan@ijs.si; Tel.: +386-1-477-3278

Received: 31 March 2020; Accepted: 27 April 2020; Published: 3 May 2020

Abstract: The paper summarizes current knowledge of growth defects in physical vapor deposition (PVD) coatings. A detailed historical overview is followed by a description of the types and evolution of growth defects. Growth defects are microscopic imperfections in the coating microstructure. They are most commonly formed by overgrowing of the topographical imperfections (pits, asperities) on the substrate surface or the foreign particles of different origins (dust, debris, flakes). Such foreign particles are not only those that remain on the substrate surface after wet cleaning procedure, but also the ones that are generated during ion etching and deposition processes. Although the origin of seed particles from external pretreatment of substrate is similar to all PVD coatings, the influence of ion etching and deposition techniques is rather different. Therefore, special emphasis is given on the description of the processes that take place during ion etching of substrates and the deposition of coating. The effect of growth defects on the functional properties of PVD coatings is described in the last section. How defects affect the quality of optical coatings, thin layers for semiconductor devices, as well as wear, corrosion, and oxidation resistant coatings is explained. The effect of growth defects on the permeation and wettability of the coatings is also shortly described.

Keywords: hard coating; growth defect; nodular defect; pinhole; flake; ion etching; focused ion beam (FIB); scanning electron microscopy (SEM); droplet

1. Introduction

The surface topography is an important characteristic of physical vapor deposition (PVD) thin films because it determines their functional performance in many applications. In general, numerous topographical imperfections on the surface degrade the quality of films; in particular cases they can even cause their catastrophic failure. The surface topography of thin films on a microscopic scale is determined by three preparation steps: (i) mechanical pretreatment of the substrate, (ii) substrate ion etching, and (iii) deposition process. Mechanical pretreatment, which usually includes grinding and polishing, can cause numerous irregularities (e.g., scratches, grooves, and ridges) on the surface of the substrate material. In the non-homogenous materials, such as tool steel, additional shallow protrusions are formed at the inclusions which are harder than the ferrous matrix (e.g., carbides and oxides) while shallow craters are formed at inclusions that are softer (e.g., MnS). Some of the protruding inclusions can be torn out during polishing due to the shear stresses and leaving pits behind. In addition, polishing residue can be incorporated into the substrate surface. All of these substrate irregularities directly affect the topography of the deposited thin film. The last step of substrate surface pretreatment normally includes cleaning by ion etching. This step is performed to remove impurities left in the previous pretreatment steps and particularly, to remove the native oxide layers and to chemically activate the substrate surface for improved film adhesion (formation of nucleation sites). However, in the case of non-homogeneous substrate material (e.g., tool steels) the ion etching can induce significant surface topography irregularities (protrusions and craters) due to different etching rates of various phases. All substrate irregularities formed during mechanical pretreatment and ion etching directly affects the topography of thin films because the growing film replicates topographical features of the substrate surface. The last part of surface imperfections comes from the deposition process itself. The physical vapor deposition (PVD) techniques, such as sputtering and evaporation, are line-of-sight processes, meaning that the material is transferred in a straight path from the source to the substrate. Due to shading (angle-of-incidence effects) of topographical irregularities on the substrate surface the thin film material is predominantly deposited in the areas which are in direct view of the sputtering or evaporation source. The result of geometrical shadowing effect is formation of growth defects (nodular defects, pinholes, pores, and other coating discontinuities). The line-of-sight deposition magnifies imperfections (e.g., foreign particles) present on the substrate surface. Even relatively small imperfections of several tens of nanometers can grow into large micrometer-sized imperfections on the surface of the thin film. All surface irregularities introduced by the PVD deposition processes cause a significant increase of surface roughness. The minor influence on the coating roughness has a columnar structure that is also a consequence of the shadowing effect.

In the PVD literature, the term "growth defects" is used (as opposed to simply "defects") in order to emphasize that the defects result from the growth process. However, the seeds that are essential for the formation of growth defects do not originate only from the deposition process but also, and in many cases predominantly, from the substrate topographical irregularities and foreign particles on the substrate. In this review, we define the growth defect as any localized imperfection in the thin film microstructure, which is in the micrometer range and forms during the film growth regardless of the seed origin. We should distinguish the growth defects from the structural defects in the crystal lattice, such as dislocations and other imperfections in the crystal structure. These types of defects are not the subject of this review.

The type of PVD technique which is used for the preparation of thin films also plays an important role in the density of generated growth defects. The plasma-based deposition process is an intensive generator of seed particles that can induce the formation of growth defects. However, PVD techniques, such as magnetron sputtering, electron beam evaporation, cathodic arc deposition, ion beam deposition, pulsed laser deposition (PLD) and others, generate substantially different types and density of seed particles and consequently different shape, size, and density of growth defects.

However, we should emphasize that the growth defects are not only limited to PVD methods but also present in thin films prepared by other deposition methods such as chemical vapor deposition (CVD) [1–3], electrodeposition [4], plasma polymerization [5], and others (the reader interested in these methods should consult references herein).

Although the growth defects were already produced by the earliest vacuum-based deposition techniques, their systematic studies could not have been undertaken until the more advanced analytical techniques with sub-micrometer spatial resolution were available. Historically, the systematic studies of the growth defects in PVD thin films can be traced to the end of the 1960s. One of the earliest studies of the growth defects in thick metallic and oxide films prepared by electron beam evaporation was reported in 1969 by Movchan and Demchishin [6]. A few years later, in 1973, Mattox and Kominiak [7] studied nodular defects in sputtered tantalum thin film. These first studies were mainly focused on the observation of growth defect geometry by scanning electron microscopy. Rigorous scientific investigation of growth defects began in the second half of the 1970s. The mechanisms of defect growth in sputtered chromium films were reported in 1979 by Patten [8]. In the same year Spalvins [9] characterized growth defects in ion-plated copper and gold films. A few years earlier, studies of Spalvins and Brainard [10] demonstrated that nodules in thick metallic films are nucleated by substrate surface imperfections arising from asperities, pits, dust particles, and flakes.

The early studies of growth defects were concentrated on metallic films, while later research focused more on the dielectric thin films, predominantly for optical applications [11]. The research of the growth defects in optical interference coatings for mirrors was stimulated by the development of high-power lasers, which require durable and highly reliable coatings. In many instances, optical coatings were damaged by very high intensity of the laser light. It was found that the damage mainly occurred due to nodular defects present in the coatings [11].

The earliest investigations of growth defects were followed by attempts to model defect growth and their shape. Leets et al. [5] and later Tench et al. [12] proposed a simple shadowing model to describe the geometry of the nodular defect. Their model predicted the parabolic shape of the nodular defects as was commonly observed in experiments. Dubost et al. [2] improved such a model by including the surface reaction probabilities of the depositing species. In addition to simple geometric models, several two-dimensional computer simulations [11–13] have been made based on a model by Dirks and Leamy [14]. Using this model, the authors reproduced the parabolic shape as well as the open boundaries between a nodular defect and the coating matrix. In the beginning of the 1990s, Liao et al. [15] and later Müller-Pheiffer et al. [16] upgraded the two-dimensional model with an atom surface diffusion and desorption.

The shape and inner structure of a growth defect was initially investigated by scanning electron microscopy (SEM) on the metallographic cross-sections and on fracture cross-sections. In this way, only a few growth defects could be analyzed since the observation of defects depended on sheer luck. Deeper knowledge on the internal structure of individual defects came with a focused ion beam (FIB) technique used together with SEM [12–17]. This technique allowed precise cross-section of an individual growth defect to examine its internal structure and to identify the seed for its formation. In recent years, this technique has been widely employed for studying the growth defects [18–22].

We should also mention the research in which the growth defects were intentionally produced by dispersing seeds, such as microspheres, on the substrate. These microspheres were subsequently coated by a single or multilayer thin film. Such studies enabled more systematic investigations under controlled growth conditions. One of the first experiments using microspheres was performed by Brett et al. [23]. They created nodular defects by coating polystyrene latex spheres of well-defined size ranges. Poulingue et al. [24] used diamond and silica seeds of micrometer size. Wei et al. [25] investigated the behavior of artificial nodules which were created from much smaller gold and SiO_2 nanoparticles. Similar investigation was performed by Mirkarimi et al. [26] using gold nanospheres of defined sizes. Recently, Cheng and Wang [27] investigated defect-driven laser-induced damages in high-reflection optical coatings using silica microsphere on substrates.

In general, the growth defects are not desired because they degrade the performance of the thin films. The detrimental effects of growth defects have been explored for a wide range of thin film functional properties. As mentioned before, one of the earliest motivations for the study of growth defects were quality problems of optical thin films and thin films in semiconductor devices. Particle contamination is especially critical in semiconductor thin film device manufacturing because it reduces the production yield of such devices as well as their performance and reliability [28,29].

There have been extensive studies on the role of growth defects on the performance of coatings for the protection against wear, erosion, corrosion, gas permeation, and others. The growth defects cause serious problems for the protective functionality of the coating because they present starting points for an environmental attack. A control over the growth defects is therefore crucial for a high-quality protective coating. When studying corrosion resistance of protective coatings, for example, Korhonen [30] demonstrated that the growth defects are the main locations for the start of pitting corrosion. In the later stages of the corrosion, such points can result in the removal of a large part or the entire coating. In the past two decades several papers have been dedicated to this topic and are described in an excellent review by Fenker et al. [31] and other papers [32–36]. Resistance of protective coatings against oxidation [37] is also a phenomenon where the growth defects act as shortcuts for oxygen diffusion. The pinholes created at pits on the substrate surface and those left by the removed nodular defects are the entrance points for molecules of the environmental species towards the substrate surface. Hence, growth defects have a very important role in gas barrier coatings too [38].

In the last two decades, growth defects have also been studied in relation to the tribological performance of the coatings [39–43]. For example, Fallqvist et al. [40] showed that the as-deposited coatings, prepared by the cathodic arc, result in a significantly higher coefficient of friction as compared to the post-polished coatings. The effect of arc-evaporated droplets on the wear behavior was investigated by Tkadletz et al. [41]. They found that droplets contribute to coating degradation by providing nucleation sites for shear cracks and by the release of abrasive fragments into the sliding contact. The particularity of the approach of our research group was that we pinpointed selected defects on the coating surface and then followed them through the tribological test with scanning electron microscopy (SEM) and a focused ion beam (FIB) microscope [42,43].

The main goal of this paper is to give an overview of the decades-long research of growth defects in thin films prepared by the PVD techniques. In this review, we cover growth defects in several different deposition techniques and their influence on various thin film applications. However, we will mainly focus on the growth defects in PVD hard coatings for protection of tools and other manufacturing components, although we will also touch upon other areas of growth defect studies, particularly in optics and microelectronics. Thermionic arc ion plating deposition system BAI 730M (Oerlikon Balzers, Balzers, Liechtenstein) [22] was used for deposition of TiN and CrN single layer and TiN/CrN double layer hard coatings. Part of the samples was coated in the cathodic arc ion plating deposition system AIPocket (Kobelco, Kobe, Japan), which is equipped with superfine cathode sources. Magnetron sputter deposition system CC800/7 (CemeCon, Wurselen, Germany) [42] was used for deposition of TiN and TiAlN hard coatings. The magnetron sputtering technique was used also for deposition of nanolayered TiAlN/CrN hard coatings, as well as for deposition of TiAlN/a-CN and TiAlN/Al2O3 double layer hard coatings (CC800/9 ML, CemeCon, Wurselen, Germany). Tool steel materials (D2, H11, L2, PM steel grade ASP30) and cemented carbide (hard metal or HM) were used as substrates. The substrates were ground and polished to a mirror-like finish with a surface roughness of S_a = 0.02 μm. Before deposition, the samples were cleaned in detergents and ultrasound, rinsed in deionized water and dried in hot air.

The paper is organized as follows. We first discuss surface irregularities induced by substrate pretreatment and deposition process, continue introducing the classification of the growth defects and then discuss their origin and morphology in detail. In the last chapter we explore the influence of growth defects on the functional properties of PVD coatings.

2. Surface Irregularities from Substrate Pretreatment

Substrate surface preparation is an integral part of any PVD film deposition process. In practice, a pretreatment of the substrate surface is always carried out before the deposition of thin films. Normally, three stages of the substrate pretreatment are included: mechanical pretreatment (grinding, blasting, polishing), wet chemical cleaning in an ultrasonic bath, and ion etching in the vacuum chamber. Mechanical pretreatment and ion etching of substrates can induce different topographical irregularities, which cause (due to the shadowing effect) the formation of numerous small- or large-scale growth defects during the deposition of coating. We will discuss here only the surface irregularities in tool steels, since these are very commonly used substrate materials in PVD production of hard coatings. Tools steels are composed of a ferrous matrix and several types of micrometer-sized carbides, which improve their mechanical properties. In addition to carbides, non-metallic inclusions (e.g., oxides, sulfides, silicates) are also present in the tool steels as inevitable impurities. Figure 1a shows a backscattered SEM top view image of the surface of D2 tool steel after middle frequency (MF) and intensive (booster) ion etching and after deposition of nanolayer nl-TiAlN/CrN hard coating (Figure 1b). Different carbide and non-metallic inclusions as well as pits of various sizes and shapes were identified by energy dispersive X-ray analysis (EDX). Their positions as well as the positions of two relatively large but pointed marks (e.g., Vickers indentations) made at opposing sites of the sample were saved in the microscope's coordinate system. The local (sample) coordinate system was uniquely fixed at these two points. By transforming the local (sample-based) coordinate system to the current instrumental coordinate system we are able to find the precise positions of selected inclusions on the substrate surface after deposition of coating (Figure 1b). EDX inspections were performed over five areas with equal dimensions of 158 μm × 118 μm on two different types of substrates (D2, powder metallurgical (PM) ASP30). Table 1 shows the average number of such inclusions per mm^2 in different types of steel. In the following sections, we will examine the influence of all substrate pre-treatment steps on the formation of large- and small-scale surface irregularities.

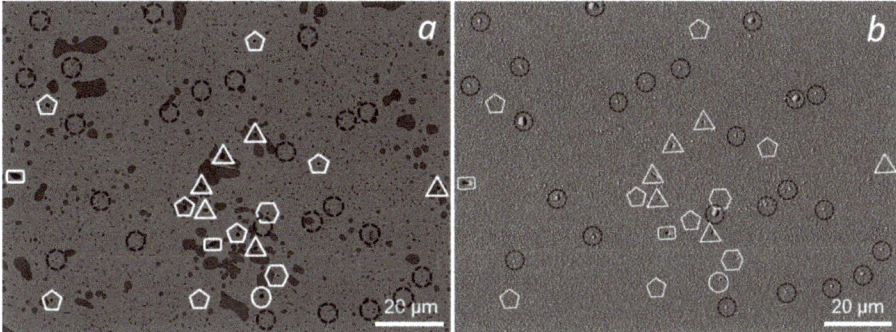

Figure 1. Scanning electron microscopy (SEM) images of D2 tool steel substrate after intensive (booster) and middle frequency (MF) ion etching (**a**) and the same surface area after sputter deposition of nanolayer nl-TiAlN/CrN hard coating (**b**). The sulfide, oxide, and other inclusions are marked with triangles, pentagons, and hexagons, respectively. Pits are designated with rectangles while sites where all other growth defects formed during the deposition process are labeled with black circles.

Table 1. Surface densities of the non-metallic inclusions in D2 and ASP30 PM tool steel substrates.

	Oxide Inclusions	Sulfide Inclusions	All Inclusions
Steel Type (AISI)	Density (mm^{-2})	Density (mm^{-2})	Density (mm^{-2})
D2	800 ± 300	460 ± 80	1400 ± 200
ASP30 PM	200 ± 120	5100 ± 500	5400 ± 300

2.1. Mechanical Pretreatment

Mechanical pretreatment of tool steels usually includes grinding [44] and polishing [45]. First, the surface is ground with progressively finer grinding papers and then it is polished with diamond paste from 15 μm down to 1 μm. There is no general recipe for polishing all types of tool materials. The grinding and polishing steps have to be slightly adjusted for each specific type of tool steel. Important parameters which determine the polishability of tool steels are the homogeneity of the microstructure, the level of purity, and the size and distribution of carbides and nonmetallic inclusions in the ferrous matrix. The biggest problems are caused by inhomogeneities. Both the purity and the homogeneity are significantly influenced by the manufacturing process of tool steel.

Although mechanical pre-treatment substantially smooths the substrate surface, it also creates numerous irregularities of the micrometer size at the same time. Mechanical pretreatment, even if performed carefully, creates various irregularities in the shape of scratches, pits, ridges, and other shapes. In addition to these relatively large topographical irregularities, smaller ones are also formed. In tool steels, slightly shallow protrusions with step-like edges are formed at the inclusions harder than the ferrous matrix (e.g., carbides and oxides) (Figure 2). At the inclusions softer than the steel matrix (e.g., MnS), inverse geometrical features are formed, known as shallow craters. The height of the protrusions and the depth of the craters depend on the hardness of the inclusions (Figure 3).

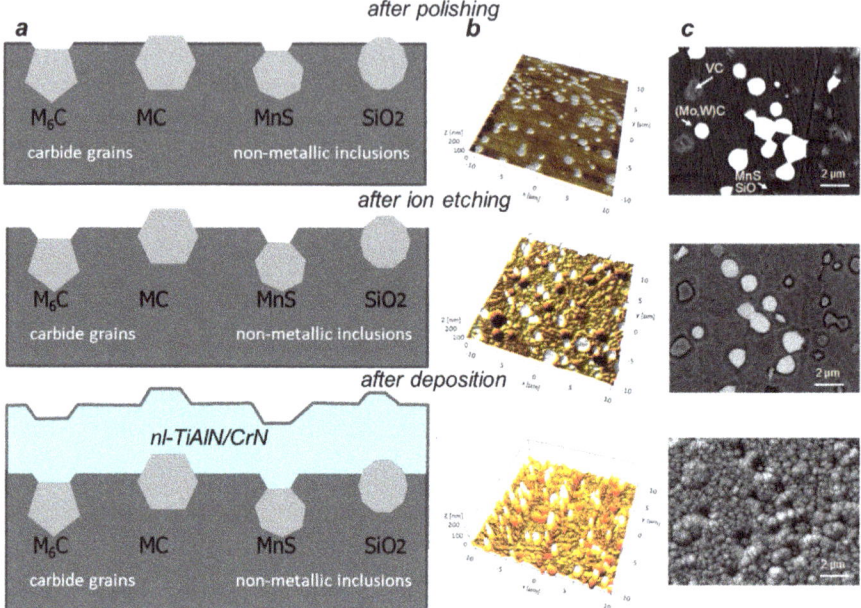

Figure 2. The schemes on the left side (**a**) illustrate the topographical changes of different types of inclusions during various steps of coating preparation (polishing, ion etching, deposition). Atomic force microscopy (AFM) (**b**) and SEM (**c**) images show the topography changes of ASP30 PM steel substrate after polishing, ion etching, and deposition of nl-TiAlN/CrN hard coatings. The SEM images (**c**) were taken at the same site on the substrate surface.

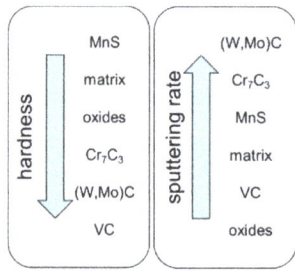

Figure 3. The total geometrical extension from the matrix level (either positive or negative) depends both on the differences in the polishing removal rate (hardness) and ion etching (sputtering) rate.

When polishing tool steels, a very common problem is over-polishing, when the polished surface gets rougher with the polishing time. Over-polishing is associated with two surface phenomena: "orange peel" and "pitting". The "orange peel" is a term used for a surface with randomly distributed smooth valleys and hills, which resembles the surface of an orange (hence the name). The formation of an orange-like surface is related to the clustered distribution of carbides in tool steel which can occur in the last polishing step if polishing is performed at too high of a pressure and prolonged time. In general, high hardness materials are less sensitive to this problem.

The most problematic issue of mechanical pretreatment is the formation of pits in the substrates (Figure 4). High pressures that are present during grinding and polishing can cause the formation of small pits (or cavities) at the positions of hard inclusions; the effect is therefore referred to as the pitting effect. The high shearing stresses also present during the polishing can tear out some of the protruding inclusions and pits with dimensions of the inclusions are left behind. The best way to avoid the orange peel and pitting effects is to keep polishing pressures constant and not too high. It is also important to use short polishing steps and apply cleaning of the substrate surface after each step.

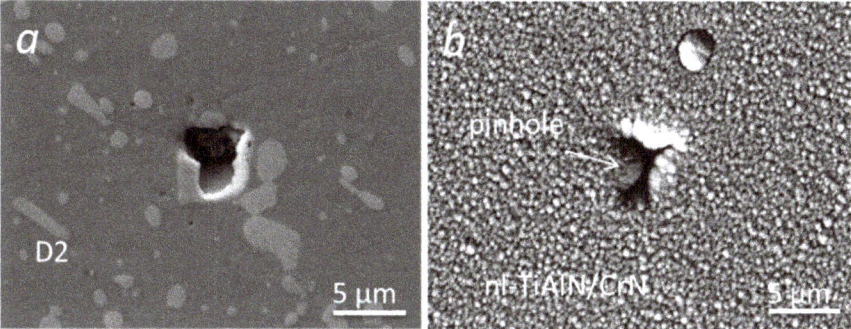

Figure 4. Pit on D2 tool steel substrate surface at the site where a carbide grain was torn out from the surface during the polishing (**a**) and the same substrate surface area after deposition of nl-TiAlN/CrN nanolayer (**b**) [46]. Close to the pinhole a nodular defect was also formed.

A part of mechanical pretreatment often includes dry or wet microblasting. In dry microblasting, abrasive media (e.g., corundum) is blown with compressed air onto the substrate to clean the substrate surface. Wet blasting is similar to dry blasting where abrasive media is mixed with water to form slurry. Both blasting techniques are used not only to clean the substrate surface, but also to alter the microtopography, hardness, and residual stresses of the substrate surface. In the pretreatment of cutting tools, wet blasting is also used for rounding the cutting edge. After such treatment the cutting edge has a more stable form, while the performance of cutting tools is significantly increased. It is clear that the impingement of micro-sized hard particles at high velocities can also cause numerous surface irregularities on the substrate surface.

2.2. Wet Chemical Cleaning

The mechanical pretreatment is always followed by the chemical cleaning of the substrates. The properties of thin films deposited by different PVD techniques depend on the cleanliness of the substrate surface on which the film is deposited [47]. Namely, any kind of contamination on the substrate surface can result in reduced adhesion of the film to the substrate, more rapid degradation of the film, greater contact resistance for electrically conducting films, and poor optical qualities for optical films. Thus, the precondition for the achievement of good adhesion of PVD coating is cleanliness of the entire substrate surface. The cleaning procedure takes place both outside (e.g., chemical cleaning) and inside (e.g., ion etching) of the vacuum chamber, just prior to the thin film deposition process.

A typical ultrasonic aqueous batch cleaning process consists of three steps: (a) ultrasonic washing in alkaline cleaning agents (pH~11); (b) ultrasonic rinsing in pure water; and (c) drying in pure hot air [47]. If we do not provide regular maintenance of water filters, air filters, and cleaning agents, then the cleaning device can also be the source of the particles, which cause the formation of growth defects. In rare cases, when the substrates are not immediately placed in the vacuum chamber for the deposition, a pitting corrosion can occur at the surface. After coating deposition growth defects are formed on the corroded area of the substrate (Figure 5).

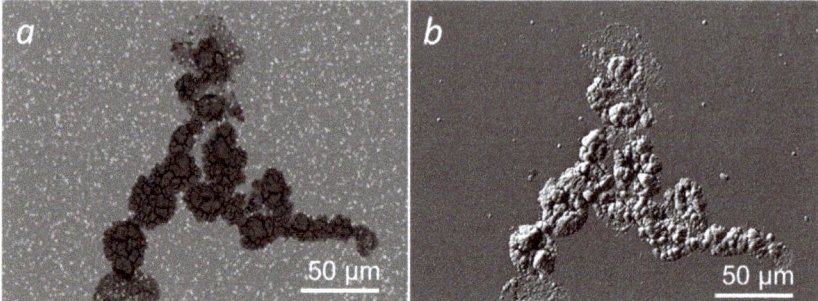

Figure 5. SEM images of a well-defined network of cracks on the corroded area of ASP30 PM tool steel substrate after wet cleaning and MF ion etching (**a**) and the same area after deposition of nl-TiAlN/CrN nanolayer hard coating (**b**).

2.3. Ion Etching

2.3.1. Basics of Ion Etching

The ion etching changes the microchemistry, surface topography, and the microstructure of the near-surface layer [47–49]. All these changes affect both the adhesion of the coating as well as its growth. However, an improved adhesion is not only a consequence of removal of surface oxides and other contaminants (which decrease the interface strength), but it is also a result of an increased density of nucleation sites and chemical activation of the surface layer. Substrate surfaces which are exposed to an ion etching erode and change topography. In general, surface topography depends on the duration of etching, density of the plasma, energy of ions, and type of ions. Different morphological features like cones, pits, hillocks, and pyramids are formed and their formation is closely related to the initial surface irregularities, impurities, and variations in the sputtering yield as a function of the angle of the ion beam incidence to the surface.

In PVD systems, there are two general concepts of ion etching procedure, which are schematically shown in Figure 6. The first and the simplest ion etching configuration is by generating plasma on the substrates themselves (Figure 6a). Such an ion etching procedure works only if high-frequency oscillatory voltage is applied to the substrates. In practice, middle frequency (MF) with several hundred kHz or radio frequency (RF) with 13.56 MHz is used. An advantage of this type of ion etching is that it forms globular plasma around the substrates and results in a more-or-less uniform etching of substrates. The disadvantage of such ion etching is that it does not allow independent control of ion densities and energies. This is determined by the voltage frequency and amplitude, which also affect the self-bias voltage.

In the second concept of ion substrate etching, an auxiliary plasma source is used. Auxiliary plasma can be generated in different ways therefore this type of ion etching is specific to a particular PVD technique. For example, plasma can be generated with the help of a hollow cathode (Figure 6b), a thermionic arc source (Figure 6c), a heated tungsten wire (Figure 6d), or some other type of plasma source. The auxiliary plasma is normally spatially confined and not all substrates are immersed in the plasma at the same time. For this reason, the substrates need to be rotated in the vacuum chamber and around their axes to achieve a more-or-less uniform ion etching. However, the main advantage of auxiliary plasma is that ion density can be controlled by the plasma source, while the ion energy is controlled by bias potential on the substrates. Substrates can be biased either by continuous pulsed or oscillatory potential, which allows an even wider control over the ion etching procedure. Although continuous substrate bias provides the most intense substrate etching, a pulsed or oscillatory substrate potential is still used in many cases because it enables the removal of native oxide and other non-conductive contaminates. Auxiliary plasmas can also be a source of metal ions. Such ion etching is available in the PVD processes that have sources of highly ionized metal plasma, such as in cathodic arc deposition [34] or high-power impulse magnetron sputtering (HIPIMS) [48]. In these cases, the source of metal ions is the cathode targets themselves, which are used during the coating deposition process. As opposed to coating deposition, the etching with metal ions is performed by applying a high negative bias potential to the substrates (typically several hundred volts); in most cases direct current (DC) bias is applied. The metal ions have a higher atomic mass than the argon ions and are more efficient in the etching of the substrate materials. The metal ions do not only etch the substrates, they are also implanted in the near-surface layer of the substrate. Such a metal ion implanted interlayer is normally beneficial since it improves adhesion of the coating. A disadvantage of etching with metal ions generated by arc discharge is the contamination of substrate surface with droplets, which reduce the adhesion of the coating and cause the formation of growth defects. This disadvantage is eliminated when HIPIMS discharge is used to generate the metal ions for the ion etching [48]. In contrast, when argon ions are used for etching, argon is also implanted in the near-surface layer, but this is normally not desired since argon is implanted in the interstitial positions of the crystal lattice, which normally increase stresses in the surface layer, while the crystallinity of the interface is completely lost.

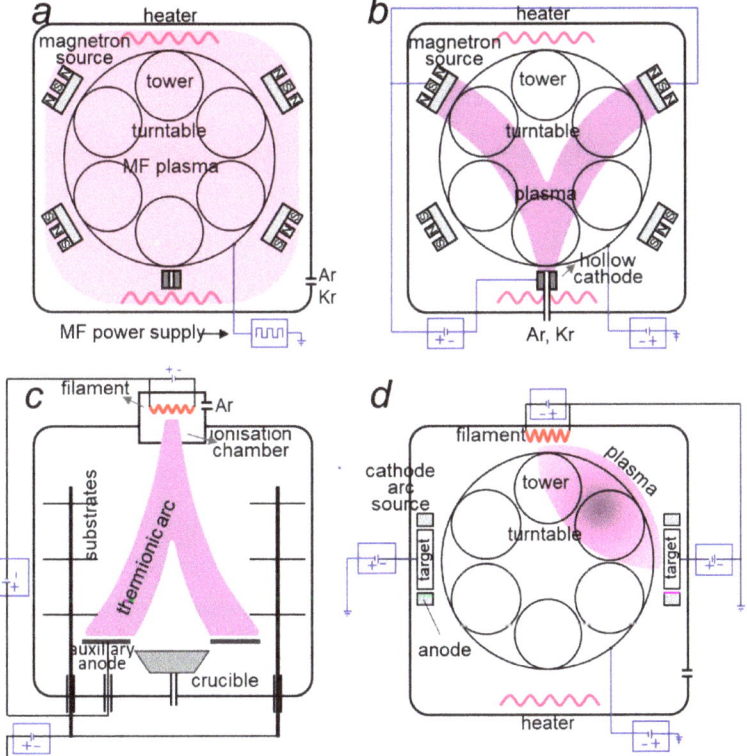

Figure 6. Examples of ion etching modes in the three different physical vapor deposition (PVD) systems we used for deposition of PVD hard coatings: (**a**) MF etching and (**b**) etching with hollow cathode plasma source (DC bias) in magnetron sputtering system CC800/9 ML; (**c**) ion etching (DC bias) in thermionic arc evaporation system BAI 730; (**d**) ion etching (DC bias) cathodic arc deposition AIPocket.

For ion etching of the substrate surface before deposition, inert ions from the broad ion beam sources (e.g., Kaufman ion source) can also be used. The advantages of such kinds of ion sources compared to competitive processes are that they generate an ion beam with a well-controlled direction, density, and energy.

2.3.2. Substrate Irregularities Induced by Ion Etching

Here we will examine substrate irregularities induced by ion etching in typical industrial PVD deposition systems we used for deposition of hard protective coatings: magnetron sputtering system CC800/9 ML, cathodic arc system AIPocket, and thermionic arc system BAI730 (see schemes in Figure 6).

If the substrate material is composed of different phases, which have different ion etching rates, then these can cause considerable geometrical irregularities on the substrate surface. An example of where such substrate irregularities form is the ASP30 PM tool steel, which is composed of several types of inclusions in the ferrous matrix. During the ion etching of tool steel material shallow craters and shallow protrusions are formed at the sites of the metal carbides and other non-metallic inclusions (Figure 2). The reason for their formation is the difference in the ion etching rate of the inclusions and the ferrous matrix. In the case of ASP30 tool steel, the sputtering rate of the M_6C carbides and MnS inclusions is higher than that of the steel matrix, while that of the MC carbides and oxide inclusions is lower. The intensity of ion etching depends on the geometry of the substrate, ion current density, ion energy, rotation mode of substrate, and plasma uniformity.

The total geometrical extension from the matrix level (either positive or negative) thus depends both on the differences in the polishing removal rate (hardness) and ion etching (sputtering) rate (Figure 3). Consequently, shallow craters (on site of M_6C, MnS) and protrusions (on site of MC and oxides) are formed. Typical height values of protrusions and depth of craters are up to a hundred nanometers.

If a foreign particle is present on the substrate before the ion etching step, then it prevents the etching of the substrate area underneath the seed. Figure 7a shows an example of an etched substrate surface with a large irregularly-shaped foreign particle. A part the substrate underneath the seed was not ion etched. As a consequence, a step-like feature formed on the substrate surface. During deposition a nodular defect is formed at the site of such a particle. The existence of the step beneath the particle proves that it was present on the substrate surface before etching (Figure 7b). If there was no step, then it reached the surface at the end of the ion etching process or immediately after it.

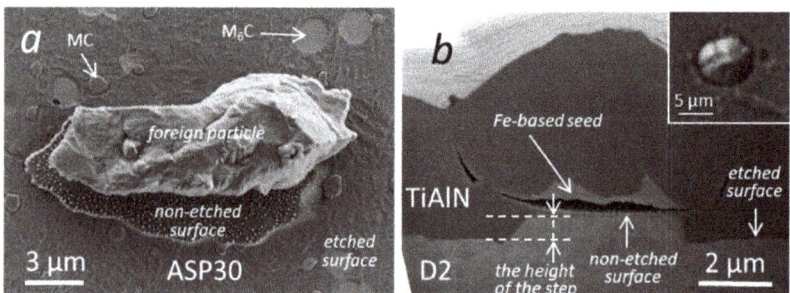

Figure 7. A foreign particle which remains on the substrate surface after the wet cleaning procedure, prevents etching of the substrate area underneath it (**a**); a step on the substrate surface beneath the iron-based seed proves that the seed particle was on the substrate surface even before etching (**b**).

If ion etching is not done properly then it can cause degradation of the substrate surface. Problems may arise due to the backscattering of the material sputtered from a substrate surface back to the surface, particularly at high pressure. Therefore, care must be taken to flush away the sputtered contaminant species. The same effect as backscattering is caused by redeposition which is defined as the return of the material sputtered from a substrate surface back to that surface. Another undesirable effect of ion etching in industrial deposition systems is the cross contamination of the substrate surface as well as target surface with batching material. During ion etching a part of the batching material is transported first from the substrate (tool) surface to the target surface. Later (in the early stage of deposition) the same material is returned back to the substrate surface forming a thin contamination film. To avoid contamination of the targets with batching material during ion etching, the target should be covered with moveable shutters.

3. Growth Defects Formed during Deposition

During deposition all morphological features of the substrate surface that are formed during its mechanical pretreatment and ion etching are transferred onto the coating surface and they are even magnified. Topographical irregularities and small foreign particles remaining on the surface of the substrate after cleaning and those which were generated during ion etching, cause the formation of growth defects in the coating due to the shadowing effect. However, a large part of the seed particles responsible for growth defect formation are generated during the coating process itself. Figure 8 shows 3D-profilometry images of a TiN coating deposited on D2 tool steel substrate by three different deposition techniques: (a) evaporation by thermionic arc (BAI730); (b) magnetron sputtering (CC800/9 ML) and (c) evaporation using cathodic arc (AIPocket). The difference in the growth defects density is evident.

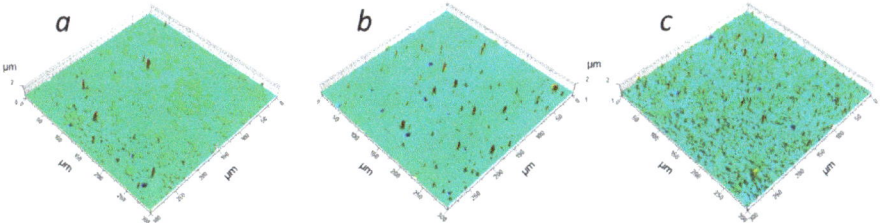

Figure 8. Three-dimensional (3D)-profilometry image of TiN coatings deposited on D2 tool steel substrate evaporation using thermionic arc (**a**), magnetron sputtering (**b**), and cathodic arc (**c**). The sharp peaks are the nodular defects while the blue dots are craters (pay attention to the strong exaggeration in z-scale).

In this section the growth defects according to their origin and shape are classified, while in the next one all potential sources of seed particles are described.

The literature is not consistent in classifying and naming the growth defects. The diverse classification comes from a wide variety of defect morphologies and their origins, and from different application fields studying growth defects. In this work, we propose to classify growth defects in the most general way. To keep things simple, we divide growth defects in two general groups: (i) the term "protrusions" is used for those defects that are above the mean surface of the film (Figure 9) and (ii) the term "holes" is used for those defects that are below the surface. We also attempt to provide a unified nomenclature for the defects with respect to their origin and morphology.

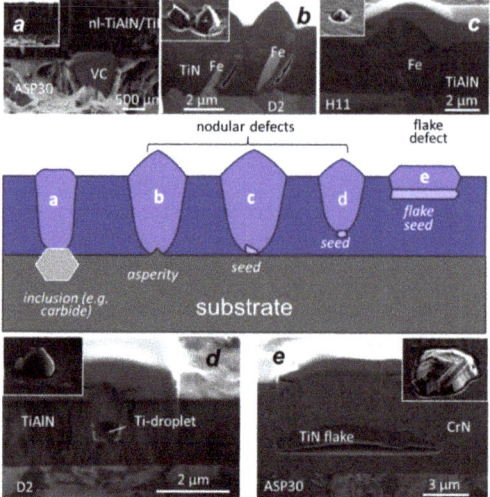

Figure 9. Schematic overview of different seeds causing the formation of protrusion defects: (**a**) carbide inclusion, (**b**) geometric irregularity, (**c**) foreign particle on the substrate, (**d**) foreign particle during growth, (**e**) flake. Focused ion beam (FIB) and fracture cross-section SEM images of typical protrusion defects in PVD hard coatings prepared by evaporation using thermionic arc (**b**,**c**), magnetron sputtering (**a**,**c**), and evaporation using cathodic arc technique (**d**) are added.

3.1. Protrusion Defects

3.1.1. Nodular Defects

Nodular defects are the most common type of growth defects [2,10,12,13] and they are present in all PVD coatings. The formation of nodular defects is caused by a seed. Seeds are usually very small particles (dust, foreign particles, particles ejected from the coating material source) or substrate protrusions. It appears that virtually any irregularities, even a minute one, may act as a seed. The nodules do not always start to grow at the substrate surface; they can also grow from a seed, which arrived on the substrate during the deposition process. The nodule starts to grow in the shape of an inverted cone that propagates through the film and forms a domed protrusion at the outer surface of the film. The nodule itself is much larger than the seed that causes it. It is not, in itself, a contaminant. It is composed of the same material as the coating but growing in a different way. Due to the shadowing of vapor flux by the seed particle and limited thermal mobility of atoms on the surface, there are discontinuous boundaries between the nodule and the surrounding coating matrix. The outer surface of the nodule is a quite sharp boundary between it and the remainder of the coating. This sharp boundary is a region of weakness and there is frequently an opening around the nodule, either partially or completely, and the nodule may sometimes be detached from the coating completely, leaving a hole behind.

As already mentioned, nodular defects originate from small seed particles and on the surface of a thin film appear in a shape of cones or domes (Figure 10a). In the literature, they are also called nodules, hillocks, peaks, or inverted cones. The part of the nodular defect that is below the film surface has a shape of an inverted cone with a parabolic cross-section, whereas the part of the defect above the film has a shape of a cone with a rounded top (similar to a dome). The base of the nodule dome (or hemi-sphere) is circular or oval in the planar projection (Figure 10b). Leets et al. [5] and later Tench et al. [12] developed a simple model, which described the geometry of the classical parabolic nodular defects observed experimentally. Their model was based on omnidirectional (isotropic) coating flux and on the assumptions that the nucleating particle is spherical and much smaller than the total coating thickness. They also assumed that the growing coating has no internal structure and that the mobility of ad-atoms may be neglected. In this case, the layers of the coating material above the seed are concentric spherical caps, while the coating is perfectly conformal (the coating thickness is assumed to be identical everywhere on the seed). In this case the topology of defect growth becomes a simple geometrical problem. The relationship between the diameter of nodular defect ($D = 2R$), the seed particle diameter (d), and the coating thickness (t) for a hemispherical seed particle on a flat substrate surface is: $D = \sqrt{8dt}$. This simple geometric model explains the shape of the nodular defects but fails to explain their size (diameter and height). It is valid when the seed particles are small, while larger particles produce more complicated structures due to shadowing effects. Dubost et al. [2] improved such a model by including proposed surface reaction probabilities of depositing species.

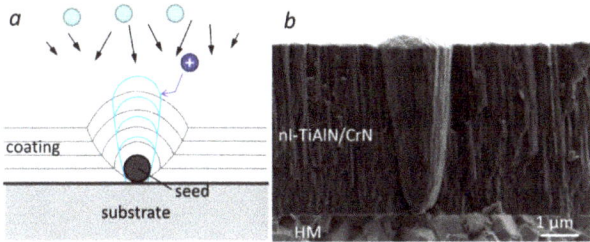

Figure 10. Scheme of a conical and parabolic nodule cross-section; the shape mainly depends on the flux distribution of the incoming atoms (**a**); fracture cross-section SEM image of a typical nodular defect with rather straight vertical walls in the nl-TiAlN/CrN hard coating prepared by magnetron sputtering (**b**).

In addition to geometric models for description of the nodular defect formation, several two-dimensional computer simulations [13] have been proposed. These simulations are based on hard disc model of Dirks and Leamy [14] in which discs (atoms) fall randomly onto the perfectly flat surface at a fixed oblique angle and then stick immediately where they land and roll into the nearest saddle position formed by the previously deposited discs. It should be emphasized that this computer simulation model is a purely kinematic one, because it does not take into account surface or particle energies, interatomic forces, crystallographic orientations, etc. These computer simulations show that nodular defects have a cylindrical symmetry with parabolic side wall structure and that the actual aspect ratio of the defects could be varied by changing deposition conditions in the model (oblique incidence, random variations of particle flux, rotating substrate). Using such a model they also reproduced the columnar microstructure and columnar tilt in the coatings, as well as opened boundaries between the nodular defect and coating matrix. However, the model does not consider complex adsorption processes that influence film-structure evolution. Liao et al. [15] and later Muller-Pheiffer et al. [16] upgraded the two-dimensional hard disk model in such a way that they included surface diffusion and desorption of arrived atoms. A more sophisticated ballistic model of coating growth, which takes into account scattering and surface diffusion of depositing species was developed by Lang and Xiuqin [50]. Their computer simulation shows that higher deposition rate, lower surface diffusivity of deposit, and higher degree of scattering of depositing particles favor the formation of nodular defects.

The deposition process can have a profound effect on the shape of the nodule. The nodule cross-section geometry mainly depends on the flux distribution of the incoming atoms (Figure 10). It is parabolic if the flux is random (isotropic) and conical if the flux is directional (narrow angular distribution of the deposition flux). A parabolic profile of the nodule is a characteristic for electron-beam deposition where a wide range of deposition angles is present. In sputtering, on the other hand, where narrower deposition angles are present, nodular defects with more straight vertical walls are formed.

The non-uniformity of incident angles of the incoming flux of the coating material and the shadowing effect cause the formation of nodules with asymmetrical boundaries (Figure 11a).

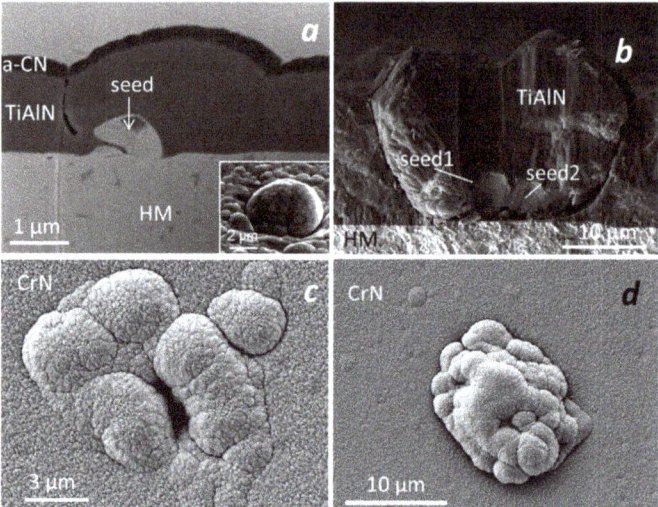

Figure 11. The material deposited on one seed forms a single nodular defect (**a**), while the material deposited on two or more close-spaced seeds forms conjoined nodular defects (**b**). Nodular defects can appear together in complex aggregates (**c**) or with a typical "cauliflower-like structure" (**d**). Nanolayer nl-TiAlN/CrN (**a**) and TiAlN (**b**) hard coatings were prepared by sputtering, while evaporation using thermionic arc was used for deposition of CrN hard coating (**c**,**d**).

Nodular defects forming on the substrates which rotate in their own plane have distinctly different shapes and sizes than those which form on stationary substrates [13]. Nodules on stationary substrates form an inverted cone with straight sides while those produced on rotating substrates form a rounded bowl-like bottom. In addition, for rotating substrates the nodule size and shape strongly depend on the angle of the vapor flux with respect to the substrate plane.

Nodular defects differ by their shape, seed depth, composition, and seed shape. The shape of the nodule and the structural characteristic of the nodule-coating interface are basically determined by the seed size and shape. If the seed is small in comparison to the coating thickness, then the shape of the nodular defect does not depend on the shape of the seed. If the seed has a smooth morphology then the nodule looks like a cone, while a seed with complex morphology results in the growth of a nodular defect with irregular surface features. Irregular shapes of nodular defects are not uncommon, particularly if they are very large. The nodular defects can appear individually (Figure 11a), form in clusters which can overlap each other (Figure 11b,c), or appear together in complex aggregates with a typical cauliflower-like structure (Figure 11d). When an irregularly shaped seed is coated, the particle flux cannot reach the area underneath the seed due to the shadowing of the particle flux and thus causes formation of voids below the particles. The coating on such kinds of seeds results in highly non-uniform coverage with a practically uncoated area underneath the seed. This can be seen in Figures 7b, 9a and 11b,c.

As the seed gets overgrown by the coating, the contour of the coating follows the shape of the seed. The nodular defect is a conformal replication of the seed particle shape. If the seed particle is overcoated by a multilayer coating, then the contours of individual layers show that the coating growth within the nodule continues in a similar way to the coating matrix (Figure 12a).

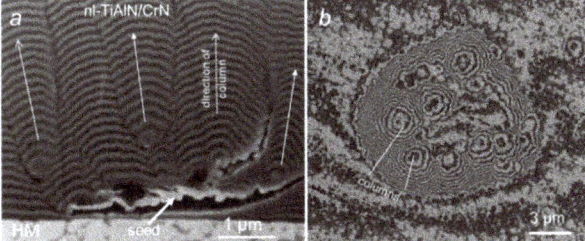

Figure 12. FIB image of a nodular defect in the nl-TiAlN/CrN nanolayer coating; several columns are formed on the surface of the seed with complex geometry (see arrows) (**a**) and SEM top view image of the ground section of the nodular defect caused by a seed with complex geometry (**b**).

Petrov et al. [51] argued that the growth of coating on seed particles is far more coarse and columnar compared to the regions around it because the top of the nodular defect grows in a regime of intensive ion bombardment while for seed regions in the shadow of the particle flux the ion bombardment is less intensive [52]. This has a large effect on the microstructure of the nodular defects which is composed of dense columnar grains, while the microstructure of the lower region of nodular defects is rough and underdense. The nodular defects typically grow in a "feather like" pattern growing from a central core.

Due to high internal or thermal stresses the internal cohesion of the nodular defect is often inferior to the cohesion at the boundary. In this case, only a part of the nodular defect is broken off, like the cases shown in Figure 13a,b. This may happen during deposition (high internal stresses) or during the cooling stage (high thermal stresses). The SEM image exposed the internal structure of nodular defects, which includes the microstructure of the coating as well as the size and shape of the seed particle.

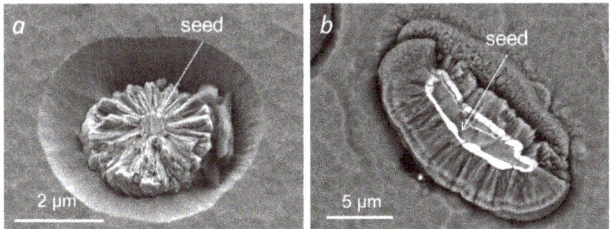

Figure 13. SEM micrographs of two broken nodular defects in TiAlN hard coating sputter-deposited on D2 tool steel substrate.

3.1.2. Flake Defects

Another type of growth defect that protrudes above the surface of a thin film is flake defects. They differ from nodular defects in the origin of the seed particles, which results in a much larger size of defects and a very different shape than nodular defects. The seed particles (see e.g., Figure 14) are large flake particles that originate from the delamination of coating that was deposited in previous batches on the fixture holders for substrate and shields within the vacuum chamber. These seed particles are delaminated due to the high thermal and internal stresses that are present in the coating prior and during the deposition. The seeds of the flake defects are typically very large, normally several tens of micrometers in the diameter and have irregular shapes. Due to very large diameter-to-thickness ratio of the flake seeds, the defects that form above the seed have flat top and step-like edges. Hence, the flake defect is a step-like projection of the overgrown seed. Flaking of the coating is also triggered by arcing on substrate fixtures (i.e., substrate turntable) during ion etching and deposition process. Overall, the surface density of flake defects is relatively low compared to the nodular defects. It mainly depends on the thickness and adhesion of coating on substrate fixtures and shields. To reduce the density of flake defects all vacuum components should be cleaned after several batches. Although the concentration of flake defects is typically low, they can be very detrimental for the performance of the coating if they extend down to the substrate and expose it to the detrimental influence of the surrounding atmosphere (oxidation, corrosion) directly.

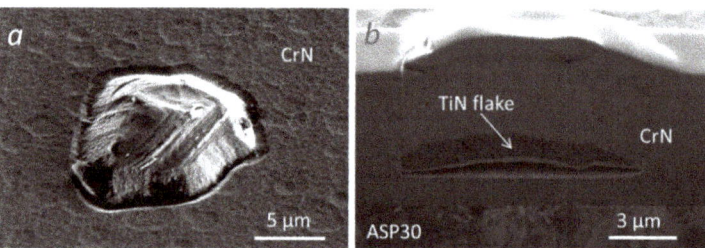

Figure 14. Top view SEM image (**a**) and FIB cross-section image (**b**) of flake defects in CrN hard coating prepared by evaporation using thermionic arc.

3.1.3. Droplet Defects

Droplet related defects are very common in the coatings deposited by cathodic arc deposition technique (Figure 15a). If metal ions are used for substrate etching, then droplets generated by arc discharge contaminate the substrate surface, which cause the formation of growth defects. While the generation of droplets continues also during deposition process the growth defects start to grow within the coating. As we will explain in Section 4.2.3 the metal droplets are produced due to the melting of the target. The liquid target material is ejected from the target as droplets. Part of these droplets can arrive on the substrate surface where the liquid material solidifies. Such droplets initiate defects in the depositing coatings.

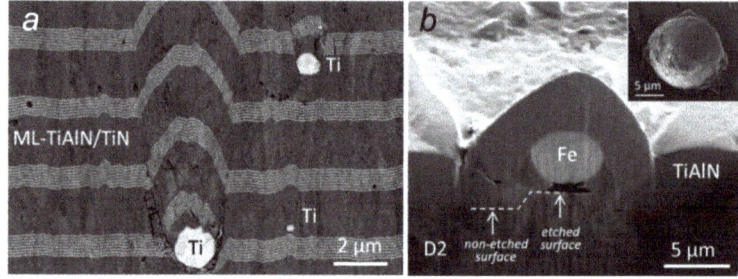

Figure 15. FIB cross-section image of a buried droplet formed in the multilayer TiAlN/TiN hard coating prepared at the company KCS Europe from Germany by cathode arc deposition technique (**a**). Fe-based droplet built in the TiAlN hard coating sputter-deposited on D2 tool steel substrate (**b**). The distinct step beneath the Fe droplet proves that it arrived on the surface of the substrate at the beginning of the ion etching.

We found that average height and average surface area of such type of growth defects are smaller in comparison with growth defects based on other types of seed particles. However, the number of droplets is more than 10-times greater. Droplets have a spherical or oval shape; therefore, the droplet related defects are of more regular shapes.

Droplets can be formed also during sputter deposition (Figure 15b), but on a much smaller scale. Their formation is caused by arcing on the substrate table and other inner components of the vacuum chamber. Therefore, composition of droplets is based mostly on iron.

3.2. Hole-Like Defects

In the literature, a great variety of terms are used in regard to the growth defects that are below the surface of the thin film. We will use the general term hole to describe any type of growth defect that is below the mean surface level of the thin film. Like protrusion defects, the hole-like defects can be distinguished with respect to their origin and shape. Figure 16 shows schematic classification of hole-like defects by their typical shape.

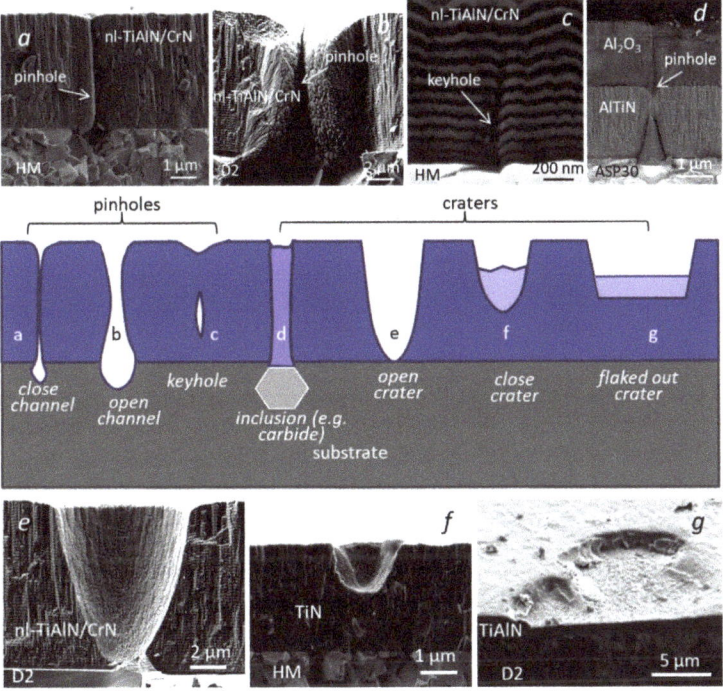

Figure 16. Schematic overview of different shapes of hole-like defects in PVD coatings: (**a**) pinhole at the site of a pit in the substrate; (**b**) open pinhole; (**c**) closed pinhole (keyhole); (**d**) pinhole that appeared at the site of a shallow crater formed during ion etching of ASP30 tool steel substrate; (**e**) crater formed due to the expulsion of a nodular defect; (**f**) a crater that does not extend through the entire coating; and (**g**) crater of irregular shape formed by detachment of flake defects. FIB and fracture cross-sectional SEM images of typical hole-like defects in PVD hard coatings prepared by magnetron sputtering (coating types: nl-TiAlN/CrN, CrN, TiAlN) and by evaporation using thermionic arc (coating type: TiN) are added.

3.2.1. Pinhole Defects

Pinholes as one of the most common growth defects in PVD thin films are discontinuities in the coating microstructure in the form of thin holes having a (sub)micron size diameter and extending from the substrate to the top surface of the coating. There are a number of causes for formation of pinholes (Figure 16). A majority of pinholes are generated at the substrate imperfections, such as cavities (pits) or shallow depressions formed on the substrate surface during its pretreatment. The usual origin of pinhole formation is geometrical: a narrow but deep cavity, where the shading effect prevents the film growth on the cavity walls. Namely, the coating is preferentially deposited on the flat front side of substrate, while the deposition rate on the sidewall of the cavity is much lower. Due to the shadowing effect, coating on the sidewalls of the cavity has a columnar, porous structure.

The angular distribution of the impinging vapor flux on the surface is the most important factor which influences the size and the number of pinholes generated by geometrical shadowing [53]. The more random the flux direction is, the smaller the number and size of the pinholes (Figure 17a,d). A random vapor flux direction is established: (i) by using substrate-holding fixtures that randomize the substrate position and angle of incidence in the vapor flux and (ii) by using extended vapor sources or several vapor sources. As compared to evaporation, conventional sputtering can provide more conformal coatings over protrusions and low aspect ratio cavities. This is because sputtering sources form a broad atom flux and atoms are ejected at wide angles too.

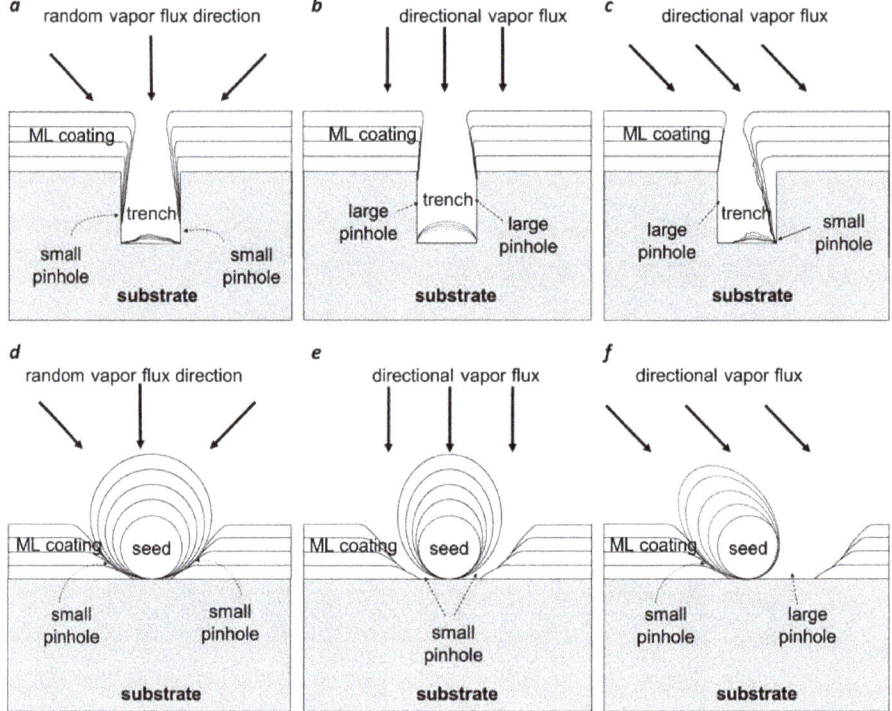

Figure 17. The schemes show the distribution of coating around two kinds of surface features (high aspect ratio trench, sphere) and formation of pinholes for a range of angles of incidence of depositing flux of atoms.

Whether a pinhole will be formed or not depends on the aspect ratio of the hole (depth/hole diameter) and not its size. In the case of a narrow but deep hole (high aspect ratio) on the substrate surface the shading effect prevents the coating growth on the hole walls. When the aspect ratio is high, the deposition starts to coat the upper sidewalls and corner of the feature, which shadows the lower area from subsequent deposition. During the following coating growth, the high aspect ratio crater is narrowing. If the PVD film is thick enough they even appear to touch and close the opening, forming an isolated pore, called a keyhole. However, a microstructural discontinuity is preserved through the growing coating, extending up to the coating surface. On the other side a laterally large but shallow hole (low aspect ratio) will not develop a pinhole. In this case an open pinhole will form. What will happen to the high aspect ratio holes on the substrate surface depends also on the thickness of the coating. If the width of the crater on the substrate surface is comparable to the final coating thickness, the pinhole will not be able to close up during the coating growth.

The pinholes are also found everywhere where there are nodular defects, which start to grow at the substrate surface. As already mentioned, the contact between the nodular defect and the undisturbed coating is poor. The contact of the seed particle with the substrate is very poor too as there is no coating at all. Therefore, the border between the nodular defect and the undisturbed coating is essentially a "circular" pinhole.

Through-porosity (pinholes) can be detected by using SEM, selective substrate liquid chemical etching, selective plasma etching, electrolytic copper decoration, or it may be measured by corrosion potentials (anodic polarization) [54].

Typically, pinholes occupy a relatively small area of the total coated surface. The significance of pinholes is highly dependent on the application. In some, pinholes are functionally insignificant, whereas in other cases, they are intolerable. Their influence on the functional properties of thin films is discussed in Section 5.

3.2.2. Crater-Like Defects

As discussed above, the bond between the nodular defect and the surrounding matrix as well as between the seed particles and the substrate is poor. Due to the buildup stresses in the growing coating, some of the nodular defects detach from the coating, leaving a crater on the coating surface. The resulting crater can be interpreted as "inverse" nodular or flake defects. Formation of such a crater may also be caused by external forces such as cleaning by ultrasonic cavitation or wiping after the coated sample is removed from the deposition chamber. Depending on the moment of the spall-off (during or after the deposition), areas of bare substrate may be found at the bottom of these holes. If the nodular defect leaves the coating during the deposition process, then the created crater is covered by the still growing coating. The microstructure of the overgrowing coating is highly columnar and porous with poor cohesion. Figure 18a shows a fracture cross-sectional SEM image of a crater-like defect formed from the nodular defect after the deposition was completed. The nodular defects probably detached during the cooling stage, where the internal stress was augmented by the thermal stress.

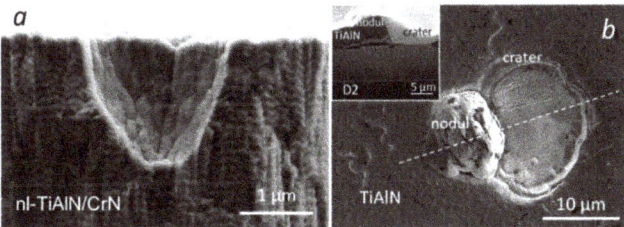

Figure 18. Crater-like defect left by the detachment of a nodule in sputter-deposited nl-TiAlN/CrN hard coating (a) and combination of flake and nodular defects in sputter-deposited TiAlN coating (b).

Similar to craters formed from the nodular defects, large craters of irregular shape are formed by detachment of large flake defects when the internal stress overcomes the adhesion (Figure 18b). In the case when this happens during the coating deposition, the remaining crater with the flat bottom is covered by the additional growing coating. The concentration of such defects is not large, but they cause large imperfections in the morphology of the thin film and can have large detrimental effects on the properties of films. This is especially the case when the craters are extending down to the substrate.

4. Origin of Seed Particles

As mentioned in previous sections, any protrusion on the substrate surface is a seed for formation of nodular defects. In general, there are several possible seed origins [18,55] which are independent of the deposition techniques. The seeds for nodular defects can form from: (a) geometric irregularities on the substrate surface after mechanical pretreatment (see Section 2.1); (b) substrate irregularities arising from ion etching (see Section 2.3); and (c) foreign particles that arrived on the substrate surface before or during the coating growth. In all these cases the consequent growth mechanisms are similar, yielding similar nodular defects, which are generally indistinguishable from a top view. The overall shape does not depend much on the seed type, nor its chemical composition.

4.1. Foreign Seed Particles

Different seed particles (e.g., dust, debris, polishing residue, impurities) remain on the substrate surface after its mechanical pretreatment, wet cleaning, drying, and batching. In order to obtain a smooth coating surface, we remove the majority of particles in the production process [55]. To approach this goal a high-quality wet cleaning procedure must be performed in an ultrasonic bath. First the steel substrates have to be demagnetized to avoid the problems in removal of ferrous debris from its surface during wet cleaning in alkaline cleaning agents. The use of ultrasonic cleaning in addition to manual cleaning has a positive impact on reduction of foreign particles. Manual scrubbing has been shown to be critical to reduce surface particulates. After rinsing in deionized water, the substrates must be dried in hot air as quickly as possible because the residual water film will stick dust particles to the surface. Fine air and water filters must be used in order to minimize the concentration of particles in deionized water and dry air. In order to minimize the re-contamination of the cleaned substrate surface before it is placed into the deposition chamber, we have to ensure a clean processing environment, a proper storage after the external cleaning, and adequate handling during batching. Seed particles can also be brought in the deposition system with fixture components. Inadequate substrate cleaning and/or inadequate cleanliness during transport and batching drastically increase the density of detrimental particles. Figure 19a, for example, shows a nodular defect that originates from a $CaCO_3$ seed particle. Such a particle is probably the residue of cleaning agent.

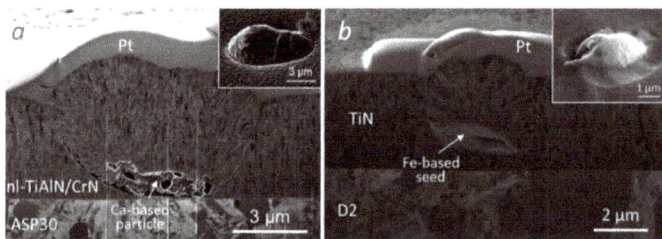

Figure 19. Top-view SEM (insets) and FIB images of nodular defects originating from Ca-based and Fe-based seeds. Nanolayer nl-TiAlN/CrN (**a**) and TiN (**b**) hard coatings were prepared by sputtering and evaporation using a thermionic arc, respectively.

An intensive generator of seed particles is the deposition system itself. Some particles may fall on the substrate surface already during rough pumping because of possible turbulent gas flow which may pick up small particles accumulated on the bottom of the vacuum chamber from previous batches. To avoid any turbulence in the gas flow, slow pump-down during rough pumping (up to 100 mbar) and "soft-venting" must be used.

The next origin of seeds is wear particles (debris) generated by all moving components in the vacuum chamber (opening and closing valves, moving parts of fixturing) (Figure 19). More motion elements in the fixture systems mean more wear particles. In particular, the problems are triggers that are often used for discontinuous rotation of substrates. Some debris also originate from maintenance and installation (e.g., wear of hand tools, insertion of bolts). Wear particles may be minimized by using appropriate non-galling materials in contact, vacuum-compatible dry lubrication of surfaces in contact, smooth surfaces and minimal contacting forces. Upward-facing samples have a higher defect density than downward- or lateral-facing samples because of particles falling on the substrate surface due to gravity. Therefore, mounting substrate surfaces facing upward should be avoided.

The seed particles may also be brought in the deposition system with processing gases. The use of high purity gasses is strongly recommended.

After several deposition runs the coating buildup on the shields and fixtures becomes too thick, and it may flake, particularly if high residual and thermal stresses are present in the coating material. The flakes, which are transferred through plasma, build up a negative charge. They are held to the substrate surface by electrostatic forces, which affect the micrometer-sized particles much stronger than the gravitational forces. One way to reduce this problem is to occasionally overcoat the brittle and poor adhered deposit of hard coatings with a softer (pure metal) material; this process is called metal layer pasting [56]. Pasting is a high-power sputtering step in metal modes that cleans up the sputter surface and also seals the re-deposited nodules with thin metal layers. If pasting is not done at recommended intervals of time, flaking of re-deposited materials from the sputtering target, shields, and fixturing is likely to cause particle formation in the deposited coatings. In any case, regular sand blasting of the deposition chamber components (e.g., fixtures, shields) is necessary to perform.

Another possible origin of dust particles is their formation in plasma [57]. Researchers in the semiconductor industry realized that sub-micrometer particles can be formed in chemically reactive plasmas (e.g., in plasma-enhanced chemical vapor deposition processes (PECVD)) by the gas-phase reaction and aggregation of atoms or molecules from etching or sputtering processes. The possibility of particle formation is smaller if the partial pressure of the reactive gas is reduced.

As mentioned above, many sources of seed particles can be eliminated using proper vacuum and substrate handling techniques. In the case of providing the cleanest conditions before coating deposition, the main source of seed particles is then the source material itself (evaporation crucible, cathodic arc or sputtering target). Each deposition technique and material combination is unique and must be studied individually. In the following text, the origin of seed particles for these three different evaporation sources is discussed in more detail.

4.2. Seed Particles Originating from Deposition Sources

4.2.1. Seeds in Electron Beam Evaporation

In addition to above described general origin of seed particles (Figure 20), specific seed particles are characteristic for electron beam evaporation. Various mechanisms of seed ejection from an electron-beam process are possible. Heating of the source material can produce seeds by several mechanisms including: (a) explosions caused by the heating of gas inclusions or micro-arcing; (b) splashing of molten material; (c) electrostatic repulsion of charged particles; (d) thermal-induced cracking; and (e) temperature-induced solid-state phase transitions [18,53].

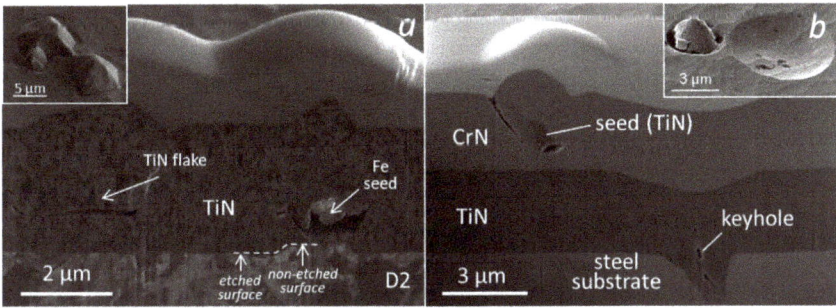

Figure 20. Typical growth defects in TiN (**a**) and TiN/CrN double layer (**b**) hard coatings prepared by thermionic arc evaporation. FIB images show the origin of defects.

In the deposition from molten source material, small droplets may be ejected from the molten pool, which lands on the substrate and is incorporated into the coating during the deposition process. This, so-called spitting phenomenon can be caused by the release of gas or vapors in the molten material during rapid heating. The source of spits can be suppressed by using pure evaporation material, preheating for degassing the evaporation material or slow heating to vaporization. In addition to these measures, the selection of deposition parameters is also very important. In the case of e-beam evaporation the key deposition parameters are e-gun voltage, e-gun emission current, and beam pattern. A low current allows the source to operate at relatively low temperatures where effects due to charging, stress relief, and phase transitions are reduced. Setting the e-gun at lower accelerating voltage keeps the e-beam heating nearer to the surface of the melt (about 25 µm in depth), thereby minimizing thermal gradients. A broad and rapidly scanning electron beam decreases the power density into the source. The splitting occurs if e-beam energy is delivered at a rate faster than the coating material can accommodate this energy by evaporation, conduction, or radiation.

4.2.2. Seeds in Magnetron Sputtering

Magnetron sputtering is one of more commonly used methods for the fabrication of PVD thin films. In magnetron sputtering, operated either in DC, RF, or pulsed mode, the plasma is generated over a large area of the cathode as opposed to the cathodic arc where plasma is localized in a small area of the arc. In the sputtering, material from the target is vaporized as individual atoms or groups of atoms and therefore, in principle, does not generate micro-droplets. In the case of magnetron sputtering the formation of specific seed particles (flakes) are caused by: (a) flaking of cones formed in the target racetrack, (b) flaking of the redeposited nodules from the target surface, and (c) by arcing [56]. In the following text, all three mechanisms are discussed in more detail.

Formation of Cones on Target Surface During Ion Bombardment

Different physical phenomena (e.g., cone formation, faceting, trenching) occur at the target surface during ion bombardment and cause its roughening. Here only cone formation as an extreme case of surface roughness is discussed. The formation of cones or micro-protrusions on target during ion sputtering in the presence of a seed material was first observed several decades ago by Wehner and Hajicek [58]. They found that cones formed by sputtering of the Cu target surface with a concurrent supply of impurity Mo ions. In this case the Mo atoms create regions with lower sputtering rates which cause a local masking effect. The substrate material around the impurity center is sputtered away faster because of a much higher sputtering yield.

Apart from these local masking effects, cones can also develop if some degree of impurity atoms is present in the target material. For example, if aluminum targets of the purity 98% and 99.99% are sputtered under identical conditions only the target of lower purity shows the cone-like features formed on its surface. The formation process of conical protrusion can be explained by considering not only the primary erosion process (i.e., different sputtering rates, the variation of sputtering yield with the ion incidence angle), but also by other effects such as ion reflection, re-deposition, and surface diffusion. The undesirable effect of cone formation is the breaking and formation of seed particles that could be built into the growing coating. The distance between the target and substrate is rather small (50–120 mm); therefore, any particle generated during sputtering has a high probability to reach the substrate surface due to electrostatic repulsion.

Flaking of Re-Deposited Nodules from the Target Surface

During magnetron sputtering in a reactive mode (oxides or nitrides) re-deposition of sputtered material occurs from the racetrack to the center and on the edges of the planar magnetron. In this region the plasma density is low and sputtering does not take place. Therefore, the material is deposited on the perimeter rather than eroded. Due to the internal stresses the redeposited material grows in the form of filaments [59,60] (Figure 21). During deposition the filaments gradually grow. Those which form near the racetrack cross the high-density regions of plasma and are resistivity heated (due to increased current flow). Heating of the filaments causes its fracture and ejection of the fragments. The ejected fragments become electrically isolated from the target and charge negatively. The charged particles are accelerated away from the target due to the repulsion effect [58,61]. Some of them arrive to the substrate surface and become seeds for the growth of nodular defects in the coating. Such a mechanism for particle formation was first observed in carbon targets used to deposit diamond-like coatings on magnetic disks and during nonreactive sputtering of TiN [60].

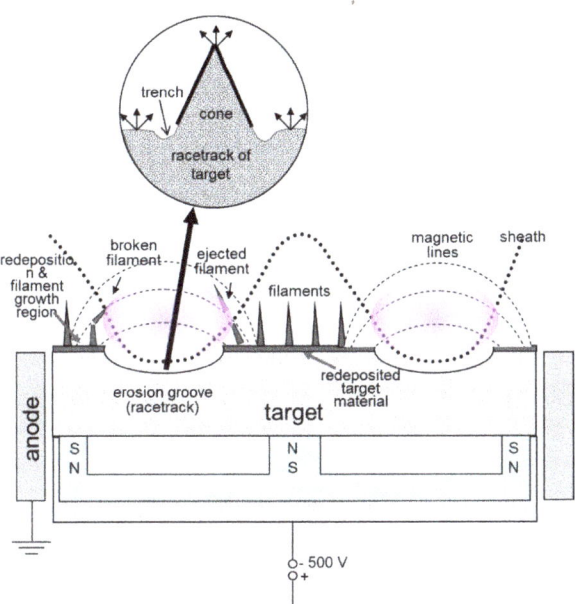

Figure 21. Scheme illustrates the formation and fracture of filaments on the perimeter and in the middle of the magnetron target surface. The cone formation on the racetrack of target is also schematically shown.

Filaments on the target surface present an important source of contamination in the sputtering process. The formation of filaments can be avoided if a cylindrical magnetron is used. On the cylindrical target a re-deposition zone does not grow, and this is the reason why such magnetron sources provide much higher process stability and cleaner coatings can be produced.

In order to prove that parts of seed particles really originate from the target surface, we sputtered deposited TiAlN coating on the test sample in a production batch together with different hard metal and high speed steel (HSS) cutting tools. After deposition we analyzed the broken nodular defects found on the coating surface by backscattered-electron (BSE) imaging (Figure 22a,b). The BSE image reveals the internal structure of nodular defects, which includes the microstructure of the coating as well as the size and shape of the seed particle. Additionally, a bright layer around the seed particles was observed. EDX analysis confirms that the seed is a TiAlN flake, while the bright layer (app. 50 nm thick) is composed of iron, tungsten, and chromium. The presence of these elements can be explained by contamination of the target surface during ion etching. As mentioned earlier, a lot of weakly bonded seed particles are present on the target surface, especially outside of the racetrack. During ion etching these particles are covered by the thin film of batching material. Immediately after starting the deposition an electrical charge may accumulate on these particles and electrostatic forces cause their self-expulsion. Some of them can be built in the coating growing on the substrate surface.

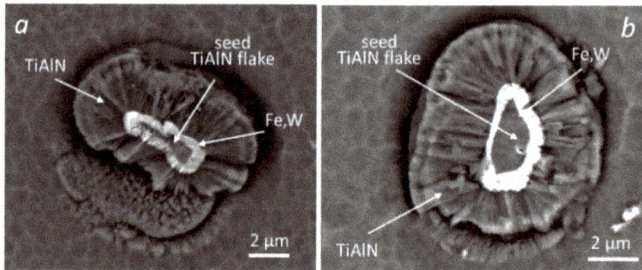

Figure 22. SEM images of the broken nodular defects in the sputter-deposited TiAlN hard coating. Seed particles are clearly visible at the fracture. The origin of these seed particles is the target of magnetron sources.

From a large number of FIB cross-sections of individual nodular defects, we found that the majority of defects in the magnetron sputtering originate from the seeds on the substrate surfaces. This finding supports the above-mentioned statement that a part of the seed particles arrives on the substrate surface just before or at the start of the deposition process.

Arcing

Arcing is a common problem in both metal and reactive magnetron sputtering because it is a significant cause of defect generation and process instabilities [62–64]. An arc appears due to the rapid accumulation of charge on a small area (e.g., nonmetallic inclusions) during the ion bombardment of the target (Figure 23). If the local electric field exceeds the dielectric strength of the insulator an electrical breakdown occurs. In the case of reactive sputtering of oxides, an isolating (or less conductive) layer may grow in the transition zone from the racetrack to the non-sputtered region, leading to the tendency of micro-arcing and thus generation of droplets (macroparticles). Native oxide and nonmetallic inclusions on the target surface are not the only cause for arcing. A similar effect occurs when the target material (e.g., powder metallurgy targets) is not fully densified or if microvoids, produced during target manufacturing are present in the target material. Pores and microvoids can trap gases, which are released during target ion etching. Locally high pressure regions of gases cause arcing and thus generation of particles.

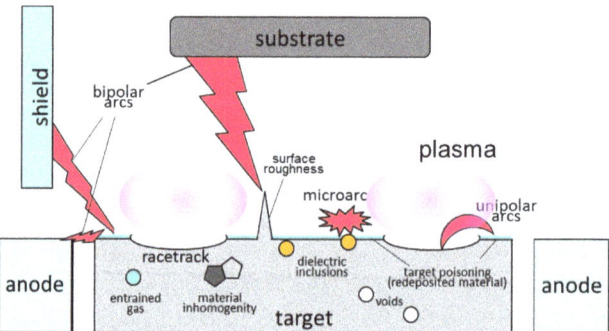

Figure 23. The scheme shows imperfections of metallurgical origin in the target material that may cause three arcing modes (unipolar arc, bipolar arc, and microarc).

There are several ways to reduce arcing on the targets. The first approach is conditioning of the targets after each deposition run. Conditioning means that the target power should be ramped up slowly in order to eliminate any native oxide or other surface contamination. Arcing can also be prevented or suppressed by using advanced power supply units. Novel power supplies are equipped with a sophisticated arc detection system which is able to detect arcs much faster (detection time is less than a microsecond), thus the amount of energy released to an arc is much smaller. The next approach is the use of the pulsed sputtering technique, developed in the mid-1990s. In this case the charge built up by ion bombardment can be compensated by electrons which arrive at the target surface during the positive half cycle of the pulse.

The problem is not only the arcing on the target but also arcing on the substrate table and other components of the vacuum chamber. Thus, the arcing on the substrate table and shields during the ion etching and coating deposition steps additionally stimulate the flaking phenomena. Due to the high energy which arises during arcing at certain points of the substrate table, anode, and other components of the vacuum chamber, locally high thermal stresses appear. The fragments of the growing coating delaminate from a spot somewhere in the deposition chamber and fall on the surface where they can get stuck. The fragment may even originate from a previous deposition of the same coating type. In general, they have the same chemical composition as the undisturbed coating. Therefore, it is practically impossible to distinguish such types of seed particles from the current growing coating in the cross-sectional fracture SEM image. Although in multilayer coatings they can be clearly visible from the shape of layer contours (Figure 24).

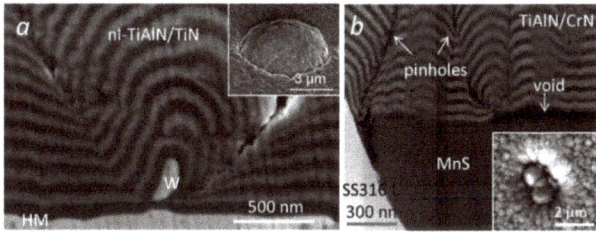

Figure 24. The seed particle in the nodular defect (**a**) or pinholes (**b**) can be clearly identified from the shape of layer contours in sputter-deposited nl-TiAlN/CrN hard coatings.

In hard coatings prepared in a sputter deposition system equipped by hollow cathodes used to assure more intensive plasma (plasma booster technology) during the etching step, we also observed rather large nodular defects with a typical diameter of about 20 µm (Figure 25a). We found that they started to grow on the substrate surface on the seeds composed of copper and tantalum. The origin of both elements is arced between the tantalum tube (nozzle) and the copper anode in the hollow cathode source, which appears occasionally. From the nodule shape we can conclude that the copper component of droplets settled on the substrate in the liquid state, while the tantalum component droplets were in the solid state. Beneath the seed (droplet) a step on the substrate surface is visible. It was formed during the ion etching process, because the droplet shaded the substrate surface. This step confirms that the Cu-Ta fragments arrived on the substrate surface during the early stage of the etching process. On the other hand, there is no step beneath the seed particle if it arrived on the substrate surface at the beginning of the deposition process.

We should also mention the ion beam sputter deposition, which is commonly used in the production of high-quality optical coatings [64]. This technique produces the lowest defect density among all PVD techniques. In this technique, target material is sputtered by an external ion source. The target is not on any electric potential; therefore, it cannot produce arcs.

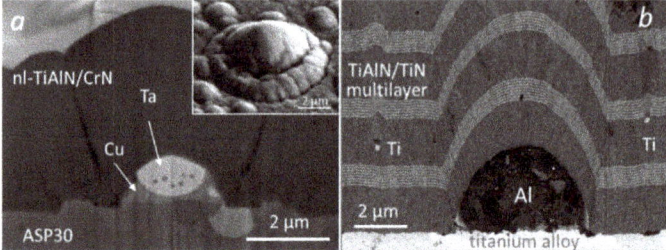

Figure 25. FIB image of a droplet-related defect in sputter-deposited nl-TiAlN/CrN hard coating. The seed composed of cooper and tantalum was formed during intensive (booster) ion etching (**a**). Two small Ti buried droplets and one large Al-based hemisphere with a flattened bottom formed during preparation of TilN/TiN multilayer using the cathode arc deposition technique (**b**). The multilayer TiAlN/TiN hard coating on the titanium alloy substrate was prepared at the company KCS Europe from Germany.

4.2.3. Seeds in Cathodic Arc Evaporation

In cathodic arc evaporation the majority of growth defects originates during the coating deposition [65,66]. The cathodic arc discharge between an anode and a cathode is localized in small spots (typically only a few micrometers in diameter) and presents an intense source of plasma with a current density of about 10^6–10^{12} A/m^2. The spot, which moves over the cathode surface, causes local melting and evaporation of the target material (cathode). The result of the arc evaporation process is not only ions and atoms from the cathode surface but also significantly larger particles (droplets). Formation of such droplets is a result of plasma pressure on the liquid cathode material. The majority of droplets have a diameter less than 1 µm. The droplets preferentially splash under a relatively shallow angle with respect to the cathode surface and since they originate during the coating deposition. they can be found at all coating depths.

On the way away from the cathode the droplets cool down. Small droplets (less than 1 µm in diameter) cool more rapidly than coarse ones. It has been shown that small droplets settle on the substrate in the solid state, while the coarse ones settle in the liquid state. Therefore, smaller droplets have a nearly spherical shape (Figures 15a and 26a,b), while the larger ones are less regular in shape (flattened droplets, Figures 15b and 25a,b). Namely those droplets which reach the substrate surface in the liquid phase, likely change the geometry upon impact. Touching the surface, these types of droplets will be deformed and quenched. Once they arrive on the surface they may stick and be incorporated into the growing film. Those droplets that become solid before impact have a high probability to reflect. While they reach the film continuously, some may be overgrown. The fact that the droplets are found at different distances from the substrate-coating interface reveals that they are generated and incorporated within the coating during the entire coating deposition process. The embedding of droplets causes the formation of nodular defects on the growing film.

The size and amount of the droplets are primarily affected by the cathode material. In general, materials with a high melting point generate less and smaller droplets. The other deposition parameters (e.g., deposition temperature, gas pressure, arc current, and power) also affect the formation of droplets. For example, increasing the reactive gas pressure in the chamber results in the generation of fewer droplets, which is due to the formation of compounds on the cathode surface. The compounds usually exhibit a higher melting point than the original target. The formation of intermetallic phases and high-melting point thin ceramic layers on top of compound cathodes may strongly influence the behavior and movement of the cathode spot and therefore the droplet generation. For instance, an increased cathode temperature leads to a higher number of generated droplets due to the larger area of the molten target material. The cathode spot movement can be used to decrease the average cathode temperature based on the reduced average arc spot rest time, equivalent to a decreased local thermal impact.

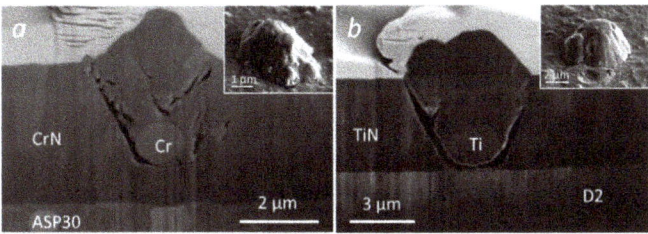

Figure 26. Typical droplet-related defects in CrN (**a**) and TiN (**b**) hard coatings prepared by cathodic arc evaporation.

The detrimental effects of droplets include local loss of coating adhesion, surface roughening, grain coarsening, nonuniform phase, and composition within the droplet [67]. The presence of droplets also leads to the creation of pores because some of the droplets are wrenched out due to high compressive stresses.

There has been a lot of research on how to minimize droplet formation or droplet incorporation into the growing film. Numerous approaches have been proposed. The simplest means to reduce the number of droplets in the vacuum arc coating synthesis are as follows [68,69]:

- increasing the arc speed on the cathode surface by using a magnetic field; in this way, the arcs are moving faster on the cathode surface, therefore they melt a smaller volume of material;
- reduction of the temperature of the cathode surface by intensive cooling;
- reduction of the arc current in order to reduce the density of ion flow;
- low-angle shielding of cathode; the majority of the droplets are emitted at angles lower than 30° with respect to the target surface;
- droplet filtering involves guiding the plasma towards the substrate using an electromagnetic field (0.01–0.1 T); in contrast to electrons and ions, the droplets are not charged and therefore will not follow the non-linear path to the substrate;
- the use of higher partial pressure of the reactive gas during deposition due to the formation of compound layers with a high melting point;
- the number of droplets can be reduced with increasing bias voltage; the latter may be attributed to the effect of the enhanced ion (re)sputtering and deflection of the negatively charged droplets.

5. The Influence of Growth Defects on Functional Properties of Thin Films

In this chapter we provide an overview on the research related to the influence of the growth defects on the functional properties of thin films and coatings. We start with the role of growth defect on the optical properties and in the semiconductor devices, which were historically studied first. Then we discuss the influence of defects on wear and friction of coatings, corrosion and oxidation resistance, surface wettability, and permeability of gas barrier coatings.

5.1. Optical Properties

Thin films for optical applications have been produced for many years and today they are an integral part of the majority of modern optical systems such as lasers, displays, lighting, mirrors, anti-reflection coatings, beam splitters and filters, decorative coating, security (antiforgery) devices, and others. Optical coatings have been traditionally deposited by evaporation (from either a crucible or an e-beam source) and sputtering, frequently assisted by ion bombardment. Optical coatings often present a critical part for the entire optical system.

The limiting aspect of optical coatings is especially their failure to perform high-power laser illumination. Several studies demonstrated that the low damage threshold of high-power laser optical components is associated with nodular defects in the optical coating. Due to the lens-like shape the nodular defects focus the light within the defect and thus light intensification is significantly greater than that in the defect-free regions. Therefore, it leads to local overheating and consequently to coating damage (crater-like pits). A few localized defects in optical interference mirror coatings can substantially increase the scattering loss to tens or hundreds of ppm [27].

Bercegol [70] showed that laser conditioning of some optical components (dielectric multilayer coatings) for high-power lasers can improve the functional laser damage resistance. This method is based on an under-threshold pre-radiation of the coating by laser, which causes a gentle ejection of nodular defects. Smooth-edged pits are left behind.

Another promising approach to reduce the density of nodular defects was proposed by Mirkarimi et al. [71,72]. Their proposal is based on integration of the thin film deposition process and direct etching of the film/substrate at normal incidence. The process consists of 50 nm silica deposition followed by ion beam etching of one half of the deposited layer. When etching a nodular defect, the sides of the defect etch faster than the top because the etching rate at normal incidence is much smaller than at high incident angles (~50°). The enhanced etching at the sides can cause the nodular defect to shrink until the defect gradually disappears (Figure 27). The geometric minimization of the coating defect significantly improves the laser resistance of the optical coating. Unfortunately, this process is not effective for the planarization of pits and scratches.

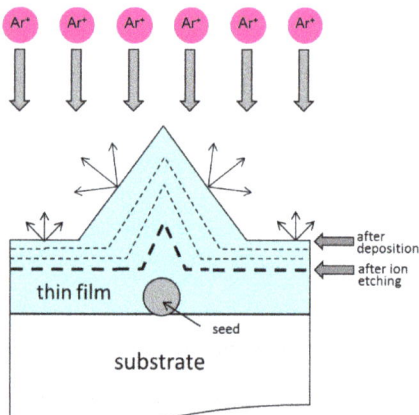

Figure 27. Illustration of the ion etching process applied for planarization of the coating surface.

5.2. Growth Defects in Semiconductor Devices

In semiconductor thin film device manufacturing (e.g., integrated circuits, flat panel displays, magnetic and optical storage media, photovoltaic devices, and other thin film devices), particle contamination and consequently growth defects are a serious problem because they cause a decrease in production yield and reliability problems during product use [29]. On magnetic storage discs, particle contamination can result in read-write errors, bad sectors, and total disc failure. In the manufacturing of complementary metal-oxide-semiconductor (CMOS) integrated circuits, particle contamination can cause pinhole formation, delamination, and interconnection shorts or opens in the metallization processes. Therefore, it is important to understand the formation of localized defects in thin films which compose the semiconductor device. In principle every process step in production of such devices (film deposition, lithography, etching) could be a source of contamination. The control of particle contamination is especially of great importance in advanced microelectronic technologies such as extreme ultraviolet lithography where defect-free masks are necessary [72]. To control these deleterious effects during semiconductor fabrication, great care is required to reduce surface contamination levels during the handling and manufacture of devices [73]. Elimination of particles that cause the so called "killer defects" during thin film processing is therefore one of the biggest challenges for semiconductor device producers.

5.3. Friction and Wear

In tribological applications the roughness of a coated component has an important influence on the friction, wear, and tendency for a material to be transferred [74]. Protruding nodular defects result in high abrasive wear of the counter material and due to the micro-ploughing and material pick-up effects, the friction coefficient increases. As long as protruding nodular defects are present on the coating surface the friction coefficient does not reach the steady state. The rougher the sample, the longer the time required to obtain the steady-state value of the friction. The period until a conformal sliding contact is formed is called the running-in phase. During the operation of tools protected by PVD hard coatings the most common reasons for their failure are coating fracture, coating delamination, and subsurface fracture (substrate failure) [75]. Such failures can be caused by cracks that initiate at different coating defects. The first contact between the tool and the workpiece material, which move relative to each other, always takes place at the highest peaks of the surfaces [76]. In addition to substrate surface asperities, such contact spots are various coating imperfections (e.g., droplets, nodular defects). Due to the small actual contact area (about 10 % of the surface area) the contact pressure at these spots is very high [77]. High pressure and shear stresses at the nodular defects cause the formation of cracks and therefore they collapse into small fragments of hard coating material. The second mechanism of wear particle generation and friction increase is caused by interlocking and breaking of all types of asperities. It all means that the nodular defects are the primary source of abrasive particles in the sliding contact.

Based on this description we can understand why wear is so intensive during the early stage, rather than later on. During this process the protruding surface asperities in the PVD hard coating and wear particles abrade the surface of the softer counter-body material. Such a plowing effect causes a high wear rate of the counter surface. During further sliding the asperities are gradually removed and both surfaces are fitted together. Consequently, the contact area increases. As soon as a smooth surface is formed the contact pressure is reduced and consequently the risk of fatigue damage is reduced. During this period the transition from the mechanical wear-dominated friction to the adhesion-dominated friction takes place [77]. Additionally, any topographical imperfections on the coated forming tool surface also present a potential initial point for the onset of galling and transfer of the workpiece material [78–80].

Only a few papers can be found in the literature that paid attention to how growth defects influence the tribological properties of PVD hard coatings. Poulingue et al. [24] designed an experiment to analyze the damage initiating from nodules through a purely mechanical approach. The artificial nodules were generated by dispersing diamond seed particles on the polished aluminum substrate before deposition. Mechanical damage was progressively induced by pulling the samples in tension in an SEM. In-situ observation showed that cracks first arise at larger nodular defects. Fallqvist et al. [40] showed that in a sliding contact, growth defects have a strong impact on the tribological behavior of the coating causing abrasive wear of the less hard counter material surface and material transfer to the coating. Both mechanisms affect friction characteristics. In order to reduce the amount of surface irregularities introduced during the coating process, they recommended post-coating polishing. In such a way it is possible to reduce the material-transfer tendencies and to stabilize the coefficient of friction in the sliding contact. Luo [39] studied the running-in period in magnetron-sputtered TiAlN/VN multilayered coatings against an alumina counter body in a large temperature range up to 700 °C. He found that during the running-in period the growth defects collapse, while the fragments are released into the wear track and form wear debris. The effect of arc-evaporated droplets on the wear behavior was investigated by Tkadletz et al. [41]. In their work special emphasis was given on the role of droplets in the performed ball-on-disk tests, where possible mechanisms triggering coating degradation were determined. Recently, we studied the influence of the surrounding atmosphere [81] and nodular defects on tribological behavior of sputter-deposited TiAlN hard coating using a new method. The novelty of our approach is based on cycle-to-cycle experiments [42,43]. We pinpointed selected defects on the coating surface and then followed them through the tribological test with a scanning electron microscope (SEM) and a focused ion beam (FIB) microscope. This approach gave us insight into the processes occurring after a certain number of sliding passes (Figure 28). The tribological tests we performed highlighted the importance not only of the nodular defects, but also of the protrusions that are located at carbide sites in different tool steels [43].

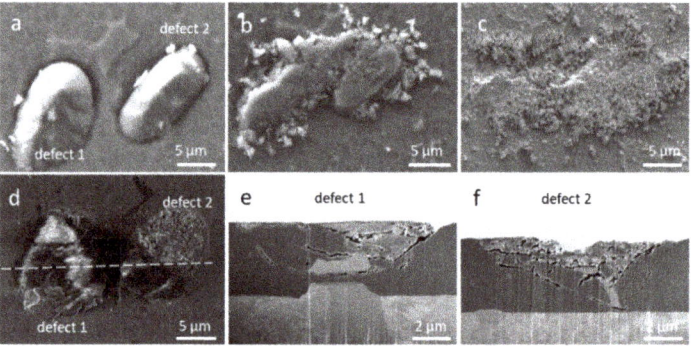

Figure 28. SEM images of the same nodular defects in the sputter-deposited TiAlN hard coating: as-deposited (**a**) and in the wear track after 1 cycle (**b**), 4 cycles (**c**), and 128 cycles (**d**) using an alumina ball (2 N, 2 Hz). FIB cross-section images (in direction marked with the dashed line) of both defects after 128 cycles (**e**,**f**).

5.4. Erosion Resistance

Limited investigations have been published on the influence of growth defects on material removal (erosion) when hard particles impact the hard coating surface [82,83]. It was found that the material removal occurred by repeated ductile indentation and cutting of the surface by impacting particles. Such solid particle erosion wear is the characteristic degradation of components in aircraft engines (operating in harsh environments), wind turbines, and power generation gas turbines. In order to enhance the higher reliability and longer lifetime of such components, many different hard protective coatings have been developed. The erosion resistance of monolithic coatings can be enhanced by using materials with both high hardness to inhibit crack initiation and high toughness to prevent crack growth. In contrast to monolithic hard coatings, various nanostructure coatings (e.g., TiN/TiAlN) are more appropriate for erosion wear protection because of their very high hardness and ability to inhibit crack propagation. Growth defects and other irregularities (scratches, pits) in the hard coating facilitate the crack initiation, which leads to premature breakup of the coating.

Wang et al. [82] performed a low-angle slurry erosive test of CrN/NbN superlattice coatings and observed the selective wear at defects. Similar wear mechanisms, like in the case of erosion with hard particles ejected from the nozzle at high pressure of air or liquid, were observed for components exposed to cavitation in different liquid media [84]. Cavitation implies the build-up and subsequent implosion of bubbles. Cavitation bubbles collapse violently either in the form of micro-jets or shock waves of high velocities, pressures, and temperatures. Due to cyclic impact of imploding cavitation bubbles on the solid surface, cavitation damage appears. Surface topography significantly influences the cavitation behavior because the implosion of bubbles is promoted especially at surface irregularities. During the cavitation erosion test, some nodular defects or droplets are removed from the surface of the coating, leaving cavities behind. Such cavities can act as crack initiation sites. Azar et al. [83] studied the cavitation erosion of TiN coating, produced by arc-PVD and found that droplet-related defects have an important influence on the cavitation erosion resistance of the coating. During the cavitation erosion test, deep cavities were formed by the detachment of conical droplets. Such localized coating damages depend on the shape, position, and depth of the droplets in the coating.

5.5. Corrosion Resistance

Not only the tribological properties, but also the corrosion properties are affected by growth defects. It is well known that any kind of porosity causes a pitting corrosion. Porosity can be either a macroporosity or a microporosity. The former arises from large growth defects such as detached droplets, nodular and flake defects. The latter is determined by the growth morphology itself (e.g., open columnar structures arising from insufficient adatom mobility).

Small microstructural defects (e.g., pinholes, pores, and cracks) formed during or after deposition of PVD hard coatings act as channels for the corrosion of the substrate [30–34,85–87]. Therefore, such coating imperfections limit its protective nature in corrosive media. When the coated substrate is immersed in a corrosive medium, the electrolyte penetrates to the substrate (driven by capillary forces) through any pinholes extending down to the substrate. This leads to the formation of local galvanic corrosion between the substrate (acting as the anode) and the coating (acting as the cathode). Thus, anodic dissolution of the exposed substrate area occurs. The corrosion attack is more intensive in the case of a less noble substrate. Transition metal nitride coatings, which are the common choice for tool wear protection, are chemically more electronegative than steel. The corrosion medium in the pores is fast due to the large ratio of the cathode to anode areas. Pits form and extend radially from the pores, resulting in the cracking and removal of the upper coating by flaking. SEM and FIB images (Figure 29) show two examples of typical pitting corrosion for TiAlN hard coating deposited on a D2 tool steel substrate. A corrosion test was performed in a chlorine solution using electrochemical impedance spectroscopy. Both samples were exposed to corrosion medium (0.5 M NaCl; pH = 3.8) for 96 h.

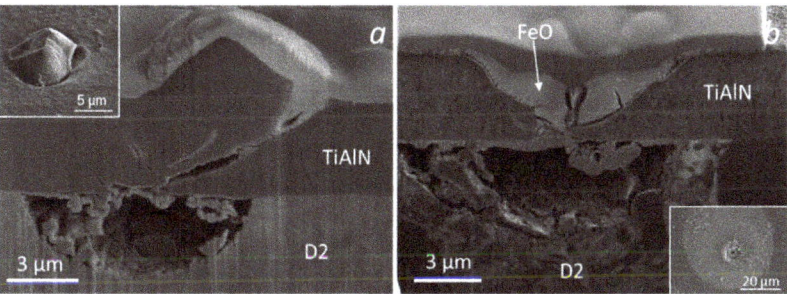

Figure 29. SEM top view (inset) and FIB images of a nodular defect (**a**) and a pinhole (**b**) in the sputter-deposited TiAlN hard coating exposed to corrosion medium (0.5 M NaCl; pH = 3.8, 96 h). An intensive pitting corrosion occurred at both sites.

Corrosion resistance of coated specimens can be improved if one could eliminate the growth defects in the coating. Although various techniques can be used to minimize the number of pinholes, they cannot be totally eliminated. Several approaches proposed to improve the corrosion resistance of PVD hard coatings were published in our previous paper [46]. One of the most promising ways is to combine the PVD coating and the thin atomic layer deposition (ALD) layer. ALD allows a deposition of conformal coating on non-even surfaces and coverage of particles, pinholes, and defects in the PVD coating.

5.6. Oxidation Resistance

High temperature oxidation resistance is one of the most important properties of PVD hard coatings for protection of cutting tools that are used for high speed, dry and hard machining. During such machining conditions the temperature at the cutting tool edge may reach more than 800 °C due to the high friction between the tool and the workpiece material. Thus, the protective coating suitable for advanced applications of cutting tools must endure extremely high thermo-mechanical loads and resist degradation in severe environments. Similar to corrosion, different coating imperfections have an important role during oxidation of PVD hard coatings. If the coated substrate is exposed to high-temperature oxidation, then the growth defects act as starting points for degradation and environmental attack (Figure 30). All the pores, voids, and gaps act as preferential diffusion paths for the oxygen transport to the inner coating/substrate interface and for metal ion transport from the substrate towards the surface.

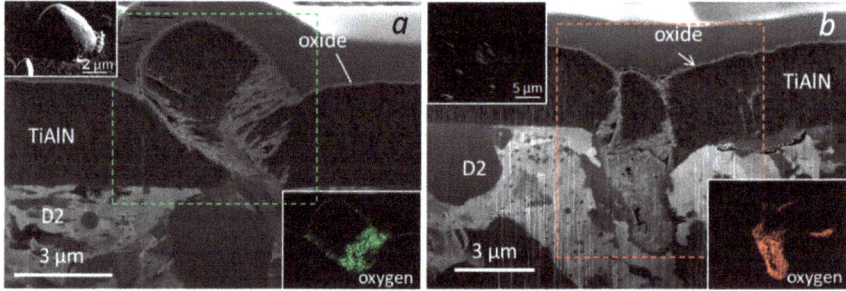

Figure 30. FIB and SEM top-view images of a nodular defect (**a**) and a pinhole (**b**) in the sputter-deposited TiAlN hard coating oxidized at 800 °C for 1 h. Oxygen elemental maps of area marked with the dashed frames are added in the inset.

There are only a few papers in the literature dealing with the role of growth defects during the oxidation process of hard coatings. The influence of defects on oxidation resistance of TiAlN is briefly mentioned in an article by McIntyre et al. [88]. The role of defects in the oxidation process is discussed in more detail by Lembe et al. [89]. They studied the influence of growth defects on the localized oxidation behavior of TiAlN/CrN superlattice coatings deposited by cathodic arc/unbalanced magnetron deposition on cemented carbide and HSS substrates. They found that oxidation behavior of the substrate material directly influences the generation of growth defects through craters caused by droplet formation. Some detached growth defects formed craters, through which oxidation products formed from the substrate material. Localized oxidation can also take place at the pores formed in the underdense region at the rim of the nodular defects. Two major kinds of oxides can emerge at the defects: oxides rich in Ti or oxides formed by substrate material. The formation of either one or the other probably depends on the depth of the defect and time of the heat treatment. Polcar and Cavalerio [90] investigated the thermal stability, oxidation resistance, and high temperature tribology of CrAlTiN coating deposited by cathodic arc evaporation on cemented carbide substrates. They found that the presence of surface defects caused oxidation of the cemented carbide substrate, which was particularly evident after the tribological tests at 700 and 800 °C. Hovsepian et al. [91] studied the influence of growth defects in deposited CrN/NbN nanostructured coatings on the high temperature corrosion resistance in a pure steam atmosphere. They found that the coating degradation mechanism when exposed to 650 °C in a pure steam environment is adverse diffusion of the substrate elements and oxygen through coating growth defects or cracks formed due to thermal expansion coefficient mismatch between the coating and the substrate. Fernades et al. [92] also demonstrated that the oxidation of TiSiVN coatings is controlled by the formation of a silicon oxide diffusion barrier which is affected by nodular defects in the as-deposited film. At nodular defects a complex oxide structure was developed. Recently our research group published the results of our investigation on high temperature oxidation of nanolayered CrVN-based coatings [93]. We found that V_2O_5 patches started to appear around the nodular defects during high temperature oxidation.

Several approaches have been proposed to increase the oxidation resistance of PVD hard coatings. All approaches are based on how to prepare the coatings with a denser microstructure. There are various ways to achieve this. More dense coatings can be prepared, for example, by the HIPIMS deposition technique. On the other hand, the oxidation resistance of PVD hard coating can be improved by doping with different elements that cause the formation of a fine grain microstructure and segregation of the doping element at grain boundaries. More grain boundaries prolong the diffusion paths of oxygen and metal atoms. In addition, nanolayer coatings are more resistant to oxidation than single-layer ones. In nanolayer coatings many interlayer interfaces act as a diffusion barrier for inward diffusion of O and the outward diffusion of metal atoms.

5.7. Gas Permeation

Gas barrier coatings were first commercially applied on polymeric foil substrates for food packaging (since the early 1970s) [38,94]. Later their use was expanded to the pharmaceutical and beverage industry. An effective barrier layer can prevent losses (e.g., aroma) from the packaged product and prevent penetration into the package (oxygen, water vapor), both of which can affect product quality and expiration date. Aluminum metallized polymeric (mainly polyester) foil substrates are widely used for this purpose. In order to fulfill additional requirements, such as product visibility and microwaveability, transparent barrier coatings based on aluminum oxide or silicon oxide have been introduced.

Thermal and electron beam evaporation are the most frequently used deposition techniques to manufacture barrier films for food packaging. These evaporation techniques allow the preparation of films with a medium barrier performance, however, at a very high productivity. On the other side, reactive sputtering and plasma-enhanced chemical vapor deposition (PECVD) assure a significantly lower water vapor permeability, but at a lower productivity.

However, the permeability of coated foil substrates is not zero. The residual permeation and the effectiveness of PVD layers as gas diffusion barriers are attributed to the presence of microscopic defects in the coating. The gas transport through the barrier layer takes place mostly at pinholes. Therefore, the density and distribution of defects in the barrier film is a critical aspect for using barrier films.

Recently the application of thin film barrier coatings has expanded to flexible electronics (e.g., organic transistors, displays, thin film solar cells, organic light emitting diodes (OLEDs)) [95–98]. Flexible electronic devices are very sensitive to a reaction with water vapor and oxygen, and therefore require an encapsulation to protect against degradation. In order to reduce the rate of permeation of gases and vapors through polymer substrates, several barrier options have been proposed and utilized. One has to be aware that not only the barrier layers need to have a low defect density, but also the substrate particles and defects need to be covered and planarized.

Commonly two approaches fulfill the specification of water vapor and oxygen permeation. One approach is the optimization of single layers by using the atomic layer deposition technique which offers the deposition of perfect, high density, uniform, and conformal barrier layers on non-even surfaces and coverage of particles and defects on the substrate surface. The other approach is the deposition of multilayer stacks [99]. In a multilayer stack two or more inorganic barrier layers (e.g., SiO_2, Si_3N_4, Al_2O_3) are combined with a polymer interlayer. The barrier layers provide a low water and oxygen permeation, while the interlayer planarizes and decouples pinhole defects as well as increases the diffusion path length of the permeating gas through the barrier layer. Magnetron processes are exclusively used to deposit the complete multilayer stack.

5.8. Wettability of Surfaces

Wettability of a liquid on a coating surface depends on the surface chemistry, surface topography (asperities, nodular defects) as well as the liquid used. Applications where solid surface wettability plays a crucial role include contact lenses, body implants, biofilm growth, super-hydrophilic surfaces, self-cleaning, and nonstick surfaces. Wettability can be estimated by measuring the contact angle of the substrate with a given liquid. The balance at the three-phase contact of solid, liquid, and vapor is given by the well-known Young equation. The Young equation assumes that the surface is chemically homogeneous and topographically smooth. However, this is not true in real surfaces like PVD coatings. Instead of one equilibrium contact angle, a range of contact angles exist, because the actual contact angle is the angle between the tangent to the liquid–vapor interface and the actual local solid surface. Therefore, in principle, the coating roughness enhances the wettability.

We investigated how the nodular defects present in the coated steel substrates influence the wettability of liquids. The static contact angle was measured by observation of a drop of a test liquid (deionized water, diiodomethane) on coated samples. It is valid that the lower the contact angle, the greater the tendency for the liquid to wet the solid surface energy. We measured the contact angles of the as-deposited nl-TiAlN/TiN hard coating and after gently polishing with diamond paste. How this changed the topography of the surface is shown the 3D-profilometer images in Figure 31. While the highest peaks (nodular defects) decreased, the lowest ones completely disappeared. The wettability measurements showed that the contact angles decreased significantly (for about 20%) if deionized water and CH_2I_2 (diiodomethane) were used, while the corresponding surface free energy increased more than 10%.

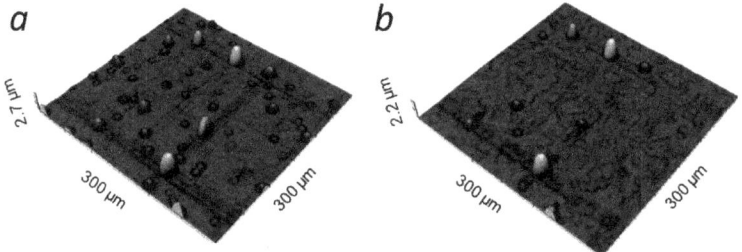

Figure 31. Three-dimensional (3D)-profilometer image of the same coating surface area before (**a**) and after polishing (**b**) with diamond paste. The sharp peaks are the nodular defects (pay attention to the strong exaggeration in z-scale).

6. Summary

We identified the origins of the growth defects found in PVD coatings prepared by three different deposition techniques. All topographic irregularities and foreign particles generated during different steps of substrate pretreatment and during the coating deposition process were systematically followed because they are seeds for the formation of growth defects. We need to be aware that contamination of substrates with foreign particles strongly depends on the quality of their wet cleaning procedure. Therefore, proper wet cleaning procedures, as well as an appropriate transport and batching, must be used in order to ensure a substrate surface that is as clean as possible.

Due to the inhomogeneity of the substrate material, certain topographic changes occur at sites of various inclusions already during polishing, as a result of the difference in their hardness in comparison with the matrix. A similar effect arises during the ion etching because the sputtering rates of various phases are also different. Whether a hole or protrusion will appear at the site of the inclusion depends on the polishing (removal) rate and etching rate. Special attention was given to the various ion etching techniques used in industrial hard coating deposition systems (DC etching, MF etching, booster etching). We were particularly interested in how they affect the substrate topography.

It was found that a large part of the seed particles responsible for growth defect formation are generated during the coating process itself. The origin of such seed particles which can take many forms are described in more detail. The emphasis is on the seed particles that are characteristic for three different evaporation techniques used in our lab: thermionic arc evaporation, magnetron sputtering, and cathodic arc evaporation. In the PVD coatings prepared by the cathodic arc deposition process, the growth defects originate mostly from droplets formed during the coating deposition. Droplet-like defects are significantly smaller than others, but their density is two orders of magnitude higher. Therefore, the roughness of such coatings is relatively high.

The primary origin of growth defects in magnetron sputtering is the seed particles formed on the target surface outside of the racetrack due to the re-deposition and poisoning effect. At the early stage of deposition all these particles are charged and therefore accelerated away from the target due to the repulsion effect. Most of them reach the substrate surface and they become the seeds for growth of nodular defects in the coating.

The lowest concentration of growth defects was found in the coatings prepared by the thermionic arc evaporation. In contrast to both deposition techniques mentioned above, in this case the evaporation source is not the origin of larger quantities of seed particles.

A common form of growth defects for all three types of coatings is also flake defects. They originate from flakes, which are formed from a thick deposit on the shields and other components of the vacuum system that break off due to the high compressive stresses. Flaking of the coating is also triggered by arcing on the substrate turntable during ion etching and deposition process.

An important part of nodular defects is formed from wear particles that are generated by all the moving parts inside the vacuum chamber. Therefore, less moving parts in the fixture systems mean less defects.

The consequences of growth defect formation on functional properties of thin film are discussed in more detail. It is explained how growth defects affect the quality of optical coatings and thin layers for the production of semiconductor devices together with their role in tribological contacts as well as in corrosion and oxidation processes. The effect on the permeation and wettability of the coatings is also briefly described. The findings of other researchers are supported with the relevant results obtained in our lab.

Author Contributions: P.P., conceptualization, methodology manuscript writing; A.D., FIB analysis of growth defects and tribological tests; P.G., SEM and FIB analysis of non-metallic inclusions; M.Č., manuscript review and project administration; M.P., critical review and participation in the writing of the first two chapters. All authors have read and agreed to the published version of the manuscript.

Funding: This work was supported by the Slovenian Research Agency (program P2 0082, projects L2-4173, L2-4239, L2-4249, L2-5470). We also acknowledge funding from the European Regional Development Funds (CENN Nanocenter, OP13.1.1.2.02.006) and the European Union Seventh Framework Programme under Grant Agreement 312483—ESTEEM2 (Integrated Infrastructure Initiative—I3).

Acknowledgments: The authors would also like to thank Tonica Bončina and Gregor Kapun for SEM and EDS analyses and Jožko Fišer for performing some laboratory tests.

Conflicts of Interest: The authors declare no conflicts of interest. The funders had no role in the design of the study; in the collection, analyses, or interpretation of data; in the writing of the manuscript, or in the decision to publish the results.

References

1. Holman, W.R.; Huegel, F.J. Interrelationships between process parameters, structure, and properties of CVD tungsten and tungsten–rhenium alloys. *J. Vac. Sci. Technol.* **1974**, *11*, 701–708. [CrossRef]
2. Dubost, L.; Rhallabi, A.; Perrin, J.; Schmitt, J. Growth of nodular defects during film deposition. *J. Appl. Phys.* **1995**, *78*, 3784–3791. [CrossRef]

3. Tait, R.N.; Smy, T.; Dew, S.K.; Brett, M.J. Nodular defects growth and structure in vapor deposited Films. *J. Electron. Mater.* **1995**, *24*, 935–940. [CrossRef]
4. Gabe, D.R. Dendritic Growth During Electrodeposition. *Metall. Mater. Technol.* **1973**, *5*, 72–77.
5. Letts, S.A.; Myers, D.W.; Witt, L.A. Ultrasmooth plasma polymerized coatings for laser fusion targets. *J. Vac. Sci. Technol.* **1981**, *19*, 739–742. [CrossRef]
6. Movchan, B.A.; Demchishin, A.V. Investigation of the structure and properties of thick vacuum condensates of nickel, titanium, tungsten and aluminum oxide. *Fiz. Met. Metalloved.* **1969**, *28*, 653–660.
7. Mattox, D.M.; Kominiak, G.J. Structure modification by ion bombardment during deposition. *J. Vac. Sci. Technol.* **1972**, *9*, 528–531. [CrossRef]
8. Patten, J.W. The influence of surface topography and angle of adatom incidence on growth structure in sputtered chromium. *Thin Solid Films* **1979**, *63*, 121–129. [CrossRef]
9. Spalvins, T. Characterization of defect growth structrures in ion-plated films by scanning electron microscopy. *Thin Solid Films* **1979**, *64*, 143–148. [CrossRef]
10. Spalvins, T.; Brainard, W.A. Nodular growth in thick-sputtered metallic coatings. *J. Vac. Sci. Technol.* **1974**, *11*, 1186–1192. [CrossRef]
11. Guenther, K.H. Nodular defects in dielectric multilayers and thick single layers. *Appl. Opt.* **1981**, *20*, 1034–1038. [CrossRef] [PubMed]
12. Tench, R.J.; Chow, R.; Kozlowski, M.R. Characterization of defect geometries in multilayer optical coatings. *J. Vac. Sci. Technol. A* **1994**, *12*, 2808–2813. [CrossRef]
13. Smith, D.J. Modeling of nodular defects in thin films for various deposition techniques. *Proc. SPIE* **1987**, *821*, 120–128.
14. Dirks, A.G.; Leamy, H.J. Columnar microstructure in vapor-deposited thin films. *Thin Solid Films* **1977**, *47*, 219–233. [CrossRef]
15. Liao, B.; Macleod, H.A. Thin film microstructure modelling. *Proc. SPIE* **1985**, *540*, 150–155.
16. Müller-Pfeiffer, S.; Anklam, H.-J. Computer simulation of hillock growth. *Vacuum* **1991**, *42*, 113–116. [CrossRef]
17. Stearns, D.G.; Mirkarimi, P.B.; Spiller, E. Localized defects in multilayer coating. *Thin Solid Films* **2004**, *446*, 37–49. [CrossRef]
18. Kozlowski, M.R.; Chow, R. The role of defects in laser damage of multilayer coatings. *Proc. SPIE* **1994**, *2114*, 640–649.
19. Stolz, C.J.; Sheehan, L.M.; Gunten, M.K.; Bevis, R.P.; Smith, D.J. The advantages of evaporation of Hafnium in a reactive environment to manufacture high damage threshold multilayer coatings by electron-beam deposition. *Proc. SPIE* **1999**, *3738*, 318–324.
20. Cheng, X.B.; Shen, Z.X.; Jiao, H.F.; Zhang, J.L.; Ma, B.; Ding, T.; Lu, J.; Wang, X.; Wang, Z. Laser damage study of nodules in electron-beam evaporated HfO_2/SiO_2 high reflectors. *Appl. Opt.* **2011**, *50*, C357–C363. [CrossRef]
21. Liu, X.F.; Li, D.W.; Zhao, Y.A.; Li, X. Characteristics of nodular defect in HfO_2/SiO_2 multilayer optical coatings. *Appl. Opt.* **2010**, *49*, 1774–1779. [CrossRef] [PubMed]
22. Panjan, P.; Merl, D.K.; Zupanič, F.; Čekada, M.; Panjan, M. SEM study of defects in PVD hard coatings using focused ion beam milling. *Surf. Coat. Technol.* **2008**, *202*, 2302–2305. [CrossRef]
23. Brett, M.J.; Tait, R.N.; Dew, S.K.; Kamasz, S.; Labun, A.H.; Smy, T. Nodular defect growth in thin films. *J. Mater. Sci. Mater. Electron.* **1992**, *3*, 64–70. [CrossRef]
24. Poulingue, M.; Dijon, J.; Ignat, M.; Leplan, H.; Pinot, B. New approach for the critical size of the nodular defects: The mechanical connection. *Proc. SPIE* **1998**, *3578*, 370–381.
25. Wei, C.Y.; Yi, K.; Fan, Z.X.; Shao, J.D. Influence of composition and seed dimension on the structure and laser damage of nodular defects in HfO_2/SiO_2 high reflectors. *Appl. Opt.* **2012**, *51*, 6781–6788. [CrossRef]
26. Mirkarimi, P.B.; Stearns, D. Investigating the growth of localized defects in thin films using gold nanospheres. *Appl. Phys. Lett.* **2000**, *77*, 2243–2245. [CrossRef]
27. Cheng, X.; Wang, Z. Defect-related properties of optical coatings. *Adv. Opt. Technol.* **2014**, *3*, 65–90. [CrossRef]
28. Selwyn, G.S.; McKillop, J.S.; Haller, K.L.; Wu, J.J. In Situ Plasma Contamination Measurements by HeNe Laser Light Scattering: A Case Study. *J. Vac. Sci. Technol.* **1990**, *8*, 1726–1731. [CrossRef]

29. Moriya, T.; Nagaike, H.; Denpoh, K.; Kawaguchi, S.; Shimada, M.; Okuyama, K. Observation and evaluation of flaked particle behavior in magnetically enhanced reactive ion etching equipment using a dipole ring magnet. *J. Vac. Sci. Technol.* **2004**, *B22*, 1688–1693. [CrossRef]
30. Korhonen, A.S. Corrosion of thin hard PVD coatings. *Vacuum* **1994**, *45*, 1031–1034. [CrossRef]
31. Fenker, M.; Balzer, M.; Kappl, H. Corrosion protection with hard coatings on steel: Past approaches and current research efforts. *Surf. Coat. Technol.* **2014**, *257*, 182–205. [CrossRef]
32. Balzer, M. Identification of the growth defects responsible for pitting corrosion on sputter-coated steel samples by Large Area High Resolution mapping. *Thin Solid Films* **2015**, *581*, 99–106. [CrossRef]
33. Balzer, M.; Fenker, M.; Kappa, H.; Müller, T.; Heyn, A.; Heiss, A.; Richter, A. Corrosion protection of steel substrates by magnetron sputtered TiMgN hard coatings: Structure, mechanical properties and growth defect related salt spray test results. *Surf. Coat. Technol.* **2018**, *349*, 82–92. [CrossRef]
34. Wang, H.W.; Stack, M.M.; Lyon, S.B.; Hovsepian, P.; Münz, W.-D. The corrosion behaviour of macroparticle defects in arc bond-sputtered CrN/NbN superlattice coatings. *Surf. Coat. Technol.* **2000**, *126*, 279–287. [CrossRef]
35. Jehn, H.A. Improvement of the corrosion resistance of PVD hard coating–substrate systems. *Surf. Coat. Technol.* **2000**, *125*, 212–217. [CrossRef]
36. Ahn, S.H.; Lee, J.H.; Kim, J.G.; Han, J.G. Localized corrosion mechanisms of the multilayered coatings related to growth defects. *Surf. Coat. Technol.* **2004**, *177–178*, 638–644. [CrossRef]
37. Braak, R.; May, U.; Onuseit, L.; Repphun, G.; Guenther, M.; Schmid, C.; Durst, K. Accelerated thermal degradation of DLC-coatings via growth defects. *Surf. Coat. Technol.* **2018**, *349*, 272–278. [CrossRef]
38. Chatham, H. Oxygen diffusion barrier properties of transparent oxide coatings on polymeric substrates. *Surf. Coat. Technol.* **1996**, *78*, 1–9. [CrossRef]
39. Luo, Q. Origin of friction in running-in sliding wear of nitride coatings. *Tribol. Lett.* **2010**, *37*, 529–539. [CrossRef]
40. Fallquist, M.; Olsson, M. The influence of surface defects on the mechanical and tribological properties of VN-based arc-evaporated coatings. *Wear* **2013**, *297*, 1111–1119. [CrossRef]
41. Tkadletz, M.; Mitterer, C.; Sartory, B.; Letofsky-Papst, I.; Czettl, C.; Michotte, C. The effect of droplets in arc evaporated TiAlTaN hard coatings on the wear behavior. *Surf. Coat. Technol.* **2014**, *257*, 95–101. [CrossRef]
42. Drnovšek, A.; Panjan, P.; Panjan, M.; Čekada, M. The influence of growth defects in sputter-deposited TiAlN hard coatings on their tribological behavior. *Surf. Coat. Technol.* **2016**, *288*, 171–178. [CrossRef]
43. Panjan, P.; Drnovšek, A.; Kovač, J. Tribological aspects related to the morphology of PVD hard coatings. *Surf. Coat. Technol.* **2018**, *343*, 138–147. [CrossRef]
44. *Grinding of Uddeholm Tool Steels*, 8th ed.; 2018. Available online: https://www.uddeholm.com/app/uploads/sites/44/2018/11/Tech-Uddeholm-Grinding-EN.pdf (accessed on 28 March 2020).
45. *Polishing of Uddeholm Mould Steel*, 6th ed.; 2016. Available online: https://www.uddeholm.com/app/uploads/sites/45/2018/02/Uddeholm_polishing-eng_t_1609_e6.pdf (accessed on 28 March 2020).
46. Panjan, P.; Drnovšek, A.; Gselman, P.; Čekada, M.; Panjan, M.; Boncina, T.; Merl, D.K. Influence of Growth Defects on the Corrosion Resistance of Sputter-Deposited TiAlN Hard Coatings. *Coatings* **2019**, *9*, 511. [CrossRef]
47. Mattox, D.M. *Handbook of Physical Vapor Deposition (PVD) Processing*, Elsevier: Amsterdam, The Netherlands, 2010.
48. Hovsepian, P.E.; Ehiasarian, A.P. Six strategies to produce application tailored nanoscale multilayer structured PVD coatings by conventional and High Power Impulse Magnetron Sputtering (HIPIMS). *Thin Solid Films* **2019**, *688*, 137409. [CrossRef]
49. Nordin, M.; Ericson, F. Growth characteristics of multilayered physical vapour deposited TiN/TaN on high speed steel substrate. *Thin Solid Films* **2001**, *385*, 174–181. [CrossRef]
50. Lang, Z.; Xiuqin, W. Formation of nodular defects as revealed by simulation of a modified ballistic model of depositional growth. *J. Mater. Sci.* **1998**, *33*, 1487–1490. [CrossRef]
51. Petrov, I.; Losbichler, P.; Bergstrom, D.; Greene, J.E.; Munz, W.D.; Hurkmans, T.; Trinh, T. Ion-assisted growth of $Ti_{1-x}Al_xN/Ti_{1-y}Nb_yN$ multilayers by combined cathodic-arc/magnetron-sputter deposition. *Thin Solid Films* **1997**, *302*, 179–192. [CrossRef]

52. Lewis, D.B.; Creasey, S.J.; Wustefeld, C.; Ehiasarian, A.P.; Hovsepian, P.E. The role of the growth defects on the corrosion resistance of CrN/NbN superlattice coatings deposited at low temperatures. *Thin Solid Films* **2006**, *503*, 143–148. [CrossRef]
53. Mattox, D.M. Atomistic Film Growth and Resulting Film Properties. *SVC Bull. Spring* **2009**, 30–31.
54. Garte, S.M. Measurement of Porosity. In *Gold Plating Technology*; Reid, F.H., Goldie, W., Eds.; Electrochemical Publications Ltd.: Ayr, Scotland, 1974.
55. Panjan, P.; Gselman, P.; Kek-Merl, D.; Čekada, M.; Panjan, M.; Dražić, G.; Bončina, T.; Zupanič, F. Growth defect density in PVD hard coatings prepared by different deposition techniques. *Surf. Coat. Technol.* **2013**, *237*, 349–356. [CrossRef]
56. Sarkar, J. Troubleshooting in Sputter Deposition. In *Sputtering Materials for VLSI and Thin Film Devices*, 2nd ed.; Elsevier: Amsterdam, The Netherlands, 2014; pp. 567–592.
57. Selwyn, G.S.; Singh, J.; Bennett, R.S. In situ laser diagnostic studies of plasma-generated particulate contamination. *J. Vac. Sci. Technol.* **1989**, *7*, 2758–2765. [CrossRef]
58. Wehner, G.K.; Hajiček, D.J. Cone formation on metal targets during sputtering. *J. Appl. Phys.* **1971**, *42*, 1145–1149. [CrossRef]
59. Heintze, M.; Luciu, I. Nodule formation on sputtering targets: Causes and their control by MF power supplies. *Surf. Coat. Technol.* **2018**, *336*, 80–83. [CrossRef]
60. Selwyn, G.S.; Weiss, C.A.; Sequedac, F.; Huang, C. Particle contamination formation in magnetron sputtering processes. *J. Vac. Sci. Technol.* **1997**, *15*, 2023–2028. [CrossRef]
61. Selwyn, G.S.; Weiss, C.A.; Sequeda, F.; Hiuang, C. In-situ analysis of particle contamination in magnetron sputtering. *Thin Solid Films* **1998**, *317*, 85–92. [CrossRef]
62. Anders, S. Physics of arcing, and implications to sputter deposition. *Thin Solid Films* **2006**, *502*, 22–28. [CrossRef]
63. Pavate, V.; Abburi, M.; Chiang, S.; Hansen, K.; Mori, G.; Narasimhan, M.; Ramaswami, S.; Nulman, J.; Restaino, D. Correlation between aluminium alloy sputtering target metalurgical characteristics, arc initiation, and in-film defect intensity. *SPIE* **1997**, *3214*, 42–47.
64. Becker, M.; Gies, M.; Polity, A.; Chatterjee, S.; Klar, P.J. Materials processing using radio-frequency ion-sources: Ion-beam sputter-deposition and surface treatment. *Rev. Sci. Instrum.* **2019**, *90*, 1–33. [CrossRef]
65. Aharonov, R.R.; Chhowalla, M.; Dhar, S.; Fontana, R.P. Factors affecting growth defect formation in cathodic arc evaporated coatings. *Surf. Coat. Technol.* **1996**, *82*, 334–343. [CrossRef]
66. Shiao, M.H.; Shieu, F.S. A Formation Mechanism for the Macroparticles in Arc Ion-Plated TiN Films. *Thin Solid Films* **2001**, *386*, 27–31. [CrossRef]
67. Munz, W.D.; Lewis, D.B.; Creasey, S.; Hurkmans, T.; Trinh, T.; Vonijzendorn, W. Defects in TiN and TiAlN coatings grown by combined cathodic arc/unbalanced magnetron technology. *Vacuum* **1995**, *46*, 323–330. [CrossRef]
68. Vereschaka, A.A.; Vereschaka, A.S.; Batako, A.D.L.; Mokritskii, B.J.; Aksenenko, A.Y.; Sitnikov, N.N. Improvement of structure and quality of nanoscale multilayered composite coatings, deposited by filtered cathodic vacuum arc deposition method. *Nanomater. Nanotechnol.* **2017**, *7*, 1–13. [CrossRef]
69. Zhirkov, I.; Petruhins, A.; Rosen, J. Effect of cathode composition and nitrogen pressure on macroparticle generation and type of arc discharge in a DC arc source with Ti–Al compound cathodes. *Surf. Coat. Technol.* **2015**, *281*, 20–26. [CrossRef]
70. Bercegol, H. What is laser conditioning? A review focused on dielectric multilayers. *Proc. SPIE* **1999**, *3578*, 421–426.
71. Mirkarimi, B.; Baker, S.L.; Stearns, D.G. Planarization of Substrate Pits and Scratches. U.S. Patent 2005/0118533A1, 2 June 2005.
72. Mirkarimi, P.B.; Spiller, E.; Baker, S.L.; Sperry, V.; Stearns, D.G.; Gullikson, E.M. Developing a viable multilayer coating process for extreme ultraviolet lithography reticles. *J. Microlith. Microfab. Microsys.* **2004**, *3*, 139–145. [CrossRef]
73. Vetter, J.; Stuber, M.; Ulrich, S. Growth defects in carbon coatings deposited by magnetron sputtering. *Surf. Coat. Technol.* **2003**, *168*, 169–178. [CrossRef]
74. Saketi, S.; Olsson, M. Influence of CVD and PVD coating micro topography on the initial material transfer of 316L stainless steel in sliding contacts—A laboratory study. *Wear* **2017**, *388–389*, 29–38. [CrossRef]

75. Podgornik, B.; Hogmark, S.; Sandberg, O. Influence of surface roughness and coating type on the galling properties of coated forming tool steel. *Surf. Coat. Technol.* **2004**, *184*, 338–348. [CrossRef]
76. Hogmark, S.; Jacobson, S.; Larsson, M. Design and evaluation of tribological coatings. *Wear* **2000**, *246*, 20–33. [CrossRef]
77. Holmberg, K.; Matthews, A. *Coating Tribology*, 2nd ed.; Tribology Series 56; Elsevier: Amsterdam, The Netherlands, 2009; p. 43.
78. Harlin, P.; Bexell, U.; Olsson, M. Influence of surface topography of arc-deposited TiN and sputter-deposited WC/C coatings on the initial material transfer tendency and friction characteristics under dry sliding contact conditions. *Surf. Coat. Technol.* **2009**, *203*, 1748–1755. [CrossRef]
79. Olofsson, J.; Gerth, J.; Nyberg, H.; Wiklund, U.; Jacobson, S. On the influence from micro topography of PVD coatings on friction behaviour, material transfer and tribofilm formation. *Wear* **2011**, *271*, 2046–2057. [CrossRef]
80. Podgornik, B.; Jerina, J. Surface topography effect on galling resistance of coated and uncoated tool steel. *Surf. Coat. Technol.* **2012**, *206*, 2792–2800. [CrossRef]
81. Drnovšek, A.; Panjan, P.; Panjan, M.; Paskvale, S.; Buh, J.; Čekada, M. The influence of surrounding atmosphere on tribological properties of hard protective coating. *Surf. Coat. Technol.* **2015**, *267*, 15–20. [CrossRef]
82. Wang, H.W.; Stack, M.M.; Lyon, S.B.; Hovsepian, P.; Munz, W.D. Wear associated with growth defects in combined cathodic arc/unbalanced magnetron sputtered CrNrNbN superlattice coatings during erosion in alkaline slurry. *Surf. Coat. Technol.* **2000**, *135*, 82–90. [CrossRef]
83. Azar, G.T.P.; Yelkarasi, C.; Ürgen, M. The role of droplets on the cavitation erosion damage of TiN coatings produced with cathodic arc physical vapor deposition. *Surf. Coat. Technol.* **2017**, *322*, 211–217. [CrossRef]
84. Krella, A.K. An approach to evaluate the resistance of hard coatings to shock loading. *Surf. Coat. Technol.* **2010**, *205*, 2687–2695. [CrossRef]
85. Biswas, B.; Purandare, Y.; Sugumaran, A.; Khan, I.; Hovsepian, P.E. Effect of chamber pressure on defect generation and their influence on corrosion and tribological properties of HIPIMS deposited CrN/NbN coatings. *Surf. Coat. Technol.* **2018**, *336*, 84–91. [CrossRef]
86. Kek, D.; Panjan, P.; Panjan, M.; Čekada, M. The role of surface defects density on corrosion resistance of PVD hard coatings. *Plasma Process. Polym.* **2007**, *4*, S613–S617.
87. Montesano, C.P.L.; Gelfi, M.; LaVecchia, G.M.; Solazzi, L. Tribological and corrosion behavior of CrN coatings: Roles of substrate and deposition defects. *Surf. Coat. Technol.* **2014**, *258*, 878–885.
88. McIntyre, D.; Greene, J.E.; Håkansson, G.; Sundgren, J.-E.; Munz, W.-D. Oxidation of metastable single-phase polycrystalline $Ti_{0.5}Al_{0.5}N$ films: Kinetics and mechanisms. *J. Appl. Phys.* **1990**, *67*, 1542–1553. [CrossRef]
89. Lembke, M.I.; Lewis, D.B.; Munz, W.D. Localised oxidation defects in TiAlN/CrN superlattice structured hard coatings grown by cathodic arc/unbalanced magnetron deposition on various substrate materials. *Surf. Coat. Technol.* **2000**, *125*, 263–268. [CrossRef]
90. Polcar, T.; Cavaleiro, A. High temperature behavior of nanolayered CrAlTiN coating: Thermal stability, oxidation, and tribological properties. *Surf. Coat. Technol.* **2014**, *257*, 70–77. [CrossRef]
91. Hovsepian, P.E.; Ehiasarian, A.P.; Purandare, Y.P.; Biswas, B.; Perez, F.J.; Lasanta, M.I.; de Miguel, M.T.; Illana, A.; Juez-Lorenzo, M.; Muelas, R.; et al. Performance of HIPIMS deposited CrN/NbN nanostructured coatings exposed to 650 C in pure steam environment. *Mater. Chem. Phys.* **2016**, *179*, 110–119. [CrossRef]
92. Fernandes, F.; Morgiel, J.; Polcar, T.; Cavalerio, A. Oxidation and diffusion processes during annealing of TiSi(V)N films. *Surf. Coat. Technol.* **2015**, *275*, 120–126. [CrossRef]
93. Panjan, P.; Drnovšek, A.; Kovač, J.; Gselman, P.; Bončina, T.; Paskvale, S.; Čekada, M.; Kek-Merl, D.; Panjan, M. Oxidation resistance of CrN/(Cr,V)N hard coatings deposited by DC magnetron sputtering. *Thin Solid Films* **2015**, *591*, 323–329. [CrossRef]
94. Siracusa, V. Food Packaging Permeability behaviour. *Int. J. Polym. Sci.* **2012**, *2012*, 1–11. [CrossRef]
95. Fahlteich, J.; Mogck, S.; Wanski, T.; Schiller, N.; Amberg-Schwab, S.; Weber, U.; Miesbauer, O.; Kucukpinar-Niarchos, E.; Noller, K.; Boeffel, C. The Role of Defects in Single- and Multi-Layer Barriers for Flexible Electronics. *SVC Bull. Fall* **2014**, *57*, 36–42.
96. Burrows, P.E.; Graff, G.L.; Gross, M.E.; Martin, P.M.; Shi, M.K.; Hall, M.; Mast, E.; Bonham, C.; Bennett, W.; Sullivan, M.B. Ultra barrier flexibe substrates for flat panel displays. *Displays* **2001**, *22*, 65–69. [CrossRef]

97. Fahlteich, J.; Fahland, M.; Schönberger, W.; Schiller, N. Permeation barrier properties of thin oxide films on flexible polymer substrates. *Thin Solid Films* **2009**, *517*, 3075–3080. [CrossRef]
98. Fahlteich, J.; Schönberger, W.; Fahland, M.; Schiller, N. Characterization of reactively sputtered permeation barrier materials on polymer substrates. *Surf. Coat. Technol.* **2011**, *205*, S141–S144. [CrossRef]
99. Michels, J.J.; Peter, M.; Salem, A.; van Remoortere, B.; van den Brand, J. A combined experimental and theoretical study on the side ingress of water into barrier adhesives for organic electronics. *Appl. J. Mater. Chem.* **2014**, *2*, 5759–5768. [CrossRef]

 © 2020 by the authors. Licensee MDPI, Basel, Switzerland. This article is an open access article distributed under the terms and conditions of the Creative Commons Attribution (CC BY) license (http://creativecommons.org/licenses/by/4.0/).

Article

Surface Topography of PVD Hard Coatings

Peter Panjan [1,*], Aljaž Drnovšek [1], Nastja Mahne [1,2], Miha Čekada [1,2] and Matjaž Panjan [1]

[1] Jožef Stefan Institute, Jamova 39, 1000 Ljubljana, Slovenia; aljaz.drnovsek@ijs.si (A.D.); nastja.mahne@ijs.si (N.M.); miha.cekada@ijs.si (M.Č.); matjaz.panjan@ijs.si (M.P.)
[2] Jožef Stefan International Postgraduate School, Jamova 39, 1000 Ljubljana, Slovenia
* Correspondence: peter.panjan@ijs.si

Abstract: The primary objective of this study was to investigate and compare the surface topography of hard coatings deposited by three different physical vapor deposition methods (PVD): low-voltage electron beam evaporation, unbalanced magnetron sputtering and cathodic arc evaporation. In these deposition systems, various ion etching techniques were applied for substrate cleaning. The paper summarizes our experience and the expertise gained during many years of development of PVD hard coatings for the protection of tools and machine components. Surface topography was investigated using scanning electron microscopy (SEM), atomic force microscopy (AFM), scanning transmission electron microscopy (STEM) and 3D stylus profilometry. Observed similarities and differences among samples deposited by various deposition methods are discussed and correlated with substrate material selection, substrate pretreatment and deposition conditions. Large variations in the surface topography were observed between selected deposition techniques, both after ion etching and deposition processes. The main features and implications of surface cleaning by ion etching are discussed and the physical phenomena involved in this process are reviewed. During a given deposition run as well as from one run to another, a large spatial variation of etching rates was observed due to the difference in substrate geometry and batching configurations. Variations related to the specific substrate rotation (i.e., temporal variations in the etching and deposition) were also observed. The etching efficiency can be explained by the influence of different process parameters, such as substrate-to-source orientation and distance, shadowing and electric field effects. The surface roughness of PVD coatings mainly originates from growth defects (droplets, nodular defects, pinholes, craters, etc.). We briefly describe the causes of their formation.

Keywords: topography; PVD coating; ion etching; sputtering; 3D stylus profilometry; atomic force microscopy; scanning electron microscopy; focused ion beam

1. Introduction

Interactions between solid surfaces depend not only on the properties of the materials in contact but also on the topography of the surfaces [1]. Therefore, the surface topography can have a considerable impact on many material functional properties (e.g., the ability to adhere to another material, optical properties, friction, wear). This applies not only to bulk material but also to all kinds of coatings, including those prepared by physical vapor deposition methods. Their topography is strongly dependent both on the substrate topography and on the topography induced by the coating deposition process. While the former depends on the substrate preparation steps before the coating process, e.g., grinding or polishing, the latter mainly depends on the ion etching step before deposition as well as on the formation of growth defects during the deposition process.

The surface topography of PVD hard coatings is an important factor influencing its tribological performance under sliding contact conditions [2,3]. Namely, the real contact area strongly depends on the roughness of the interacting surfaces. The reduction in contact area in the case of a rough surface affects friction, adhesion, wear and other tribological phenomena. However, the real area of contact is not only a function of surface roughness, but also a function of the applied normal load. Thus, in the case of an elastic contact, it increases with normal load. A sliding of two surfaces does not always generate constant friction force, but the motion changes periodically between adhesion and sliding (i.e., the stick-slip phenomenon). Rougher coating surfaces cause higher friction and wear because of abrasive and ploughing effects. Because of a smaller real contact area in the case of a rough surface, a high surface pressure at contact spots also increases the tendency for crack initiation and the risk of fatigue-related damage. On the other hand, a very smooth coating surface contact may increase the adhesion between the surfaces, and therefore, the material transfer in the contact will be more pronounced [4]. However, a certain degree of roughness can also be useful; for example, when oil is used, it allows oil retention in the sliding contact area [5,6]. Particularly, surface irregularities (e.g., pores, crevices) on rough sliding surfaces can store lubricant and supply it to the interface. At these sites, the wear particles can also be trapped, which can significantly reduce abrasive wear. Thus, the challenge is to find the optimal surface roughness for the contacting surfaces to achieve optimal tribological performance of the coating.

Wettability is the next example of the coating surface property, which highly depends not only on its chemical composition but also on its topography. Applications where solid surface wettability plays a crucial role include contact lenses, super-hydrophilic surfaces, self-cleaning (lotus effect), nonstick surfaces, biofilm growth and body implants. Generally, the wettability of solid surfaces is evaluated by contact angles at the liquid droplet/solid surface/air interfaces. The balance at the three-phase contact is given by the Young equation, where it is assumed that the surface is chemically homogeneous and topographically ideally smooth. However, this is not true in the case of a real surface. Instead of one equilibrium contact angle, a range of contact angles exist, because the actual contact angle is the angle between the tangent to the liquid–vapor interface and the actual local solid surface. When using water as a liquid phase, a surface is defined as hydrophilic for a contact angle less than 90° (the liquid will subsequently spread over the surface) and hydrophobic for a contact angle greater than 90°.

The surface topography and surface roughness of the orthopedic and dental implants (in addition to the chemical composition) have a decisive influence on their integration and biological response in soft and hard tissues [7–10]. Several biocompatible wear-resistant and antimicrobial PVD coatings have been developed, which can improve selected surface properties (e.g., hardness, wettability, elastic strain, friction coefficient, wear) of implants (e.g., hip and knee prostheses) [11]. It was established that surfaces containing complex, rough, textured, and porous topographical features stimulate the protein adsorption and subsequent cell adhesion on implant materials and the formation of the extracellular matrix [10]. This is mainly because complex topographical features on the surfaces provide, in comparison with smooth ones, more surface area for interaction with the proteins and surrounding physiological environment. In recent years, surface topography has also been found to substantially influence the interaction between bacteria and surfaces [9]. In general, a large surface area with a rough surface promotes bacterial adhesion, while surfaces with the specific micro- or nanoscale surface features present enhanced antibacterial properties, either by preventing bacterial adhesion or by inactivating (killing) the adherent bacteria [9]. However, smooth surfaces are desired for medical devices that are directly exposed to blood, such as cardiovascular stents, to minimize clotting and restenosis. Recently, a lot of attention has been paid to surface topography modifications of the implant surface, particularly in the nanoscale regime, and the use of biocompatible coatings, in order to mimic the surrounding biological environment as well as to prevent the infection of tissue [10].

Optical coatings are yet another example of PVD coatings, where the topography of the coating surface has a major effect on their functionality [12]. The light reflected from an imperfect optical surface consists of a specular reflected component and a diffuse reflected component. The most significant sources of light scattering are the surface roughness and growth defects (e.g., nodular defects, pinholes). In general, the scattered light is affected only by those features that are of the same size or larger than the wavelength of light. Diffuse scattered light degrades the performance of high-precision optical systems for several reasons: (i) it reduces optical throughput (since some of the scattered light will not even reach the focal plane), (ii) it causes a reduction in contrast (signal-to-noise ratio) and (iii) the small-angle scatter causes a decrease in resolution (image blur) [13].

Although the topography of coatings strongly affects the performance of PVD coatings, literature systematically addressing this subject is scarce. Authors most often cite only some general data, such as surface roughness, while a deeper view of these issues is missing. Only a few studies are devoted to the influence of mechanical pretreatment and ion etching on the topography of PVD coatings [14,15]. In this paper, we limit ourselves to the topography of PVD hard coatings deposited on different substrates by various PVD methods used in industrial production. There are several significant differences between these methods both in the ion etching and the deposition steps. Therefore, we describe and analyze all coating preparation steps that can affect the topography of the coating. This includes selection of substrate and coating material, mechanical and chemical pretreatment, ion etching, deposition methods and deposition parameters. Such studies are particularly important for the reduction in the surface density of artefacts such as growth defects.

2. Materials and Methods

2.1. PVD Processes

All coatings were prepared in industrial batch-type deposition systems. A more extensive description of the four deposition processes is given below, while the essential process parameters and schematic drawings for each deposition method are given in Table 1 and Figure 1.

Low-voltage electron beam evaporation (or thermionic arc evaporation) system BAI 730 (BAI, Balzers, Vaduz, Liechtenstein) was used for deposition of TiN single layer coatings (Figure 1a). This system consists of a thermo-emissive cathode (filament), a crucible (evaporation source) connected to a low-voltage supply, and an auxiliary anode (around the target) [16]. The plasma (thermionic arc discharge), created between the ionization chamber and auxiliary anode, is used for the heating, etching and evaporation steps. A great advantage of this method is that the substrates can be immersed in a very dense plasma. Prior to deposition, the substrates are heated by applying a positive voltage to the substrate table, attracting electrons from the plasma and heating the substrates up to 450 °C. Then, the surface of substrates is cleaned by ion etching for 15 min. Argon ions are drawn out from the arc discharge and accelerated towards the substrates using a voltage of −200 V. To regulate the etching intensity, the substrate voltage can be adjusted independently. The low inert gas pressure (e.g., 0.1 Pa) ensures that the mean-free-path length is much longer (e.g., 5 cm) than the structural details of the substrates, thus providing excellent penetration. The plasma boundaries perfectly follow the contours of complex-shaped substrates because, according to the Child–Langmuir law, the width of the dark space is less than 1 mm in a high-density plasma (e.g., >1 mA/cm^2). Control of the etching process is easy and there are fewer problems with mixed loads. After the etching step, the crucible is made the anode of the arc discharge. The evaporation material held in the water-cooled copper anodic crucible is heated by electrons impinging on the crucible from the thermionic arc. During deposition, the bias voltage on the substrates is −125 V. The ion flux to the substrates is approximately proportional to the film deposition rate because of the high amount of coating material vapor ionizes in the plasma. In order to improve the coating uniformity, the rotating cylindrical substrate holders are placed concentrically around the crucible. Additionally, the crucible moves vertically during deposition to further improve the coating uniformity. The deposition rate of TiN hard coating is about 50 nm/min (for 2-fold rotation of the substrate).

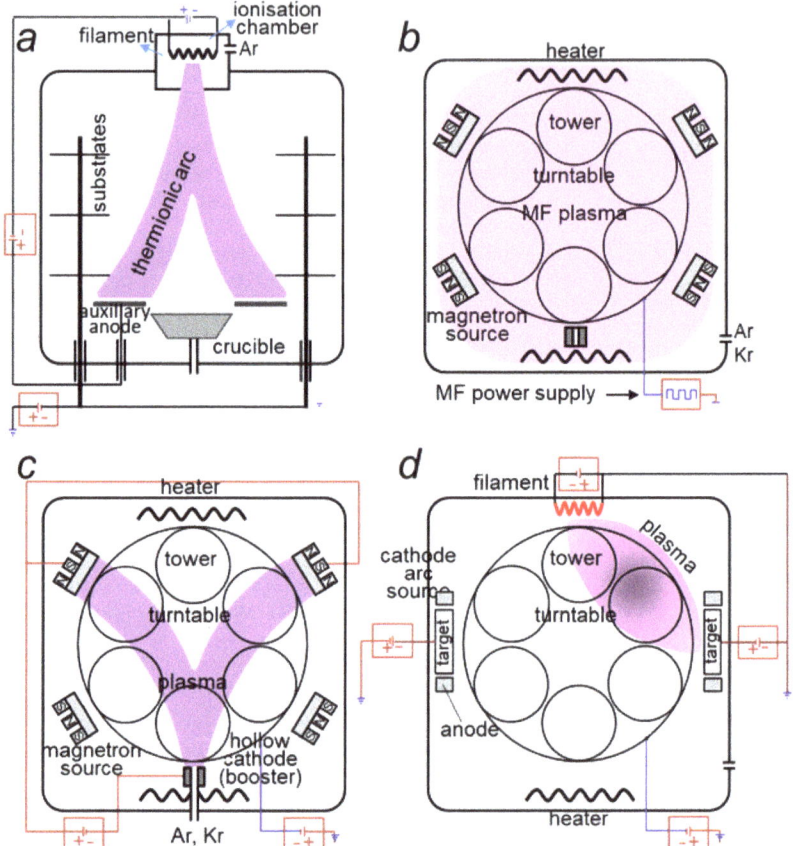

Figure 1. Schematic drawings of different deposition methods and ion etching modes used for the preparation of PVD hard coatings (the approximate area of plasma during the ion etching is marked by purple color) [17]: (**a**) side view of ion etching (DC bias) in low-voltage electron beam evaporation system BAI 730 (BAI); (**b**) top view of RF etching in magnetron sputtering system CC800/7 (CC7) and MF etching in CC800/9 sinOx ML (CC9); (**c**) top view of etching with hollow cathode plasma source (DC bias) in magnetron sputtering system CC9; (**d**) top view of ion etching (DC bias) cathodic arc deposition AIPocket (AIP).

A modified *magnetron sputter deposition system CC800/7* (CC7, CemeCon, Würselen, Germany) was used for depositions of TiAlN and TiN hard coatings (Figure 1b). This system is equipped with four 8 kW unbalanced magnetron sources, arranged in the corners of a chamber and operating in DC mode. The dimensions of an individual source are 200 mm × 88 mm. They are not positioned at the same height but at different vertical positions, which improves coating uniformity. This system utilizes infrared heaters for heating the substrates prior to the ion etching. During substrate cleaning, an RF bias is applied (maximum RF power 2 kW), while the etching time is 85 min. During deposition, the total operating pressure is maintained at 0.75 Pa, with the flow rates of nitrogen, argon and krypton being 100, 160 and 110 sccm, respectively. High-purity gases (99.998%) are used. Additional plasma is generated between the magnetron and the auxiliary anode–hollow cathode (such "booster" module is used during the deposition to enhance the plasma density). A DC bias of −95 V is applied to the substrates. The deposition time is 135 min. After this time, the deposition process is interrupted for another intermediate ion etching (for 55 min at the same conditions as in the initial substrate etching step). This is followed by an additional deposition of a coating (deposition time is 30 min). The intermediate etching creates new nucleation sites for the subsequently deposited nitride coating, resulting in a fine-grained and less porous microstructure of the top layer.

Table 1. Deposition methods and process parameters used for the preparation of PVD hard coatings. The deposition rates and ion current density are averages since all deposition techniques exhibited both spatial and temporal variations during a given deposition run as well as from one run to another run (depending on batching material).

	Deposition System	BAI 730	CC800/7	CC800/9 sinOx ML	AIPocket
preheating	heating method	Electron bombardment	Infrared heating	infrared heating	infrared heating
	preheating temperature (°C)	450	450	450	450
etching	etching mode	DC	RF	MF/booster	pulsed DC
	type of ions	Ar	Ar + Kr	Ar + Kr	Ar
	negative substrate etching voltage (V)	200	200	650/200	300/400
	etching time (min)	15	85 + 55 **	15/60	45
deposition	deposition method	low-voltage electron beam evaporation	Magnetron sputtering	magnetron sputtering	cathodic arc evaporation
	temperature (°C)	450	450	450	450
	working gas	Ar + N_2	Ar + Kr + N_2	Ar + Kr + N_2	N_2
	pressure of working gas (Pa)	0.2	0.75	0.66	4
	deposition time (min)	80	165	200	45
	negative substrate bias voltage (V)	125	95	90	70
	average deposition rate * (nm/min)	50	20	20	85
	average substrate current density (mA/cm^2)	3–5	~2	~2.5	-

* For twofold rotation of substrates, ** intermediate etching.

The magnetron sputter deposition system CC800/9 sinOx ML (CC9, Cemecon, Würselen, Germany) was used for the deposition of different nanostructure (nl-nanolayer, nc-nanocomposite) hard coatings (nl-TiAlN/CrN, nl-TiAlN/TiN, nl-AlTiN/TiN, nc-TiAlN/TiSiN/TiAlSiN, nc-TiAlCrSiN) as well as for the deposition of TiAlN/a-CN and TiAlN/Al$_2$O$_3$ double layer hard coatings (Figure 1c). In this system, the chamber is equipped with four rectangular magnetron sputtering cathodes (500 mm × 88 mm) which can operate in DC (TiAlN, TiN, CrN, AlTiN, nc-TiSiN, nl-TiAlSiN, nc-TiAlCrSiN, a-CN) or pulsed DC modes (Al$_2$O$_3$). The turntable can provide up to threefold rotation of the substrates. Prior to the depositions, the chamber is evacuated to a base pressure of 3 mPa, and heated to around 450 °C. In the next step, the substrates are ion-etched for 55 min in a mid-frequency plasma (Ar and Kr gas mixture, 240 kHz, duty cycle 1600 ns), with a bias voltage of 650 V, applied to the substrate table. An additional etching option, primarily used for round cutting tools, is so-called "booster" etching, where the working gas is injected through upper and lower "booster" etch nozzles (i.e., the hollow cathode), where intensive ionization of the working gas (Ar, Kr) occurs. The plasma is created between the magnetron cathodes that are facing opposite to the "booster" (see Figure 1c). Such additional discharge enhances the plasma density and thus the intensity of the etching process. Nitride-based coatings are deposited in a mixture of argon (160 sccm), krypton (110 sccm) and nitrogen (80 sccm), at a total pressure of 0.66 Pa. A DC bias of −90 V is applied to the substrates. The "booster", which is active also during the deposition process, enhances the potential gradient in the vacuum chamber so that the ionization degree of the sputtered atoms is increased to more than 50% [18]. The drawback of the "booster" etching is associated with shadowing, where some parts of the tool with complex geometries can be less exposed to ion etching.

Another series of the coatings was deposited by *cathodic arc evaporation deposition system AIPocket* (AIP, KCS Europe GmbH, Monschau, Germany). This system uses a technology called "super fine cathode" (SFC) [19]. SFC is a magnetically controlled cathodic arc source, which enables the deposition of low-stress, relatively smooth and thick coatings. In this system, prior to the deposition, the substrates are heated up to 450 °C and then they are cleaned by argon ion etching process. The ion etching in the AIP deposition system is also based on the auxiliary plasma (Filament-Assisted Pulse Etching or FAPE™ technology). In this system, the plasma is generated between a resistance-heated tungsten filament and two nearby cathodes (Figure 1d). Argon ions are drawn out from the arc discharge and accelerated to the substrate using a (pulsed) voltage of −400 V. The voltage and the current between the filament and the cathodes are 65 V and 6 A, respectively. For the coating deposition process, four targets are employed, and a power of 3 kW (20 V/150 A) is applied to each of the four targets. The deposition pressure is 4 Pa, while a bias voltage of 70 V is applied to the substrate table. The deposition time is 45 min, which results in 4 µm thick coatings on substrates mounted on a turntable that has the possibility of threefold rotation.

2.2. Substrate Materials

Six different substrate materials were used in this work: conventional tool steels (M2, D2, H11), powder metallurgical (PM) steel (ASP30), stainless steel (SS 316L) and cemented carbide (HM). The round substrates (3 mm in thickness, 18 mm in diameter) were ground and polished using 1 µm diamond paste for the final polish. Before deposition, the samples were degreased and cleaned in an industrial sized automated ultrasonic cleaning line. The first step was ultrasonic degreasing in de-ionized water with a degreaser (pH ~11) to remove surface impurities (cleaning time 15 min), followed by ultrasonic rinsing in deionized water and then drying in pure hot air. The coatings were deposited on all six different substrate materials with 1-fold, 2-fold and 3-fold rotations. Most substrates were mounted in the middle vertical position, while some of them were positioned at various heights in the vacuum chamber. Most substrates were mounted in the middle of the vertical position, while some of them were positioned at the bottom and top positions of the substrate holder.

2.3. Analytical Methods

Different analytical techniques were used to characterize the surface morphology and surface roughness as well as to look at individual features on the surface. Surface topography was observed with an optical microscope (OM) Axio CSM 700 (Carl Zeiss Microscopy, Oberkochen, Germany). However, the lateral and depth resolution of the optical microscope is not good enough for detailed analysis of micrometer-sized defects. Additionally, in the optical image, all types of growth defects are seen as black dots and therefore, it is impossible to distinguish between protrusions and craters. The most common technique for the study of substrate and coating morphology is scanning electron microscopy (SEM). In our study, a JEOL JSM-7600F SEM microscope (JEOL, Tokyo, Japan) was used. The applicability of SEM microscopy is limited due to relatively low-depth resolution. Therefore, atomic force microscope Solver PRO (AFM,) (NT-MDT Spectrum Instruments, Moscow, Russia) was employed to study the surface topography on a nano level. The use of an AFM microscope is limited due to the small scan area (of about 50 µm × 50 µm) and is therefore not appropriate for the study of growth defects, because its density is rather low (a few hundred defects/mm^2). The most suitable method for this purpose is a 3D stylus profilometry [20]. In the scanning mode, the stylus profilometer can give a 3D image of the surface on a large scan area (from several hundreds of micrometers square to several millimeters square) with all the micrometer-sized details. The Bruker Dektak XT stylus profilometer was used in this study.

The microstructure and the coating morphology were studied using fracture cross-sections examined in the SEM microscope. Cross-sections for SEM investigations were also prepared by focused ion beam techniques (FIB) using a FIB source integrated into the Helios Nanolab 650i field emission scanning electron microscope (FEI, Amsterdam, The Netherlands). Detailed microstructural analysis was performed by JEOL ARM 200 CF transmission electron microscope (TEM). The specimen for TEM characterization was prepared with a Helios NanoLab 600i focused ion beam system using the standard lift-out technique. SEM and TEM images were recorded using the ion beam and the electron beam.

3. Results and Discussion

In general, the topography of a PVD coating may originate from three different contributions [21]: (a) topography of the substrate surface, (b) intrinsic coating micro-topography or (c) growth defects forming during the deposition process.

3.1. Topography of the Substrate Surface

Substrate surface preparation is an integral part of any PVD film deposition process. The topography of the surface of an engineering material is determined by the method of its processing and the nature of the material itself. In general, the substrate surface topography may originate from mechanical pretreatment (grinding, blasting, polishing) and ion etching [3,14].

3.1.1. Substrate Irregularities Induced by Mechanical Pretreatment

Mechanical pretreatment of tool materials (e.g., tool steels, cemented carbides) usually includes grinding, blasting and polishing. Such pre-treatment significantly smooths the substrate surface, but it also creates different micrometer-size irregularities [22]. Thus, grinding, even if performed carefully, creates various irregularities of different shapes (such as scratches, grooves, ridges, pits) that are produced by the abrasive SiC particles in the grinding paper. The influence of mechanical preparation on the topography of the substrate surface is described in more detail in our recent review paper [17].

During polishing, which is the next step in the mechanical treatment of the substrates, protrusions are generated. These are formed due to hardness difference between the different phases in the substrate material. Namely, the removal rate depends on the hardness of the individual phase. Therefore, the harder phases that are more resistant against polishing will protrude from the surface (Figure 2) [14]. In such cases, the surface topography reflects the microstructure of the substrate material. As it can be seen in Figure 2, both the ASP30 and D2 tool steel substrates have higher roughness in comparison to H11 steel. After polishing the D2 and ASP30 tool steel substrates, shallow protrusions (about 8 nm in height) appeared at the sites of carbide inclusions, while no such protrusions are observed in the case of H11 tool steel. Namely, the amount and type of carbides depend on the carbon content and the quantity of carbide-forming elements (chromium, molybdenum, vanadium and tungsten, for example). The carbon content in H11 steel is much lower than in ASP30 and D2 steel. Therefore, the carbides in H11 steel are also much smaller and distributed uniformly in the steel matrix. The homogeneity of the microstructure, the level of purity, the size and distribution of carbides and other hard constituents in the steel matrix are the parameters that influence the polishability of tool steel substrates [22].

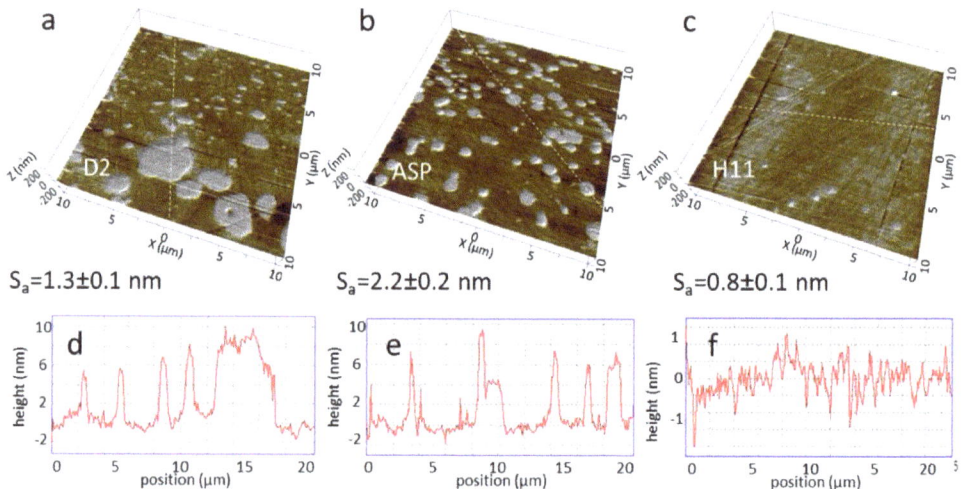

Figure 2. AFM images and surface roughness (S_a) of bare D2 (**a**), ASP30 (**b**) and H11 (**c**) tool steel substrates after polishing. The scan area was 20 μm × 20 μm, while the z-scale is 200 nm. Figures (**d**–**f**) show typical roughness profiles along the dashed lines in figures (**a**–**c**). Pay attention to the different z-scales.

3.1.2. Substrate Irregularities Induced by Ion Etching

In this paper, we examine the substrate irregularities induced by ion etching in typical industrial PVD deposition systems. The ion etching procedures are in part described in the experimental section (see schemes in Figure 1). Here, we provide additional details for the four deposition systems used in this work: magnetron sputtering systems (CC7, CC9), cathodic arc evaporation system (AIP) and the low-voltage electron beam evaporation system (BAI). In these four deposition systems, two different concepts of inert gas (Ar or Ar + Kr) ion etching are used. The first concept of ion substrate etching is based on the plasma generated by applying oscillating voltage between the substrate and the electrically grounded vacuum chamber. Such ion etching can only operate if high-frequency oscillatory voltage is applied. A radio frequency (RF) with 13.56 MHz and middle frequency (MF) with 240 kHz are used in CC7 and CC9 deposition systems, respectively. An advantage of this type of ion etching is that it results in a more-or-less uniform etching of all substrates in the deposition system. However, the disadvantage is that it does not allow independent control of ion densities and energies. The etching efficiency depends on the voltage frequency and amplitude.

In the second concept of substrate ion etching, an auxiliary plasma source is used (BAI, "booster" etching in CC9, AIP). The auxiliary plasma is normally spatially confined and therefore, not all substrates are immersed in the plasma at the same time. In order to achieve a uniform etching, the substrates need to rotate past the dense plasma. However, the main advantage of auxiliary plasma is that the ion density can be controlled by the plasma source, while the ion energy is controlled by the bias potential on the substrates. Furthermore, substrates can be biased either by continuous, pulsed, or oscillatory potential, which offers even greater control over the etching parameters. Although continuous substrate bias provides the most intense substrate etching, a pulsed or oscillatory substrate potential is still used in many cases because it enables the removal of native oxide and other non-conductive contaminates.

The above-mentioned concepts of ion etching are based on the use of inert gas ions. In this case, implanted ions can cause tensile stresses or even the formation of an amorphous zone [23]. This is not the case if metal ions are used. Metal ions can be generated in the deposition system with cathodic arc or high-power impulse magnetron sputter (HIPIMS) sources. However, a disadvantage in the case of arc discharge is the contamination of the substrate surface with the droplets of pure metal emitted by the cathode spots. This problem can be completely eliminated if HIPIMS discharge is used to generate the metal ions.

Ion bombardment changes the microchemistry of the substrate, its surface topography and microstructure of the near-surface layer. The ion etching results in the removal of native oxides and other contaminants, increased density of nucleation sites and chemical activation of the surface. All these changes affect not only the adhesion of the coating but also its growth. The etching efficiency depends on the etching method (Figure 3), etching time, the geometry of the substrate, substrate material (Figures 4–6), current density, ion energy, rotation mode of the substrate (Figure 7), plasma homogeneity and other parameters. The efficiency of ion etching can be determined by observing the selected area on the substrate surface under optical, SEM and AFM microscopes or by making a 3D profile image (Figure 8).

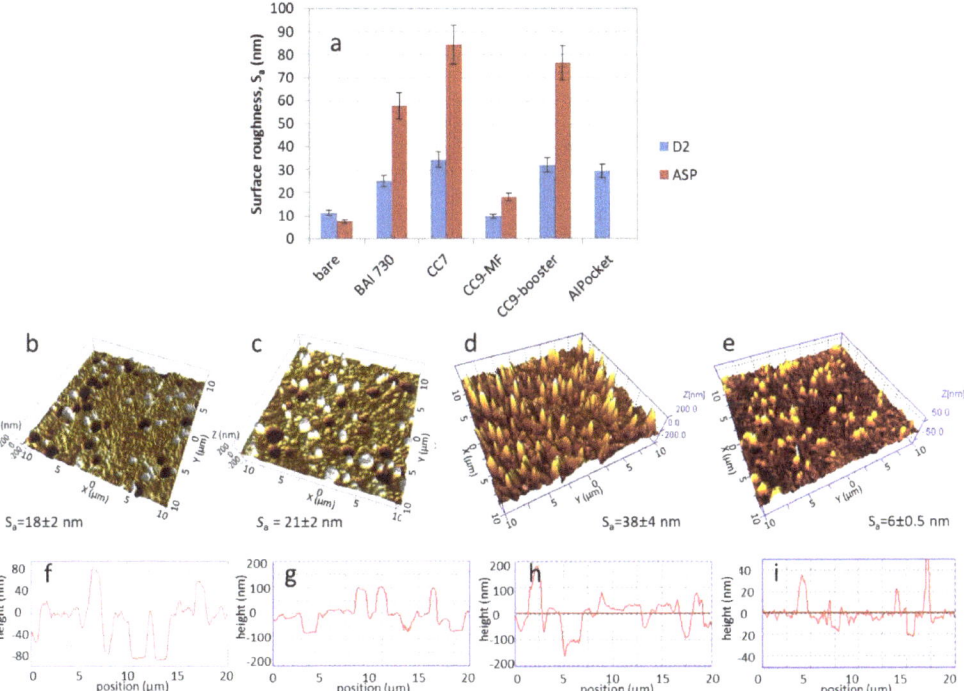

Figure 3. (a) 3D surface roughness (S_a) (scan area was 300 μm × 300 μm) of D2 and ASP tool steel substrates after ion etching in different industrial deposition systems. AFM images (**b–e**), surface roughness (S_a) and line profiles (**f–i**) of the ASP substrates (the scan area was 20 μm × 20 μm) after ion etching in: BAI 730 (**b,f**), CC7 (**c,g**), MF + booster etching in CC9 (**d,h**) and MF etching in CC9 (**e,i**). Pay attention to the different z-scales on AFM images and height scale on line profiles.

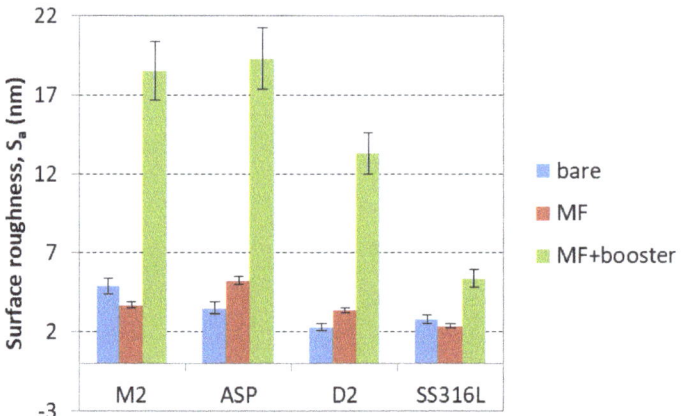

Figure 4. AFM surface roughness of different substrates (M2, ASP, D2, SS 316L) before and after MF and MF + booster ion etching. The scan area was 20 μm × 20 μm.

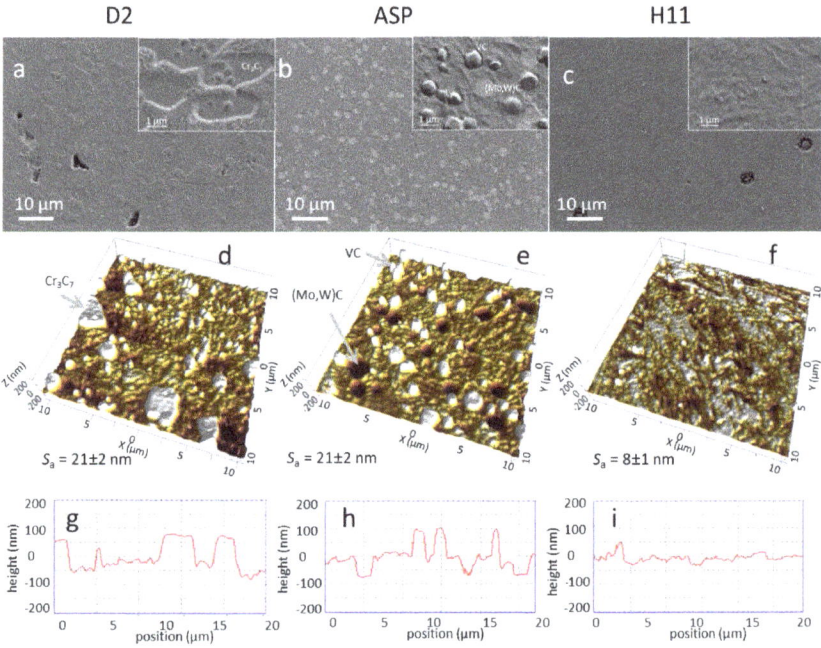

Figure 5. SEM images (**a**–**c**), AFM images (**d**–**f**), line profiles (**g**–**i**) and surface roughness (S_a) of D2, ASP30 and H11 tool steel substrates after RF ion etching in CC7 deposition system. The scan area and z-scale are 20 μm × 20 μm and 200 nm, respectively.

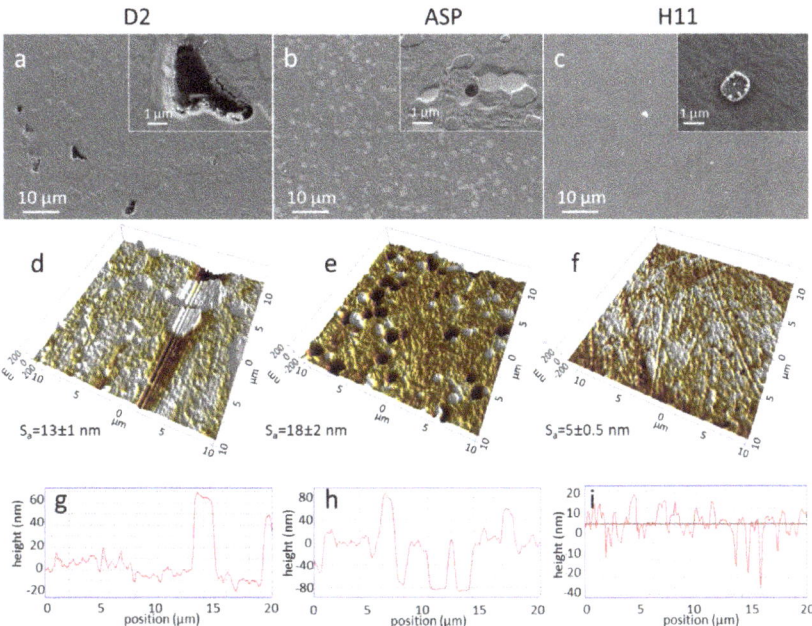

Figure 6. SEM images (**a**–**c**), AFM images (**d**–**f**), line profiles (**g**–**i**) and surface roughness (S_a) of D2, ASP30 and H11 tool steel substrates after DC ion etching in BAI deposition system. The scan area and z-scale are 20 µm × 20 µm and 200 nm, respectively.

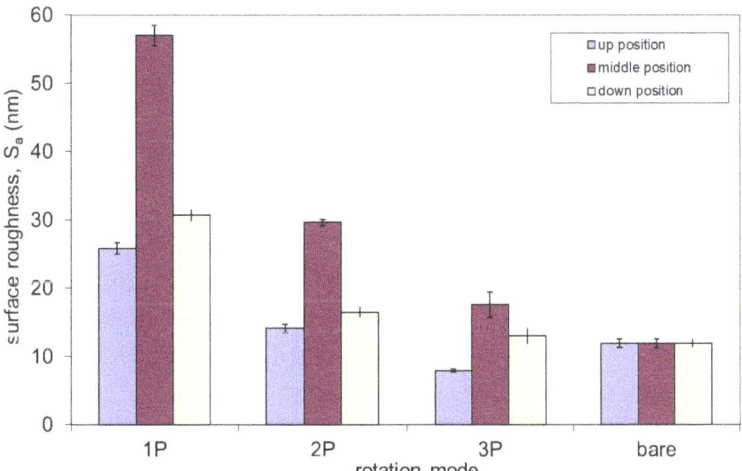

Figure 7. The 3D surface roughness (S_a) after filament-assisted pulse etching in AIP deposition system. The etching procedure was performed in the same batch using onefold, twofold and threefold rotation of ASP substrates, while their vertical position was also different. The scan area was 300 µm × 300 µm.

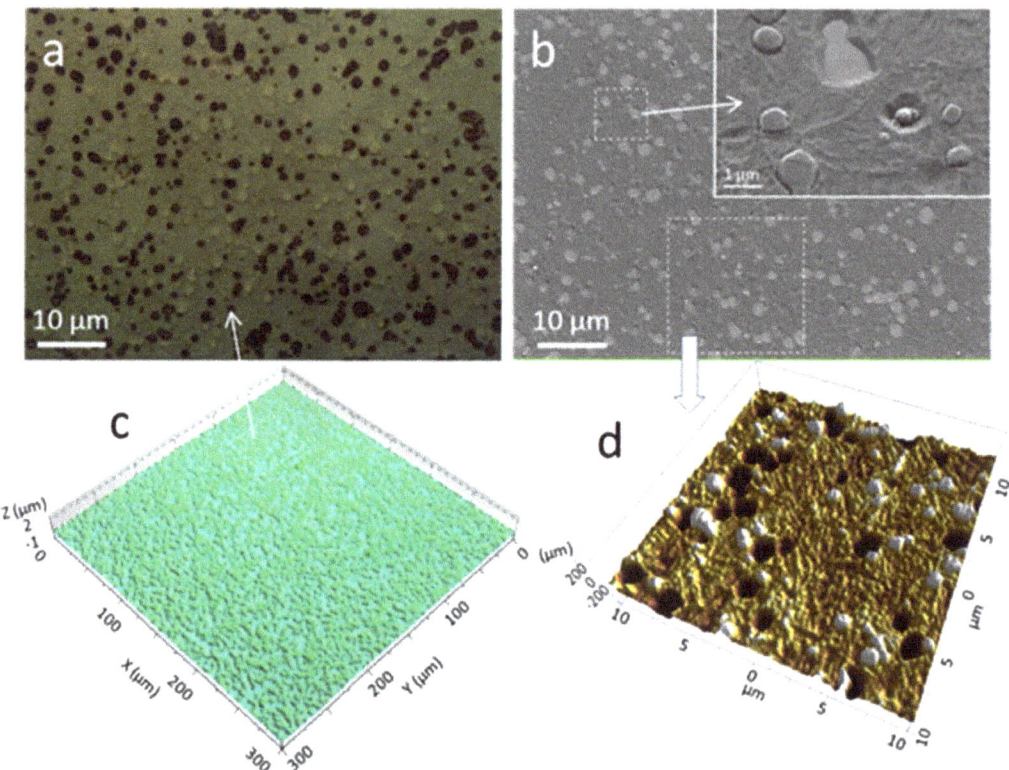

Figure 8. OM image (**a**); SEM images at lower (**b**) and higher magnification (see inset); 3D profile image (**c**); AFM image (**d**) of the ASP tool steel substrate surface after DC ion etching in BAI deposition system.

Etching of a surface by ion bombardment does not remove the contaminated surface layer without side effects. These side effects are implantation of bombarded ions, radiation damage, surface topography development, ion beam mixing, preferential sputtering and related changes in surface stoichiometry [24]. Furthermore, ion bombardment causes surface micro-roughening and the formation of various micro-sized surface features (Figure 9). Cones, pyramids, pits, hillocks, steps, etc., have been observed, and their formation is closely related to the initial surface irregularities, impurities, intrinsic or ion-beam-induced defects and variations in the sputtering yield as a function of the angle of ion beam incidence to the surface [25,26]. Especially large topographic changes occur at high fluencies of ions. Cones are the most common surface features produced by sputtering. They can be formed by different mechanisms. In 1971, Wehner and Hajicek [27] published a paper where they demonstrated that cones on etched substrates can arise as a result of a very small amount of certain foreign atoms (Figure 10), which are present as impurity atoms in the substrate material or are supplied during sputtering from another source. The islands of backscattered materials or contaminants act as local etch masks, as their sputtering rate is different in comparison with the surrounding material (matrix). This can cause the formation of various surface structures.

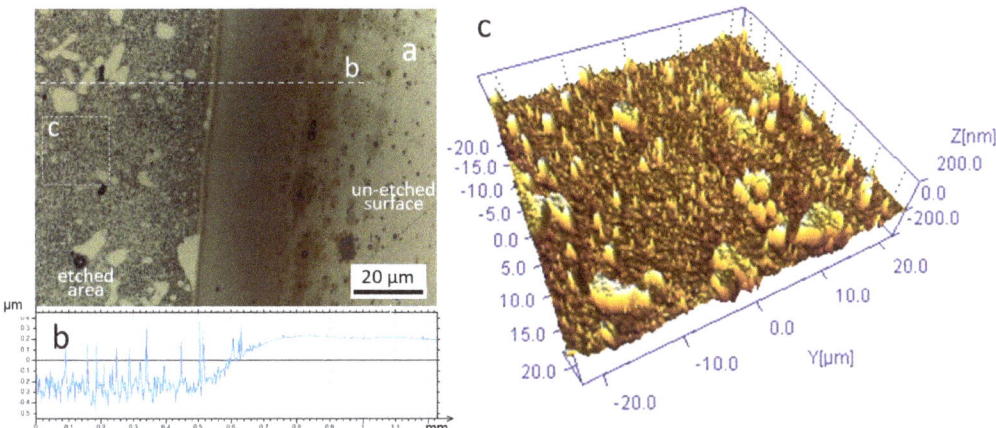

Figure 9. Optical image (a) and roughness profile measured along the dashed line (b) show the transition from the region exposed to ion etching to the unexposed one (this part of the substrate was covered with a stainless steel foil). Ion etching of D2 substrate was performed in the CC9 deposition system (MF + booster). The etching rate was quantified by measuring the step height between the etched and un-etched surface area (b). AFM image (c) of ion-etched area of D2 tool steel substrate (scan area was 50 μm × 50 μm).

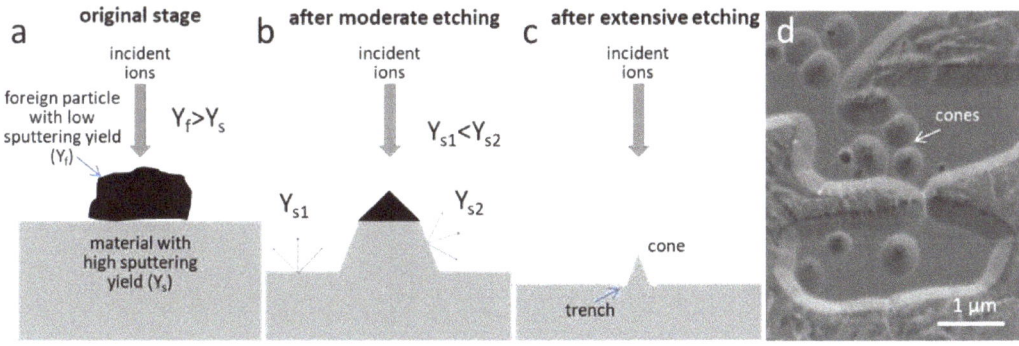

Figure 10. Scheme showing stages of cone formation during ion etching of substrate covered by a micrometer-sized foreign particle. If the foreign particle is located on the substrate surface, then it prevents the etching of the substrate area underneath the seed (a). Therefore, a step will be formed under the edge of the particle (b). As the particle itself shrinks, the step will be inclined. When the shielding particle is eroded away, a sharp cone of matrix material remains (c). Top view SEM image (d) shows the cones formed on the surface of carbides in D2 substrates during RF ion etching in the CC7 deposition system.

Cone-like features can also form underneath a foreign particle, which screens the underlying material from the ions. Figure 10 shows the scheme of the cone formation stages during ion etching of the substrate where a micrometer-sized foreign particle (with low sputtering yield) is present. As the sputtering proceeds, the area of this particle decreases, and the surrounding substrate material is sputtered away at a higher rate (Figure 10b). When the shielding particle is eroded away, a sharp cone of matrix material remains (Figure 10c). The size and shape of the cone depends on the ratio of particle shrinkage rate to the removal rate of bulk material. As the bombardment continues, the cone decreases in size, and it eventually disappears.

Cone-like structures can also form due to the initial micro-topography of the substrate surface (e.g., asperities). The variation in sputtering yield with the angle of ion incidence causes the surface to erode more rapidly where the angle of incidence is higher. This is demonstrated in Figure 11, where we calculated sputtering yields as a function of Ar⁺ incidence angle for several substrate materials investigated in this work. We used SRIM (Stopping and Range of Ions in Matter) software, which is based on the binary collision approximation to simulate the sputtering [28] process and calculate the sputtering yields and other sputtering parameters. The calculations were made for 5000 argon ions bombarding the surface with energy of 200 eV. It can be seen that sputtering yield strongly depends on the angle of ion incidence. In some cases, the sputtering yield at higher incidence angles can be twice or three times as high as the sputtering yield for perpendicular ion bombardment. Hence, the surfaces that are exposed to the ion flux under higher angles with respect to the surface normal will be etched much faster than those etched in the direction of the surface normal. These results in the preferential etching of inclined surfaces and promotes formation of faceted surfaces. It should be mentioned that the plasma sheath around the substrates is relatively large (around 1 mm) compared to the height of the protrusions (typically less than 0.5 µm). The plasma sheath, therefore, does not follow the topography of protrusions. This means that the ion flux at the protrusion arrives under various angles, while on the flat parts of the substrate (e.g., iron matrix of the steel), the ions bombard the surface predominantly in the perpendicular direction.

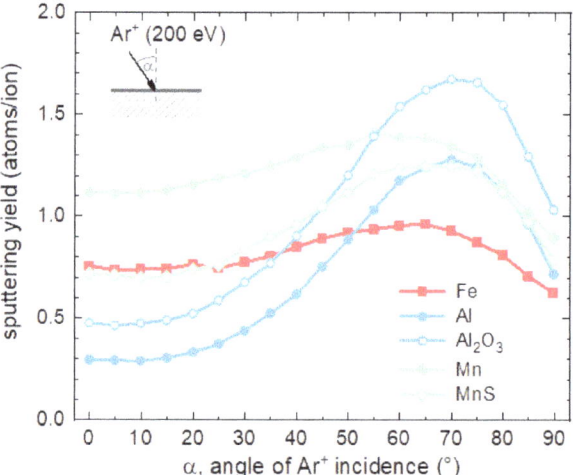

Figure 11. Sputtering yields as a function of Ar⁺ incidence angle for several substrate materials investigated in this work. Calculations were performed by SRIM program.

The sputtering yield changes (decreases and also increases) when the surface of the substrates is faceted or roughened. Faceting usually starts at corners of protrusions (Figure 12), which always have some rounding, and therefore present a variety of incidence angles to the incoming ions (ranging from glancing to normal) [29]. This causes more rapid etching at the corner than on the flat surface, and simultaneously, the sputtering yield at the corner increases. Therefore, with increasing etching time, the corner is increasingly faceted. The angle of the faceted corner coincides with the angle at which the sputtering rate has the highest value. Thus, the angular dependence of the sputtering yield of the cone material largely determines the final shape of the cone. The enhanced etching rate at the sidewall causes the formation of cones at any protrusions on the substrate surface. The cones continue to sharpen as etching proceeds. Due to the higher sputtering (etching) rate at the sidewalls, the cones will eventually disappear after a longer period of etching. However, the substrate surface will at the same time expose new second-phase regions or impurity particles and new cones will start to form. Thus, a steady-state surface topography forms.

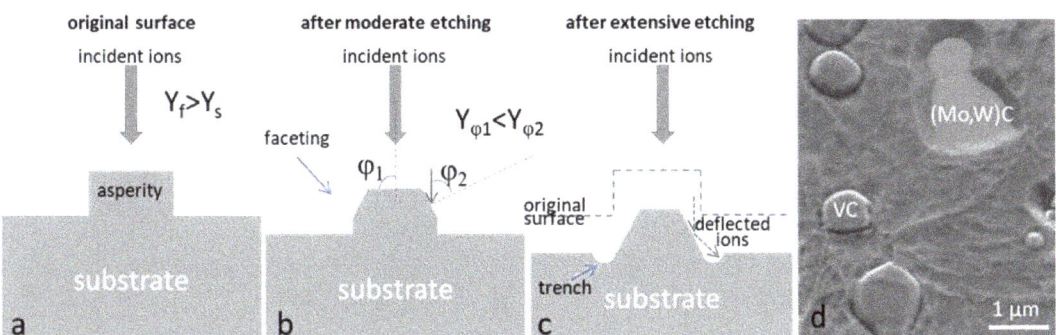

Figure 12. Schematic diagram outlining the mechanism of faceting and trench formation during ion etching of a protrusion on the substrate surface (**a**–**c**). Top view SEM image of ASP30 substrate after DC ion etching in BAI deposition system (**d**). In the SEM picture, we can see the faceting of carbides and the formation of trenches.

In general, the picture described above is more complex if other phenomena, such as surface diffusion of atoms on the substrate surface, are considered. Re-deposition and cross-contamination of sputtered materials, as well as sputtering by reflected ions, lead to additional changes in the sidewall shape. Cones are usually surrounded by shallow trenches which are created at the bottom of etched sidewalls (Figure 12c). Typically, trenches are formed due to enhanced erosion by incident ions reflected off the sidewall by energetic sputtered atoms from the cone sidewall. As a result, the cones develop grooves around them (Figure 12d).

The sputtering rate also depends on the texture of the coating or crystal orientation of individual grains and inclusions. Thus, when the polycrystalline substrate is ion etched, there will be a preferential sputtering due to the presence of individual grains and therefore, a stepped surface topography usually develops.

If the substrate material is composed of different phases of material, shallow craters and protrusions are generated during ion etching due to the difference in the sputtering rate between the different phases and matrix. For example, in ASP30 tool steel substrate, the sputtering rate of the M_6C and the MC carbides is higher and lower than the martensitic matrix, respectively [30]. Thus, the consequence of ion etching is the formation of a large number of pits ((Mo,W)C carbides) and hillocks (VC carbides) that are evenly distributed over the substrate surface, as seen in Figure 13. Shallow craters and protrusions also form at sites of nonmetallic inclusions that have a higher (e.g., MnS) or lower (e.g., oxides) etching rate than the martensitic matrix (Figure 14). According to SRIM calculations shown in Figure 11, the sputtering yield of Al_2O_3 is lower than the sputtering yield of Fe for Ar^+ incidence angles below 40° but increases dramatically at higher angles. The sputtering yield of MnS is similar to the sputtering yield of Fe for lower incidence angles but increases substantially at angles above 30°. Hence, predicting the final surface topography of a material such as tool steel, which has many nonmetallic inclusions, is a difficult task since it depends on many local parameters as well as on the specifics of the etching procedure and substrate rotation. Nevertheless, our observations show that substrate roughness increases with the etching time.

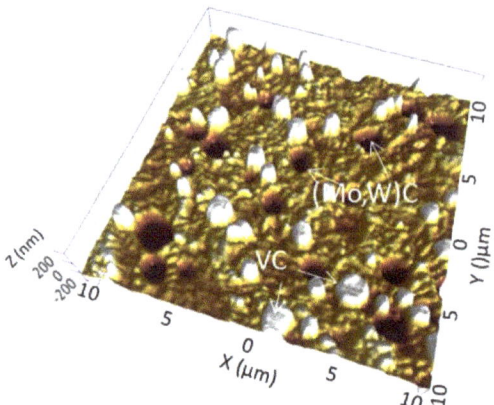 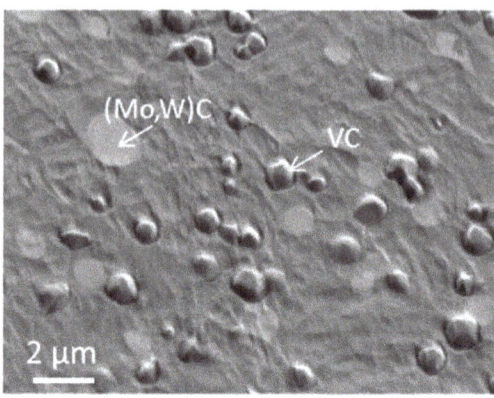

Figure 13. AFM and SEM images of the shallow protrusions and shallow craters formed by the RF ion etching of ASP30 tool steel substrate in the CC7 deposition system. The SEM image is a side view from about 45° inclined direction. Carbide inclusions do not retain their original profiles; they develop facets on the sidewalls and trenches at the bottom. The VC carbides become more protruding from the matrix and sharper after longer etching times.

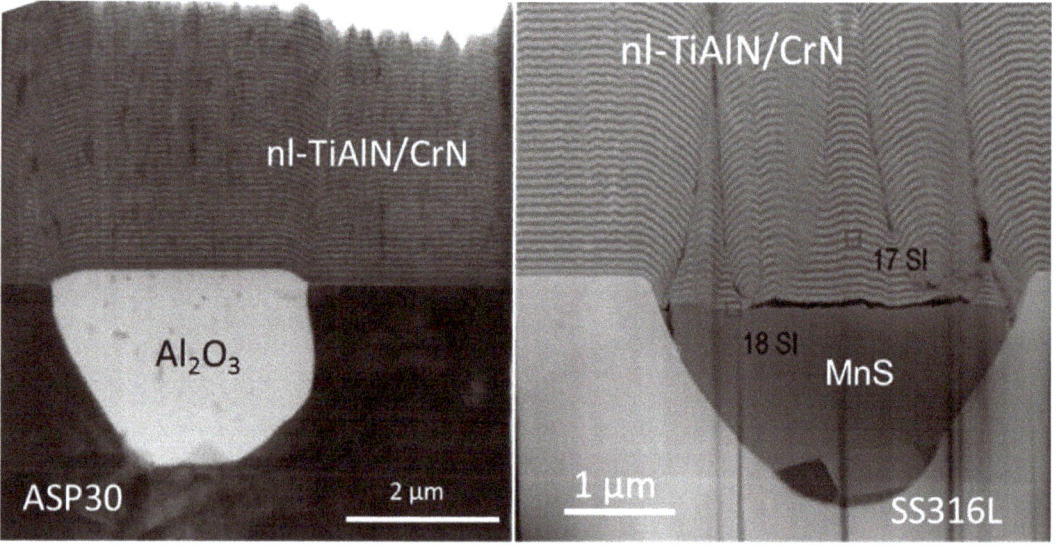

Figure 14. STEM image of FIB cross-section of the nanolayered TiAlN/CrN hard coating at sites of Al_2O_3 (ASP30 substrate) and MnS (SS316L substrate) nonmetallic inclusions.

As mentioned above, the etching efficiency depends not only on batching material and etching conditions, but also on the geometry of the substrates (tools), their positions in the vacuum chamber and the rotation mode [31,32]. On the industrial scale, PVD coatings are commonly deposited on substrates with complex three-dimensional shapes that have sharp edges (e.g., cutting tools), holes and slots. Ion etching on such substrates is nonhomogeneous due to varying ion currents arriving at the substrates. During a given deposition run as well as from one run to another, the etching rate is not only spatially dependent due to the different batching configurations but also temporally dependent due to multiple rotations of substrates. The effects of sample shape, orientation and distance have only scarcely been studied. Most of the available literature refers only to the etching of flat substrates. Therefore, it is very important to study more in detail the differences in etching between 3D substrates and flat substrates that are usually used in tests.

A typical job-coating batch is filled with tools of different geometry, size and height. We can expect a highly non-uniform electric field distribution, not only around the different tools positioned at various heights within the batch but also on different parts of the same tool. For this reason, it is very difficult to predict the degree of etching on different parts of the tool. The knowledge on the etching efficiency within a batch therefore needs to be acquired empirically. An example of differences in the surface topography for the similar etching and coating conditions is shown in Figure 15. The figure shows 3D surface topography images of two D2 tool steel substrates after TiN deposition in the BAI deposition system. The individual samples were prepared in two different batches for the same etching and deposition parameters. The only difference between both samples was the batching configuration.

Since all topographic irregularities formed during ion etching are transferred to the coating surface, they can also be observed on the coated tools under an optical microscope. Under the microscope, the contours of carbides are clearly visible and more or less pronounced, depending on the etching efficiency (Figure 15a,b). It is evident that the TiN at the sites of carbide grains protrudes out of the surface much more on the sample in Figure 15a than on the sample in Figure 15b.

The etching efficiency can be quantified from roughness profiles (Figure 15c,d). The highest peaks in the 3D surface profile image (Figure 15e,f), which belong to the nodular defects formed during deposition, must be eliminated in the analysis. The height of steps at the sites of carbides is the measure of etching rate. Thus, for two different batching configurations, presented in Figure 15, the step heights are 165 nm and 94 nm, respectively. From the photos of both batches, we can conclude that the first sample was less shaded by other tools, so the ion flux density and consequently the etching rate was higher. Based on these and other similar observations, we conclude that the degree of etching is highly dependent on the batching configuration.

On tools with sharp edges and corners, the applied negative bias leads to non-uniform electric fields and non-uniform plasma densities over the different faces of the substrate surface and thus to the non-uniform bombardment of overall substrate surface. The current density and consequently the erosion rate are higher at the corners than on the flat surface. On the other hand, ion etching of holes and slits is lower due to shading effects and possibly lower electric field (Figure 16) [33]. This non-uniform ion bombardment leads to variations in surface topography over the substrate surface. Therefore, a major problem in using ion etching for substrate cleaning is how to obtain a uniform and controlled ion etching over a surface. The uniformity of the electric field is further worsened by the fixturing system. As a general rule, the best plasma system design is the one that is geometrically symmetric, which is, however, difficult to achieve in many cases. Thus, for example, in the case of a magnetron deposition system, the magnetic field confines electrons and increases the local plasma density in one region, which leads to a decrease in plasma density in another region [34]. In order to achieve a more uniform etching, multiple rotations of substrates are necessary. However, even multiple rotations of the substrates do not completely solve the problem of uniform etching. For example, an end mill can be etched quite uniformly around the main flank, but in the flute (on the rake face), the etching efficiency will be lower. The disadvantage of complex movement is that the etching efficiency of the substrate decreases with the increased degree of rotation.

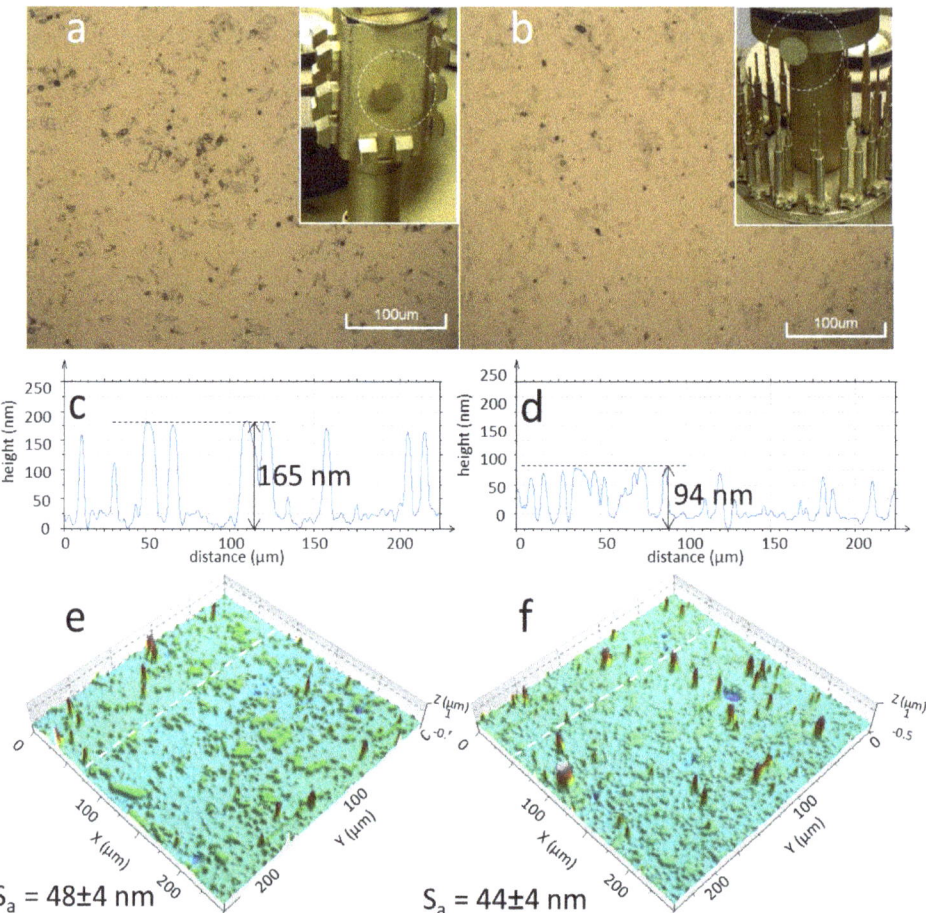

Figure 15. (a,b) OM images, (e,f) 3D surface profile images and (c,d) line profiles (along dashed lines in the Figure (e,f)) of TiN hard coating deposited on D2 substrate in the BAI deposition system at the same etching and deposition parameters. It is evident that the D2 substrate on the left was exposed to much more intensive ion etching (the step height at the carbide grain is 165 nm) than that right one (the step height at the carbide grain is 94 nm). The reason is the difference in batching configuration (see insets). The surface roughness, however, differs only slightly, due to the higher growth defect density on the right sample (406 defects/mm^2) in comparison with the left one (236 defects/mm^2). The nodular defects are the sharp peaks.

The erosion rate is higher at the corner than on the flat surface of the sample. This is attributed to the concentration of the electrical field lines around the corners and consequently, a higher ion current and erosion rate (Figure 16). Thus, we noticed that in the case of threefold rotation of the round plate substrate, the etching of the plate edge is much more pronounced than on the flat surface. This effect is much less pronounced for twofold rotation and the least for onefold rotation, where the electrical field lines are perpendicular to the substrate surface all the time. In a certain phase of threefold sample rotation, however, the sharp edge is highly exposed to intense ion bombardment. For example, Figure 17 shows the SEM images of D2 round substrates (5 mm in thickness, 16 mm in diameter) coated in the CC9 deposition system with nanostructured hard coating using threefold rotation. The ion etching is much more pronounced at the edge of the round sample (S_a = 195 nm) than on the flat surface in the middle (S_a = 54 nm) of the sample.

Due to variation in the electrical field at the substrate surface, the distribution of ion current density during etching and deposition is highly non-uniform, and at sharp edges, it may be many times higher than in the middle. Therefore, we can expect different coating properties and performance compared to the coating on the flat part of the sample. The majority of scientific papers concentrate on ion etching and subsequent coating deposition on a flat substrate. Only a few studies reported the differences between coatings deposited on the curved and flat substrates [15,35]. However, care must be taken when applying the observations made on the flat test samples to the tools with complex geometry.

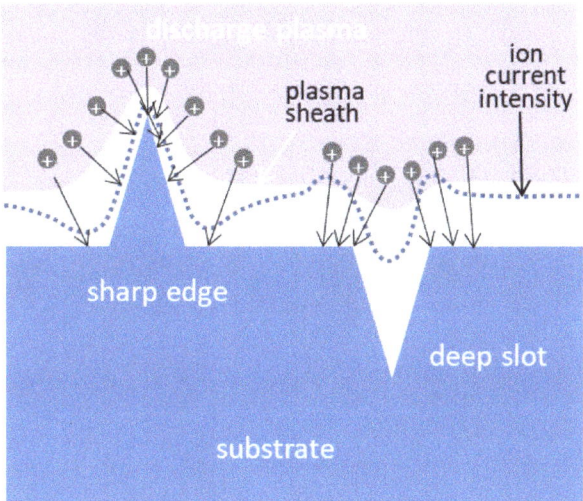

Figure 16. The sheath between homogeneous plasma and negatively biased substrate with complex geometric shape (sharp edges, slots) and the corresponding ion current density.

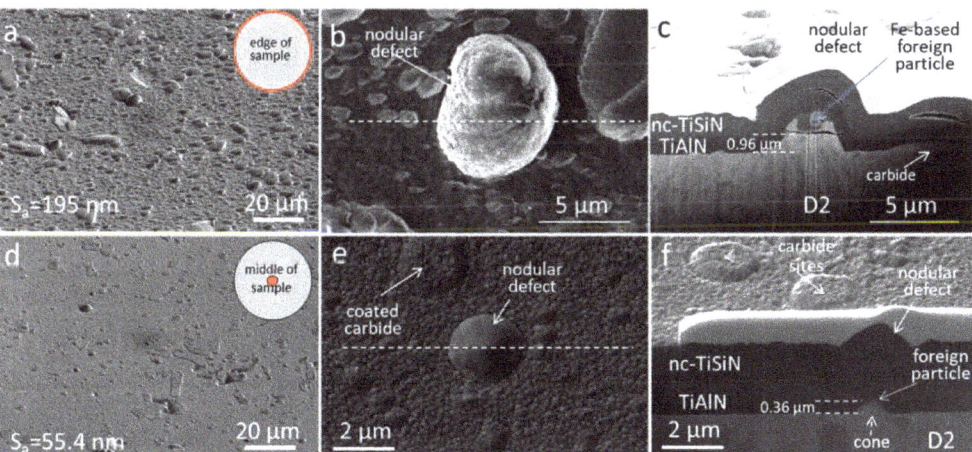

Figure 17. SEM images of the nanostructured hard coating TiAlN/TiSiN/TiAlSiN deposited in the CC9 deposition system: (**a**–**c**) at the edge and in (**d**–**f**) the middle of D2 round substrate. The SEM images (**a**,**d**) are the side view from about 45° inclined direction. Top view SEM images in the middle show selected nodular defects (**b**,**e**) and their FIB cross section (**c**,**f**). A step formed under the seed particles (marked with dashed lines) shows that a 3-times thicker layer was removed on the edge during ion etching in comparison with the middle of the substrate.

3.2. Micro Topography of Coating Surfaces

As mentioned above, all topographical irregularities on the substrate surface formed during pretreatment are transferred through the coating. Therefore, the intensity of ion etching can also be estimated after coating deposition, either using optical, AFM and SEM microscopes or a 3D profilometer. However, after coating deposition, additional topographic changes occur related to the intrinsic micro- and nanomorphology of the coating itself and growth defects (Figure 18). Low magnification observations (Figure 18a) show a smooth coating surface interrupted by unevenly distributed growth defects (nodular defects, pinholes, craters) of various shapes and sizes (see inset b in Figure 18a). At a high magnification of the area between growth defects, columns with dome-shaped tops can be observed. At even higher magnifications, the nanoscale sub-cells on the top of every columnar grain are visible (see inset d in Figure 18c).

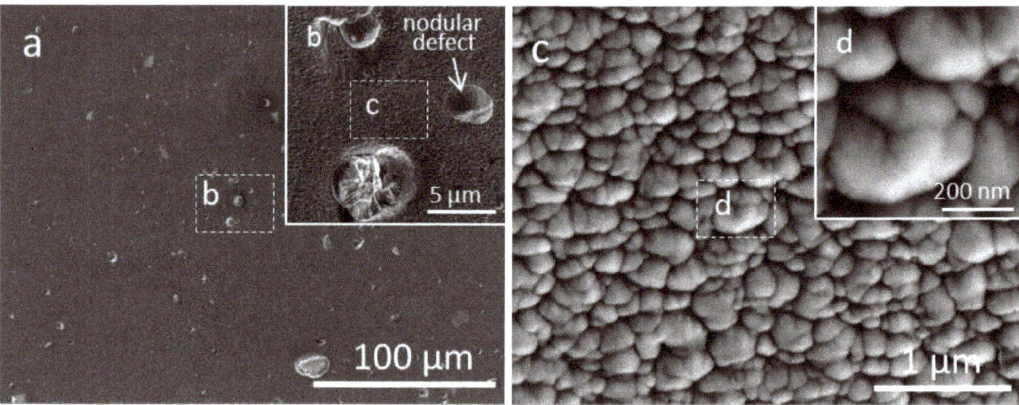

Figure 18. SEM observations of (**a**) macro-, (**c**) micro- and (**d**) nano-morphology of a magnetron sputtered nanolayer and nanocomposite coating TiAlN/TiSiN/TiAlSiN deposited on D2 tool steel substrate. Inset (**b**) shows the typical growth defects formed during deposition process.

The micro morphology of the depositing film is determined by the surface roughness and the surface mobility of the depositing atoms. If the substrate surface is rough, the variation in the angle of incidence causes the formation of a less dense coating with a more complex morphology. The peaks of protrusions receive the flux of depositing atoms from all directions and, if their surface mobility is low, the peaks grow faster than the valleys. The self-shadowing effect is even greater if the flux of depositing atoms is off-normal because the valleys are in "deeper shadows" than when the flux is normal to the surface [36]. The consequence of such film growth is the formation of a columnar microstructure with dome-shaped surface topography (Figures 19 and 20). The roughness of interfaces that are rather smooth near the substrate gradually increases towards the top of the coating (Figure 20). Rounded columns result in higher cumulative surface roughness of the coating, which increases with the coating thickness. The columnar morphology forms regardless if the material is crystalline or amorphous [37]. The driving force for columnar grain growth is surface energy minimization. Additionally, the self-shadowing effect increases the degree of roughening and causes a voided columnar structure.

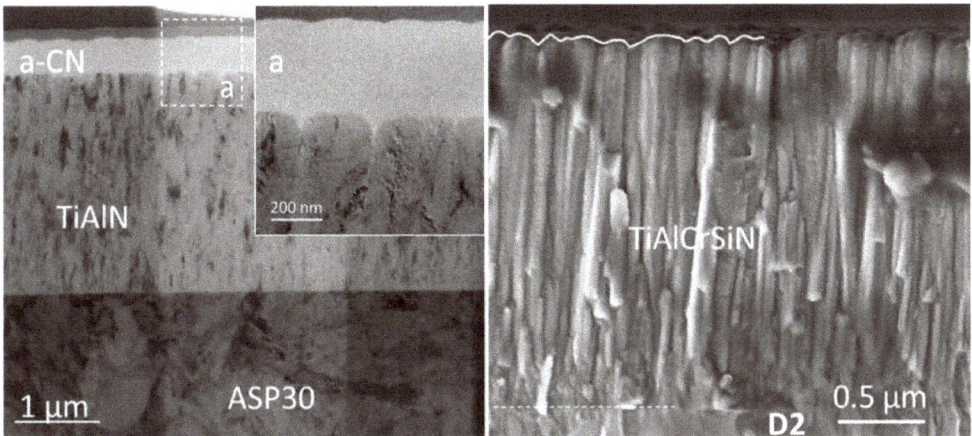

Figure 19. TEM cross-sectional image of a-CN/TiAlN and fracture cross-section SEM image of a TiAlCrSiN hard coating show how columnar microstructure changes the surface topography. It is evident that the surface roughness of the coating has its origin in the columnar microstructure. Both coatings were sputter deposited in the CC7 deposition system on tool steel substrates (ASP30, D2).

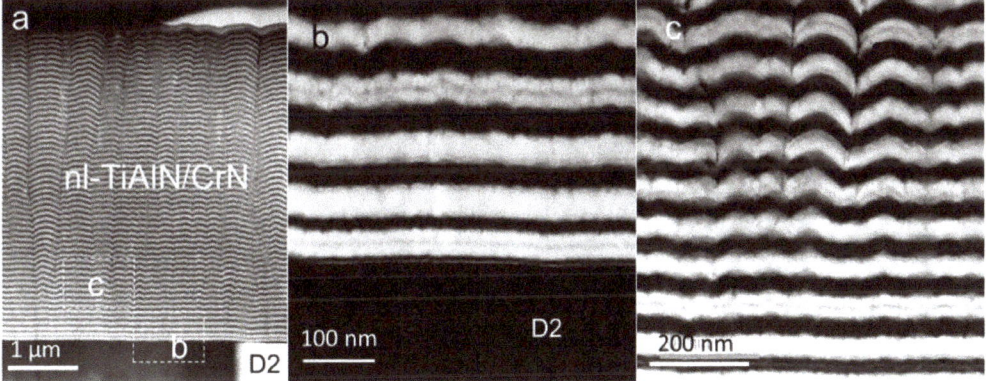

Figure 20. Bright-field STEM cross-sectional images of TiAlN/CrN nanolayered coating sputter deposited in CC7 deposited on D2 tool steel substrate (**a**). The roughness of the TiAlN/CrN interfaces gradually increases up to the top of the coating (**b**,**c**).

The surface roughness can be reduced by using high mobility depositing atoms. Their mobility depends on their energy, substrate temperature and interactions with the substrate surface (i.e., chemical bonding). For example, the most common way to increase the surface mobility of atoms is by low-energy ion bombardment during the deposition.

This simultaneously promotes chemical reactions and introduces new nucleation sites. The columnar growth can be disrupted by intense ion bombardment of the growing coating. For example, transition metal–nitride coatings produced by magnetron sputtering typically exhibit a pronounced columnar microstructure (Figures 19b and 21). However, if the deposition is interrupted and the coating is then exposed to ion bombardment for a certain time, new nucleation sites for the subsequently deposited nitride coating are created. Such intermediate ion etching causes an interruption of the coating columnar growth and eliminates the porosity along the columns. This is reflected in a fine-grained and less porous microstructure of the top layer. A change in the surface topography from dome-shaped column tops (Figure 21b) to a dimpled surface occurs (Figure 21c). After intermediate ion etching, the roughness of the coating surface decreases because the sides of coating protrusions etch faster than the top. This consequently leads to the shrinking of the protrusions and even the elimination of some smaller cones [38].

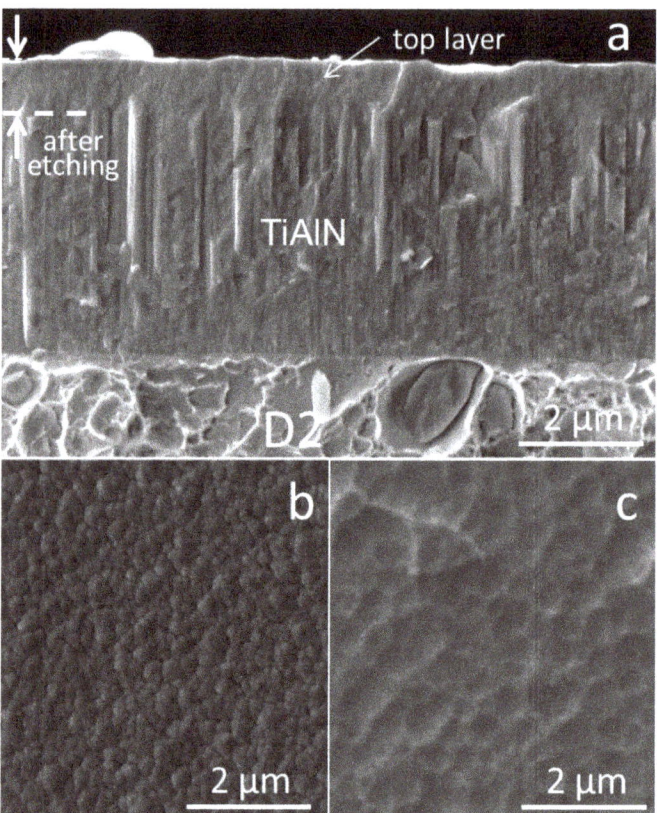

Figure 21. (**a**) Fracture cross-sectional SEM image of TiAlN coating sputter deposited in CC7 on D2 tool steel substrate; (**b**) top-view SEM image of the TiAlN coating prepared without (**b**) and with (**c**) an intermediate ion etching step.

The effect of ion bombardment during deposition strongly depends on the ratio between the deposition rate and the ion current density to the substrate, which should be as low as possible. For the selected deposition method, this ratio is influenced by the target-to-substrate distance, the substrate geometry, substrate rotation mode and batching configuration. At a certain bias voltage on the substrates, this ratio depends mostly on the rotation mode and is the highest for onefold rotation. We also have to consider that both the angle-of-incidence of the depositing atom flux and the deposition rate change constantly during the deposition process. Therefore, it is practically impossible to obtain a uniform and controlled ion bombardment over the surface of the complete batching material. The use of a substrate bias, however, leads to problems when depositions are made onto complex and irregularly shaped tools, with sharp edges and corners. In this case, an increased ion current density around the edges causes either an increased rate of re-sputtering of the deposited material [39] (e.g., magnetron sputtering) or a thickness increase (cathodic arc evaporation) [40]. This effect can be eliminated by proper selection of deposition parameters (e.g., bias voltage, deposition rate, batching configuration, the target-to-substrate distance).

Hard coatings prepared by various deposition techniques and deposition parameters exhibit a wide variety of microstructures which are reflected in different sizes of grains, textures and morphology. For example, Figure 22 shows three different coatings prepared by different deposition methods on the same type of substrate material (ASP30). There is a significant difference in surface morphology between the different coatings. It seems that the initial nucleation of the coating follows the carbide structure in the steel substrates. Thus, on the TiAlN/Al$_2$O$_3$ coating sputter deposited in CC9, small pits appear at sites of (Mo,W)C carbides in the ASP30 tool steel substrate (Figure 22a). The top Al$_2$O$_3$ layer in this coating was deposited by the pulsed magnetron sputtering technique. Very small protrusions at sites of VC carbides can also be observed. On the other hand, pronounced protrusions are formed on the nl-TiAlN/TiN coating prepared in the CC9 deposition system at sites of VC carbides (Figure 22b). The surface topography is completely different when a TiN coating is prepared on ASP30 substrate in the BAI deposition system. It is characterized by dimpled surface topography (Figure 22c). The fracture cross section SEM images also reveal that at the sites of selected VC carbides, epitaxial growth of TiN coating occurs (see inset in Figure 22b,c). This phenomenon, however, does not occur at (Mo,W)C carbide sites and martensitic matrix, where a dense columnar morphology can be observed. All these phenomena show a decisive impact of the microstructure and the micro topography of the substrate surface, formed during ion etching, on the coating topography.

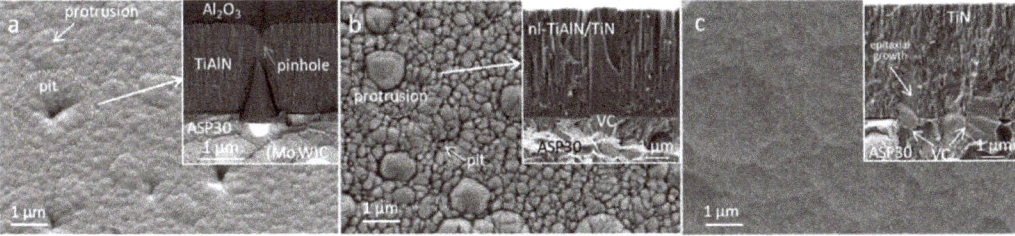

Figure 22. Top view SEM images of TiAlN/Al$_2$O$_3$ (**a**), nanolayered TiAlN/TiN (**b**) and TiN (**c**) hard coatings deposited on ASP30 tool steel substrates in CC7 (**a**,**b**) and BAI deposition systems. Fracture cross sections of the individual coatings are shown in the insets.

SEM images (Figure 23) present the observations of macro- (a–c) and micromorphology (d,e) of TiN hard coatings deposited on D2 tool steel substrates by three different deposition methods: low-voltage arc evaporation (a,d), magnetron sputtering (b,c) and cathodic arc evaporation (c,f). There are obvious differences in coating topography. Differences in topography are not only the result of different ion etching modes, but they also depend on the ratio of the deposition rate and the ion current density to the substrates, which is specific for the selected deposition method.

In the first of these two experiments, we analyzed the morphology of different hard coatings which were deposited by different methods on the same type of substrate (ASP). In the second one, we analyzed the TiN coating deposited on D2 substrates by three different methods. In the following, however, we will address the morphology of TiAlN hard coating, sputter-deposited in the same batch of the CC7 deposition system, on three different types of substrates. Coating topography was analyzed at macro and micro levels using 3D profilometry, AFM and SEM microscopes (Figure 24). At the micro-level (see SEM images), no significant difference in topography can be observed. For all three substrate materials, the dimpled surface topography of the TiAlN coating is characteristic. AFM images that were recorded at a scanning area of 20 μm × 20 μm, reveal a similar surface topography for all three tool steel substrates. Additionally, on coated ASP and D2 substrates, protrusions are visible at sites of carbides. The difference in surface roughness is not very large. It is the largest for D2 (S_a = 30 nm) and the lowest for H11 (S_a = 22 nm) substrates. On the other hand, the surface roughness determined by the 3D profilometer is much larger because the measurements that were performed at a scanning area of 300 μm × 300 μm, and also include the contribution of growth defects. The line profiles were recorded on the defect-free areas. We can see that the surface roughness of TiAlN coatings deposited on ASP and D2 substrates is comparable and much higher than that deposited on the H11 substrate.

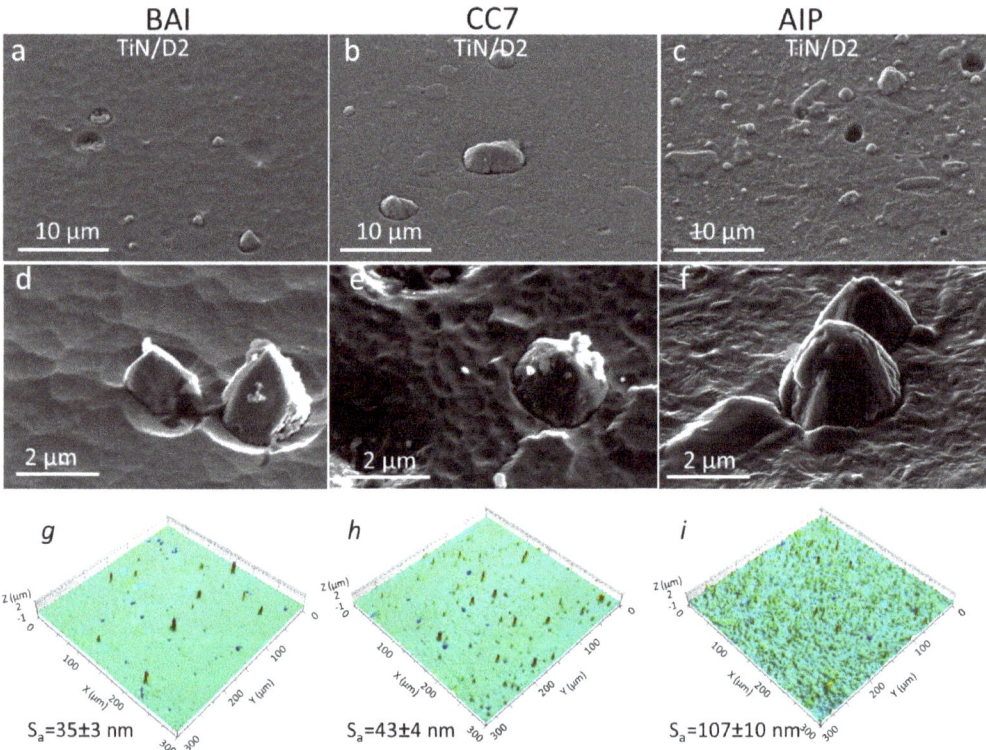

Figure 23. Top view SEM images at lower (**a**–**c**) and higher (**d**–**f**) magnifications and 3D surface profile images (**g**–**i**) of TiN coating deposited on D2 tool steel substrate using BAI, CC7 and AIP deposition systems. The SEM images are the side view from about 45° inclined direction. The sharp peaks in 3D surface profile images are the nodular defects while the blue dots are craters.

In the next test, the nl-TiAlN/TiN hard coating, prepared by magnetron sputtering in the CC9 deposition system, was applied on four different substrates (ASP30, D2, H11, SS 316L). One-third of the substrates were cleaned with MF ion etching and one-third with a combination of MF and "booster" ion etching, while one-third of the samples were not exposed to etching. SEM observations of surface micro topography, presented in Figure 25, show that there are only small differences in the intrinsic morphology of all coatings. All samples are characterized by a dome-shaped morphology of the coating surface in the area without growth defects. The average diameter of the columns is comparable for all samples, regardless of the type of substrate and the method of etching. Similar surface morphology of the coating also occurs on the surface of the carbides. This could mean that the geometrical shadowing effect plays a decisive role in the formation of a columnar microstructure. We assume that the etching of the martensitic matrix in individual steel was similar.

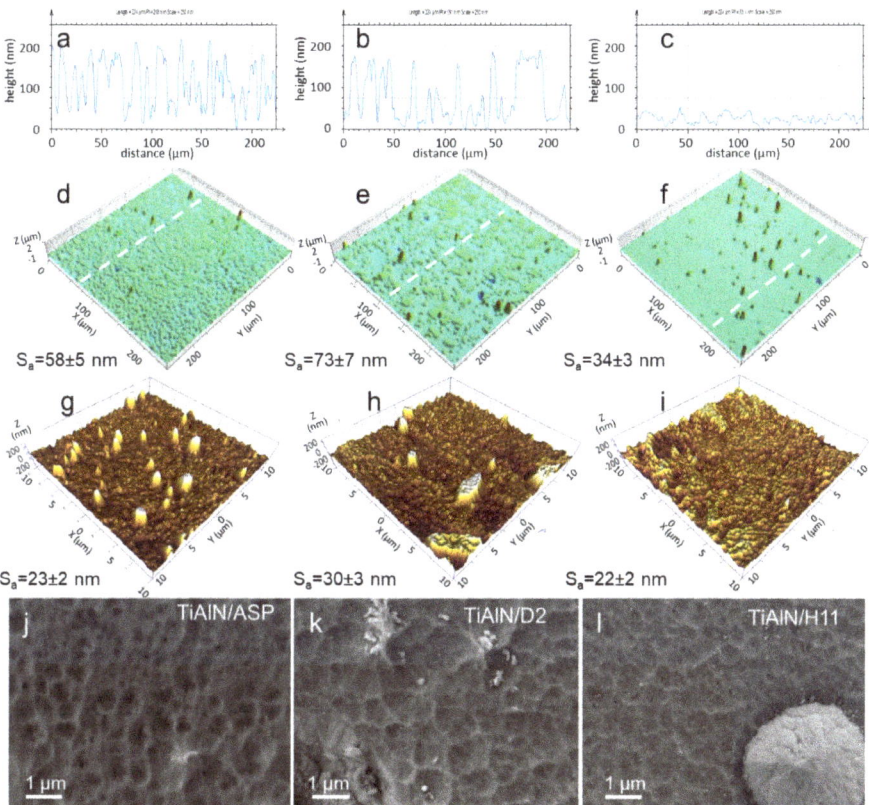

Figure 24. The 3D surface profile images (**d–f**), AFM images (**g–i**) and top-view and SEM (**j–l**) images of TiAlN hard coating sputter deposited in CC7 system on three different tool steel substrates (ASP30, D2, H11). The sharp peaks in 3D surface profile images are the nodular defects while the blue dots are craters. The line profiles (**a–c**) were recorded on the defect-free areas of 3D profile images (**d–f**).

In another test, we deposited different hard coatings on D2 tool steel substrates in the deposition system CC9. The top-view SEM images of different coatings are given in Figure 26. We can see that four different coatings have a similar micro- and nano-morphology, which is similar to cauliflower. This test indicates that the chemical composition of the coating has no effect on its morphology.

The analysis of microstructure and topography shows that there are considerable differences between the coatings deposited in the same batch but using different modes of substrate rotation (Figures 27 and 28). In the case of onefold rotation, the microstructure is characterized by a well-developed columnar microstructure with pores between the individual columns and rough topography. In two- and threefold rotations, the distance to the target and orientation changes significantly during the deposition process. The first consequence is that the incidence angle of the metallic atoms and ions changes all the time, which works strongly against the directional columnar growth. The second consequence arises when the sample is facing away from the target (this applies mainly for threefold rotation) but is still exposed to the bombardment of ions. Namely, the ions from bulk plasma follow the electric field, which is confined in the plasma sheath of the substrate (a few mm). If an ion arrives near the edge of the sheath, it is accelerated toward the substrate. This can also occur in the areas in shadow of the particle flux from the target. In the threefold and partly in the twofold rotations, the periods of growth are followed by periods of densification, which ensures the formation of a compact, dense film [41]. The individual columns are much more pronounced in 1-fold rotation than 2-fold and 3-fold rotations. The microstructural differences between the samples prepared by two- and threefold rotations are relatively modest.

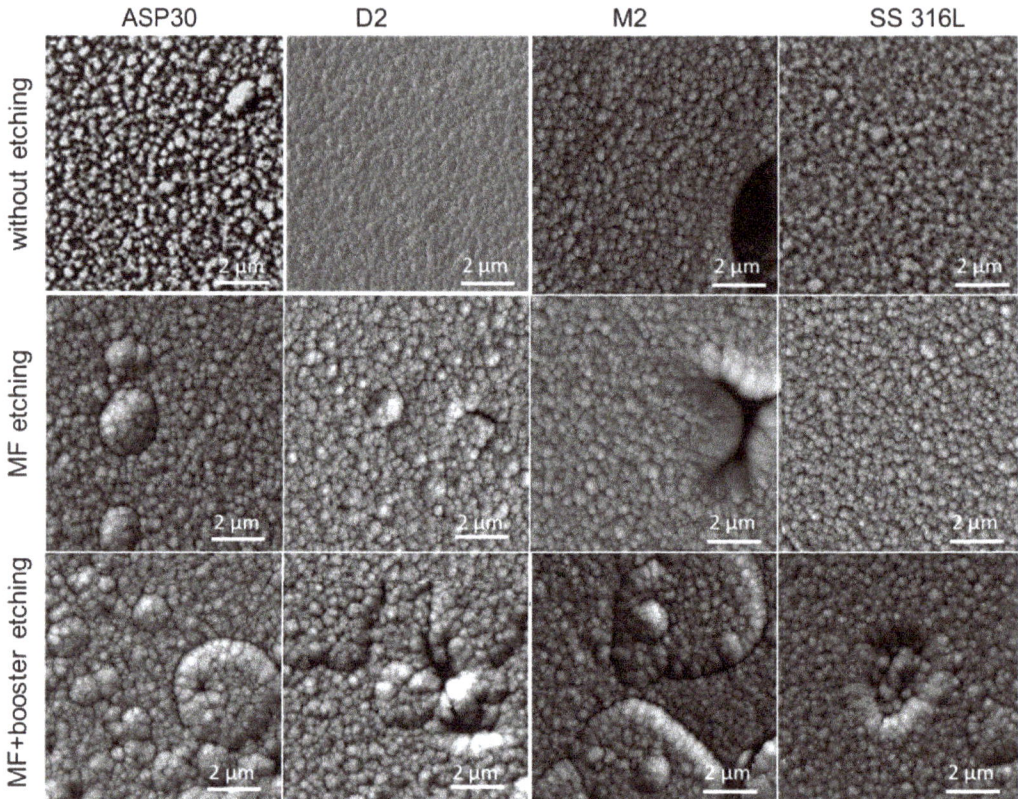

Figure 25. Top-view SEM images of surface topography of nanolayer TiAlN/CrN hard coatings deposited on un-treated, MF and MF + booster etching tool steel (PM ASP30, M2, D2) and stainless steel SS3016L substrates. There are no significant differences in topography.

The effect of ion bombardment on the morphology and topography strongly depends on the type of ion species bombarding the growing coating. There are two sources of ions: working gas ions (i.e., Ar^+ and Kr^+) or metal ions from the target material. In the low-voltage electron beam evaporation (BAI system) and cathodic arc evaporation (AIP system), a large share of the evaporated target material is ionized. This is not the case with the conventional magnetron sputtering process, where the ionization degree of the sputtered metal is typically low. Both sputter deposition systems used in this study (CC7, CC9) are equipped with a high ionization module (HIS), which can improve the ionization of the sputtered metal atoms to more than 50% [18]. The highest ionization degree and charge state of the target material is achieved in the cathodic arc processes (AIP). Differences in the method of evaporation of the target material, the degree of ionization, the type of ions, the ion flux density on the substrates and the deposition rate are also reflected in the microstructure and surface morphology of the coating prepared by different deposition methods. In general, the higher degree of ionization and higher energy of metal atoms means higher coating density.

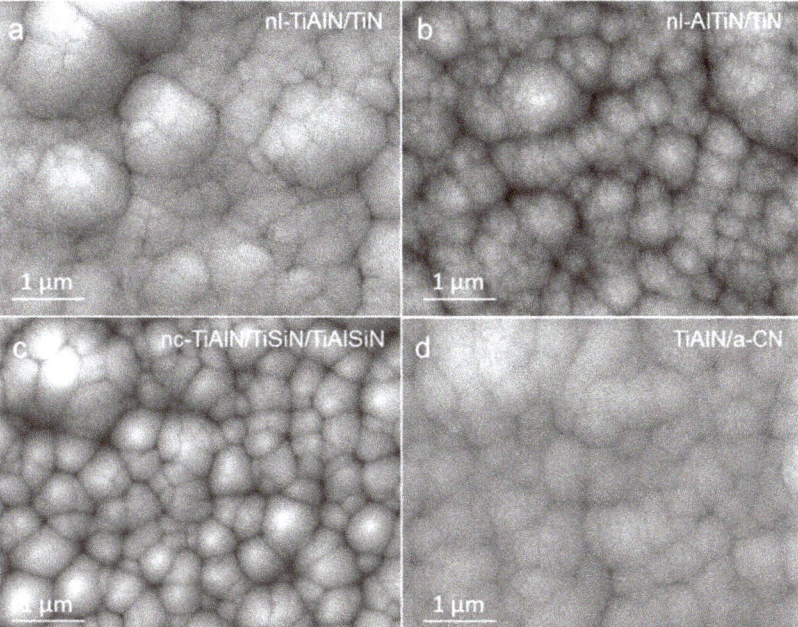

Figure 26. Surface morphology of four different hard coatings: nl-TiAlN/TiN (**a**), nl-AlTiN/TiN (**b**), ns-TiAlN/TiSiN/TiAlSiN (**c**) and TiAlN/a-CN (**d**). They were deposited on D2 tool steel substrates in the CC9 system using 1-fold rotation.

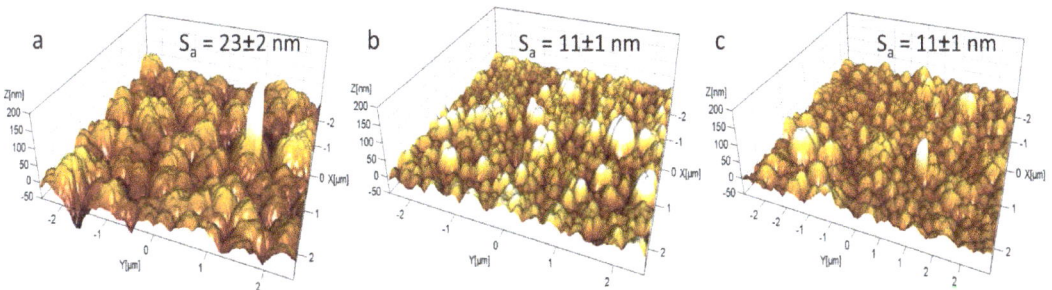

Figure 27. AFM images (scan area 5 μm × 5 μm) of double layer TiAlN/a-CN hard coating sputter deposited on D2 tool steel substrates in the CC9 system. The samples were prepared in the same batch but rotated around a different number of axes: (**a**) onefold, (**b**) twofold, (**c**) threefold rotation.

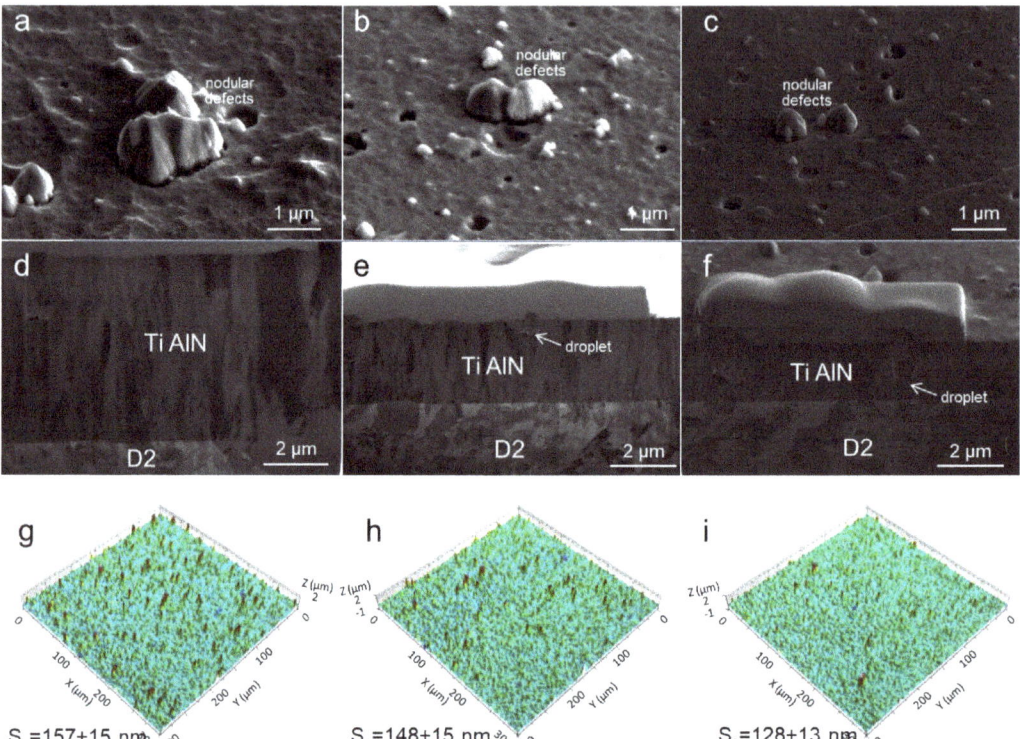

Figure 28. Top-view SEM images (**a**–**c**), SEM images of FIB cross-section (**d**–**e**) and 3D surface profile images (**g**–**i**) of TiAlN hard coatings deposited on D2 tool steel substrate in the AIP deposition system using 1-fold (**a**,**d**,**g**), 2-fold (**b**,**e**,**h**) and 3-fold rotation (**c**,**f**,**i**). The magnification scales are identical. The top SEM images were recorded from the side view at about 45° direction.

3.3. Influence of Growth Defects on Coating Topography

The growth of the coating is also strongly affected by small foreign particles (e.g., dust particles, flakes) remaining on the substrate surface after the cleaning procedure and those generated during the etching and deposition. All topographical irregularities on the substrate surface are transferred through the coating and even magnified due to the geometrical shadowing effect. The shadowing effect is a geometric phenomenon related to the line-of-sight impingement of arriving atoms. Therefore, even relatively small imperfections with the size of several tens of nanometers cause the growth of micrometer-sized imperfections (growth defects) on the coating surface (Figure 29). Growth defects are generally both larger and more protruding than the carbide-induced coating topography. Growth defects of various shapes and sizes are unevenly distributed, with low numbers and low density, while their size varies from one defect to another (depends on the size and geometry of a seed). Recently, we published a review paper where the reader can find more details related to the growth defects in PVD coatings [17].

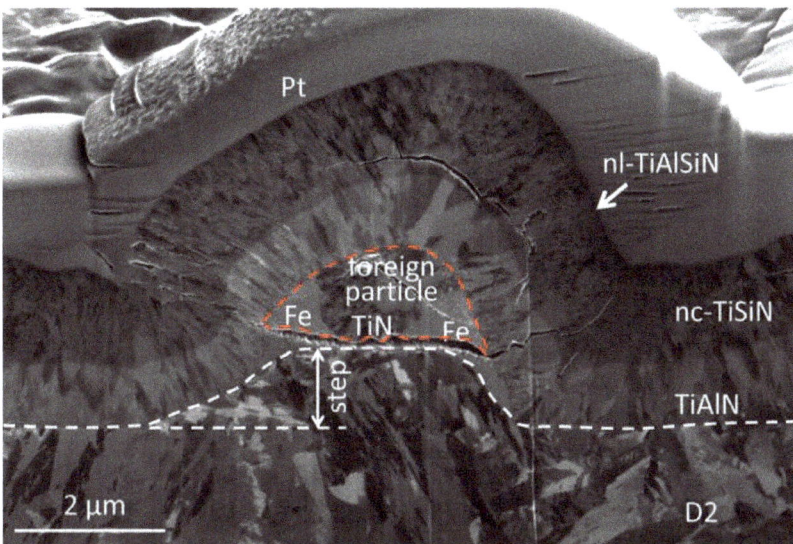

Figure 29. The SEM image was taken at the FIB cross-section of the nodular defect in a nanostructured TiAlN/TiSiN/TiAlSiN hard coating. The nodular defect was formed at the site of a foreign particle. A step on the substrate surface beneath this particle proves that it was on the substrate surface even before etching. The substrate beneath the foreign particle is about 0.8 µm thicker than surrounding area, exposed to ion etching.

Growth defects cause a significant increase in the surface roughness. In the area between the growth defects, the coating exhibits a relatively smooth surface with a roughness that is only slightly larger than that of the uncoated substrate. All AFM images of coated substrates presented in this study were taken at the area between the growth defects. Consequently, the AFM surface roughness is much lower than that obtained by the 3D profilometry on a much larger scanning area (Figure 24).

4. Conclusions

In this study, we analyzed the surface topography of different PVD hard coatings prepared in four batch-type industrial deposition systems, which differ significantly in both the method of ion etching and deposition. We showed that the coating topography originates from the topography of the substrate surface, intrinsic coating micro-topography and growth defects forming during the deposition process. The substrate surface topography is affected by mechanical pretreatment and in particular by the ion etching method. We found that the etching efficiency depends not only on the etching method and etching parameters but also on the batching configuration and substrate geometry. It is challenging to obtain uniform and controlled ion etching over a large surface because the etching rate varies spatially due to the different batching configurations and complex substrate geometry but also temporally due to multiple rotations of substrates. We also discussed in more detail the influence of ion bombardment during the etching procedure on the surface micro-roughening and the formation of various micro-sized surface features (e.g., cones, pits, steps, trenches).

Hard coatings prepared by various deposition techniques and at different discharge parameters exhibited a wide variety of micro-topographies. The coating morphology is determined by the substrate surface roughness (due to geometrical shadowing effect) and the surface mobility of the depositing atoms. There were considerable differences between the coatings deposited in the same batch using different substrate rotation modes. Specific rotation mode determines the ratio between the deposition rate and the ion current density to the substrate. Additionally, in the case of two- and threefold rotations, the orientation of the substrate changed significantly and non-periodically, which strongly affected the ion etching and coating deposition. In general, the intensive ion bombardment of the growing coating disrupted its columnar microstructure and changes its topography. Based on these observations, the use of the same deposition protocol for deposition of PVD coatings on the one-, two- and threefold rotating substrates in the same batch is not suitable; a rotation-specific protocol offers a better solution. Batching configuration is also of great importance, both in terms of ion etching and coating deposition. One should avoid batching tools of very different sizes and geometries in the same deposition process.

The biggest changes in the topography of the surface of the coatings were caused by the growth defects. Investigation of the as-deposited coatings showed that there was a significant amount of surface growth defects, mainly nodular defects and shallow craters, both formed during the coating deposition process. These growth defects contributed the most to the roughness of the substrate.

Surface topography analysis of PVD coatings prepared in different industrial deposition systems is a necessary tool to better understand of the relationship between the coating topography and the process parameters. Such analysis should be performed regularly to produce coatings with reproducible performance.

Author Contributions: Conceptualization, manuscript writing, P.P.; AFM, SEM and FIB analysis, manuscript review, A.D.; manuscript review, SEM analysis, M.Č.; AFM and TEM analysis, manuscript writing, M.P.; analysis, manuscript writing, SRIM simulation, N.M. All authors have read and agreed to the published version of the manuscript.

Funding: This work was supported by Slovenian Research Agency (program P2-0082). We also acknowledge funding from the European Regional Development Funds (CENN Nanocenter, OP13.1.1.2.02.006) and European Union Seventh Framework Programme under grant agreement 312483–ESTEEM2 (Integrated Infrastructure Initiative—I3).

Institutional Review Board Statement: Not applicable.

Informed Consent Statement: Not applicable.

Data Availability Statement: Not applicable.

Acknowledgments: The authors would also like to thank Peter Gselman, Tonica Bončina and Gregor Kapun for SEM analyses and Jožko Fišer for performing some laboratory tests.

Conflicts of Interest: The authors declare no conflict of interest.

References

1. Assender, A.; Bliznyuk, V.; Porfyrakis, K. How Surface Topography Relates to Materials' Properties. *Science* **2002**, *297*, 973–976. [CrossRef] [PubMed]
2. Harlin, P.; Carlsson, P.; Bexell, U.; MOlsson, M. Influence of surface roughness of PVD coatings on tribological performance in sliding contacts. *Surf. Coat. Technol.* **2006**, *201*, 4253. [CrossRef]
3. Panjan, P.; Drnovšek, A.; Kovač, J. Tribological aspects related to the morphology of PVD hard coatings. *Surf. Coat. Technol.* **2018**, *343*, 138–147. [CrossRef]
4. Olofsson, J.; Gerth, J.; Nyberg, H.; Wiklund, U.; Jacobson, S. On the influence from micro topography of PVD coatings on friction behaviour, material transfer and tribofilm formation. *Wear* **2011**, *271*, 2046–2057. [CrossRef]
5. Etsion, I. State of the Art in Laser Surface Texturing. *J. Tribol.* **2005**, *127*, 249. [CrossRef]
6. Vilhena, L.M.; Sedlaček, M.; Podgornik, B.; Vižintin, J.; Babnik, A.; Možina, J. Surface texturing by pulsed Nd:YAG laser. *Tribol. Int.* **2009**, *42*, 1496–1504. [CrossRef]
7. Kearns, V.R.; McMurray, R.J.; Dalby, M.J. Biomaterial surface topography to control cellular response: Technologies, cell behaviour and biomedical applications. In *Surface Modification of Biomaterials*; Williams, R., Ed.; Woodhead Publishing Limited: Cambridge, UK, 2011; pp. 169–201.
8. Curtis, A.; Wilkinson, C. Topographical control of cells. *Biomaterials* **1997**, *18*, 1573. [CrossRef]
9. Wu, S.; Zhang, B.; Liu, Y.; Suo, X.; Li, H. Influence of surface topography on bacterial adhesion. *Biointerphases* **2018**, *13*, 060801. [CrossRef]

10. Damiati, L.; Eales, M.G.; Nobbs, A.H.; Su, B.; Tsimbouri, P.M.; Salmeron-Sanchez, M.; Dalby, M.J. Impact of surface topography and coating on osteogenesis and bacterial attachment on titanium implants. *J. Tissue Eng.* **2018**, *9*, 1–16. [CrossRef]
11. Oshida, Y.; Guven, Y. Biocompatible coatings for metallic biomaterials. In *Surface Coating and Modification of Metallic Biomaterials*; Wen, C., Ed.; Woodhead Publishing: Cambridge, UK, 2015; pp. 287–343.
12. Liu, X.; Li, D.; Zhao, Y.; Li, X.; Shao, J. Characteristics of nodular defect in HfO_2/SiO_2 multilayer optical coatings. *Appl. Surf. Sci.* **2010**, *256*, 3783–3788. [CrossRef]
13. Piegari, A.; Flory, F. *Optical Thin Films and Coatings: From Materials to Applications*; Woodhead Publishing Limited: Cambridge, UK, 2018.
14. Rebeggiani, S.; Rosen, B.G.; Sandberg, A. A quantitative method to estimate high gloss polished tool steel surfaces. *J. Phys. Conf. Ser.* **2011**, *311*, 012004. [CrossRef]
15. Gassner, M.; Schalk, N.; Sartory, B.; Pohler, M.; Czettl, C.; Mitterer, C. Influence of Ar ion etching on the surface topography of cemented carbide cutting inserts. *Int. J. Refract. Metals Hard. Mater.* **2017**, *69*, 234–239. [CrossRef]
16. Moll, E.; Daxinger, H. Method and Apparatus for Evaporating Materials in a Vacuum Coating Plant. U.S. Patent 4,197,175, 8 April 1980.
17. Panjan, P.; Drnovšek, A.; Gselman, P.; Čekada, M.; Panjan, M. Review of Growth Defects in Thin Films Prepared by PVD Techniques. *Coatings* **2020**, *10*, 447. [CrossRef]
18. Erkens, G. New approaches to plasma enhanced sputtering of advanced hard coatings. *Surf. Coat. Technol.* **2007**, *201*, 4806–4812. [CrossRef]
19. Yamamoto, Y.; Sato, T.; Takahar, K.; Hanaguri, K. Properties of (Ti,Cr,Al)N coatings with high Al content deposited by new plasma enhanced arc-cathode. *Surf. Coat. Technol.* **2003**, *174–175*, 620–626. [CrossRef]
20. Sousa, V.F.C.; Silva, F.J.G.; Lopes, H.; Casais, R.C.B.; Baptista, A.; Pinto, G.; Alexandre, R. Wear behavior and machining performance of TiAlSiN-coate tools obtained by dc MS and HiPIMS: A comparative study. *Materials* **2021**, *14*, 5122. [CrossRef]
21. Saketi, S.; Östby, J.; Olsson, M. Influence of tool surface topography on the material transfer tendency and tool wear in the turning of 316L stainless s steel. *Wear* **2016**, *368–369*, 239–252. [CrossRef]
22. Klocke, F.; Dambon, O.; Behrens, B. Analysis of defect mechanisms in polishing of tool steels. *Prod. Eng.* **2011**, *5*, 475–483. [CrossRef]
23. Hovsepian, P.E.; Ehiasarian, A.P. Six strategies to produce application tailored nanoscale multilayer structured PVD coatings by conventional and High Power Impulse Magnetron Sputtering (HIPIMS). *Thin Solid Films* **2019**, *688*, 137409. [CrossRef]
24. Taglauer, E. Surface Cleaning Using Sputtering. *Appl. Phys. A* **1990**, *51*, 238–251. [CrossRef]
25. Ghose, D.; Karmohapatro, B. Topography of Solid Surfaces Modified by Fast Ion Bombardment. *Adv. Electron. Electron Phys.* **1990**, *79*, 73–154.
26. Navinšek, B. Sputtering—surface changes induced by ion bombardment. *Prog. Surf. Sci.* **1976**, *7*, 49–70. [CrossRef]
27. Wehner, G.K.; Hajicek, D.J. Cone Formation on Metal Targets during Sputtering. *J. Appl. Phys.* **1971**, *42*, 1145. [CrossRef]
28. Ziegler, J.; Biersack, J.P.; Ziegler, M.D. *SRIM-The Stopping and Ranges of Ions in Solids*; SRIM Co.: Chester, UK, 2008.
29. Vossen, J.L. The preparation of substrates for film deposition using glow discharge techniques. *J. Phys. E Sci. Instrum.* **1979**, *12*, 159. [CrossRef]
30. Nordin, M.; Ericson, F. Growth characteristics of multilayered physical vapour deposited TiN/TaN_x on high speed steel substrate. *Thin Solid Films* **2001**, *385*, 174–181. [CrossRef]
31. Baptista, A.; Silva, F.; Porteiro, J.; Miguez, J.; Pinto, G. Sputtering Physical Vapour Deposition (PVD) Coatings: A Critical Review on Process Improvement andMarket Trend Demands. *Coatings* **2018**, *8*, 402. [CrossRef]
32. Terek, V.; Miletić, A.; Kovačević, L.; Kukuruzović, D.; Škorić, B.; Panjan, P.; Terek, P. Surface Topography and Grain Morphology of Nanolayer TiAlN/TiSiN Coating Governed by Substrate Material and Rotation during Deposition. *Mater. Proc.* **2020**, *2*, 32.
33. Grigoriev, S.; Metel, A. Plasma- and beam-assisted deposition methods. In *Nanostructured Thin Films and Nanodispersion Strengthened Coatings*; Voevodin, A.A., Shtansky, D.V., Levashov, E.A., Moore, J.J., Eds.; Kluwer Academic Publishers: Dordrecht, The Netherlands, 2004; pp. 147–154.
34. Mattox, D.M. Particle bombardment effects on thin film deposition: A review. *J. Vac. Sci. Technol. A* **1989**, *7*, 1105. [CrossRef]
35. Macak, E.B.; Munz, W.D.; Rodenburg, J.M. Electron microscopy studies of hard coatings deposited on sharp edges by combined cathodic arc unbalanced magnetron PVD. *Surf. Coat. Technol.* **2002**, *151–152*, 349–354. [CrossRef]
36. Mattox, D.M. Atomistic Film Growth and Resulting Film Properties. *SVC Bull. Spring* **2009**, 30–31.
37. Petrov, I.; Barna, P.B.; Hultman, L.; Green, J.E. Microstructural evolution during film growth. *J. Vac. Sci. Technol. A* **2003**, *21*, S117–S128. [CrossRef]
38. Mirkarimi, B.; Baker, S.L.; Stearns, D.G. Planarization of Substrate Pits and Scratches. U.S. Patent 2005/0118533 A1, 2 June 2005.
39. Johansson, B.O.; Sundgren, J.E.; Hentzell, H.T.G.; Karlsson, S.E. Influence of substrate shape on TiN films prepaared by reactive sputtering. *Thin Solid Films* **1984**, *111*, 313–322. [CrossRef]
40. Čekada, M.; Panjan, P.; Drnovšek, A.; Drobnič, M. Increase of coating thickness on sharp edges, deposited by cathodic arc evaporation. *Surf. Coat. Technol.* **2021**, *405*, 126691. [CrossRef]
41. Panjan, M.; Čekada, M.; Panjan, P.; Zupanič, F.; Kölker, W. Dependence of microstructure and hardness of TiAlN/VN hard coatings on the type of substrate rotation. *Vacuum* **2012**, *86*, 699–702. [CrossRef]

Article

Microstructure and Surface Topography Study of Nanolayered TiAlN/CrN Hard Coating

Peter Panjan [1,*], Peter Gselman [1,2], Matjaž Panjan [1], Tonica Bončina [3], Aljaž Drnovšek [1], Mihaela Albu [4], Miha Čekada [1] and Franc Zupanič [3]

1. Jožef Stefan Institute, Jamova Cesta 39, 1000 Ljubljana, Slovenia
2. Interkorn d.o.o, Gančani 94, 9231 Beltinci, Slovenia
3. Faculty of Mechanical Engineering, University of Maribor, Smetanova 17, 2000 Maribor, Slovenia
4. Graz Centre for Electron Microscopy, Steyrergasse 17, A-8010 Graz, Austria
* Correspondence: peter.panjan@ijs.si; Tel.: +386-1-477-3278

Abstract: The microstructure and surface topography of PVD hard coatings are among the most important properties, as they significantly determine their mechanical, tribological and other properties. In this study, we systematically analyzed the microstructure and topography of a TiAlN/CrN nanolayer coating (NL-TiAlN/CrN), not only because such coatings possess better mechanical and tribological properties than TiAlN and CrN monolayer coatings, mainly because the contours of the individual layers, in the cross-sectional STEM or SEM images of such coatings, make it easier to follow topographic and microstructural changes that occurred during its growth. We investigated the effects of the substrate rotation modes on the microstructure and surface topography of the NL-TiAlN/CrN coating, as well as on the periodicity of the nanolayer structure. The influence of the substrate material and the ion etching methods were also studied, while special attention was given to the interlayer roughness and influence of non-metallic inclusions in the steel substrates on the growth of the coating. The topographical features of the NL-TiAlN/CrN coating surface are correlated with the observations from the cross-sectional TEM and FIB analysis. Selected non-metallic inclusions, covered by the NL-TiAlN/CrN coating, were prepared for SEM and STEM analyses by the focused ion beam. The same inclusions were analyzed prior to and after deposition. We found that substrate rotation modes substantially influence the microstructure, surface topography and periodicity of the NL-TiAlN/CrN layer. Non-metallic inclusions in the substrates cause the formation of shallow craters or protrusions, depending on their net removal rates during the substrate pretreatment (polishing and ion etching), as compared to the matrix.

Keywords: magnetron sputtering; nanolayer hard coatings; growth defects; surface topography; interlayer roughness; non-metallic inclusion; focused ion beam (FIB); scanning electron microscopy (SEM); scanning transmission electron microscopy (STEM)

1. Introduction

In order to provide efficient wear protection for tools with PVD hard coatings, besides high hardness, fracture toughness and oxidation resistance, they must have a smooth surface as well. The surface topography of PVD hard coatings is of key importance, especially for their tribological properties [1–6]. In the case of the cutting process, for example, a smooth, hard coating surface reduces mechanical interaction with the chip; therefore, the friction and material transfer are also reduced.

- In our previous works, we discussed, in more detail, the coating surface topography [6–9]. We have shown that the topography of a PVD coating surface originates from: A substrate surface topography formed during the substrate mechanical pretreatment (grinding, blasting and polishing) and during the substrate cleaning by ion etching prior to the coating deposition;
- Intrinsic coating features such as grain size, phase composition, texture, etc.;

- Growth defects (e.g., nodulus and pinholes) formed during the coating growth process.

We have also shown that the multilayered coatings are more suitable for the study of growth defects because the contours of individual layers in the cross-sectional SEM and TEM images reveal their internal structure. That is also the reason why, in this study, we analyzed the surface topography of the TiAlN/CrN nanolayer coating (NL-TiAlN/CrN) in more detail.

On a macroscale, the surface roughness mainly originates from growth defects in a size range above 3 μm. The inspection of the coating surface at lower magnification shows a large number of unevenly distributed growth defects of various shapes and sizes. The growth defect density depends on the type of ion etching and deposition processes and deposition parameters, as well as on the type of substrate material and its mechanical and ion etching pretreatments. In order to change the micro-topography and to reduce the surface roughness of the coated tool, post-coating treatments (polishing and blasting) are often used.

On the microscale, the surface roughness mainly depends on the columnar microstructure and the crystallinity of the coating. Depending on the deposition conditions, the coating microstructure can be coarse columnar, fine columnar or amorphous. The formation of the columnar microstructure on a rough substrate surface is a result of the geometrical shadowing effect. Because the sputtered atoms travel from the source to the substrate in a straight line, the substrate surface morphology affects their angle of incidence; therefore, the surface coverage decreases. However, the growth of columns depends not only on the substrate surface roughness but also on the angle of incidence of the arriving atoms, the deposition temperature and the intensity of energetic particle bombardment [8]. While the first two mechanisms increase the roughness of the coating surface, the other two increase the surface mobility of the depositing atoms, which consequently decreases its roughness or even disrupts the columnar microstructure. A columnar microstructure can also develop on a smooth substrate surface but only if the pronounced preferential growth of crystal planes occurs. The coating surface morphology and, thus, its roughness are determined by the diameter of the columns and by the faceted column tops. Both contributions increase with the coating thickness. Columnar growth can also be observed in the multilayered hard coatings. In such coatings, the growth of some columnar grains stops at the layer interfaces, while others start growing, as the layer interfaces provide more opportunities for the nucleation of new crystal grains. These processes result in a refinement of crystal grains.

During the coating deposition, the surface roughness is increased not only due to columnar growth but also because crystal grains with different orientations grow faster than others. Thus, the growth of new columns with more preferred growth orientations or phases dominates [10]. In a multilayer structure, the last deposited layer can provide a template for the growth of the next layer. If the two materials in the multilayer structure have the same crystal structure and if their lattice mismatch is small enough (no more than several percent), this can result in epitaxial growth [11–13]. Epitaxial growth can extend through the whole layer if its thickness is kept small enough (a few nm) [10].

In this paper, we focus on the microstructure, surface topography, interlayer roughness and formation of growth defects in a NL-TiAlN/CrN hard coating. This coating was chosen because the growth defect formation and other coating surface irregularities are easier to observe in a multilayer structure. In addition, the NL-TiAlN/CrN coating possesses enhanced mechanical and tribological properties as compared to TiAlN and CrN monolayer coatings. The stresses formed at the interfaces due to different lattice constants significantly contribute to higher hardness of the nanolayer structure coating in comparison with the corresponding monolayers [14]. The improvement of the coating properties is contributed to the layer interfaces in such a structure, which act as a barrier to dislocation motion, crack propagation and elemental diffusion. In the literature, the NL-TiAlN/CrN coating has been investigated by several groups [13,15–18]. It was demonstrated that such a coating could be deposited with clear superlattice characteristics. We should also mention that the properties (e.g., morphology, microhardness, fracture toughness and residual stresses) of such nanolayer coatings could be tailored in a controlled manner by varying their modulation period [19].

2. Materials and Methods

The NL-TiAlN/CrN coatings were prepared in the industrial deposition system CC800/9 (Cemecon, Würselen, Germany). The deposition system was, in detail, described elsewhere [8]. The unit was equipped with four unbalanced magnetron sources placed in the corners of a rectangular vacuum chamber. For the deposition of the NL-TiAlN/CrN coatings, two segmental TiAl targets were positioned at one side of the chamber and two Cr targets at the opposite side. The dimension of the sources was 500 mm × 88 mm. The TiAl target was made of a titanium base with 48 cylindrical aluminum plugs within the racetrack. The coating composition was Ti 23 at.%, Al 27 at.%, and N 50 at.%N. In order to form a nanolayer structure, the substrates were mounted on a planetary substrate holder system that enabled up to three-fold rotation. A detailed description of the rotation geometry is given in [20,21]. During the preparation of nanolayer coatings, all the cathodes were operated simultaneously, while the nanolayer structure resulted from the substrate rotation.

As substrates, we used test plates made of cold work tool steel (AISI D2, Ravne steel factory, Ravne, Slovenia), conventional high-speed steel (HSS, AISI M2, Ravne steel factory, Ravne, Slovenia), powder metallurgical tool steel ASP30 (AISI M3:2+Co, Uddeholm, Hagfos, Sweden),stainless steel (AISI 316L, Acroni, Jesenice, Slovenia) and cemented carbide (WC-Co, Mecut, Ceranesi, Italy). Substrates were first ground and polished to a mirror finish, corresponding to surface roughness S_a values of around 12 nm. Before deposition, they were cleaned with detergents and ultrasound, rinsed in deionized water and dried in hot air.

The coating preparation was performed in a standard way, which included substrate heating by resistive heaters up to about 450 °C, etching by argon and krypton ions and DC sputter deposition. In a separate batch, one set of the substrates was cleaned by mid-frequency (MF) ion etching (240 kHz, duty cycle 1600 ns) conducted for 60 min with a peak power of 1400 W. The MF ion etching was performed in mixed argon (flow rate 180 mL/min) and krypton (flow rate 50 mL/min) atmosphere under the pressure of 0.35 Pa, while the bias voltage applied on the turntable was 650 V. The next set of substrates was cleaned by MF ion etching and, additionally, with the so-called "booster" ion etching. During the booster etching, the working gas was injected through the upper and lower booster etch nozzles (i.e., a hollow cathode), where intensive ionization of the working gas (Ar, Kr) occurs. Such additional discharge enhances the plasma density and, thus, the intensity of the etching process.

After ion etching, the sample surfaces were examined with a scanning electron microscope (SEM), energy-dispersive X-ray spectroscopy analysis (EDX), atomic force microscope (AFM) and 3D stylus profilometer. After that, all the substrates were put again into the deposition system where they were once more heated to ~450 °C and cleaned by MF ion etching for just 10 min to remove any possible impurities formed during the previous analysis.

The NL-TiAlN/CrN coating was deposited in a mixture of argon (flow rate 150 mL/min), krypton (flow rate 100 mL/min) and nitrogen (flow rate 70 mL/min) at a total pressure of 620 mPa. A DC bias of −90 V was applied to the substrates. The power on the Cr and TiAl targets was 4.5 kW and 9.5 kW, respectively. Coating thickness was measured by the ball crater method (Calotest), as well as from the SEM images of the sample fracture. For one-, two- and three-fold rotations, the thicknesses were approximately 13.5 µm, 7.5 µm and 3 µm, respectively.

Surface topography of precoated and coated substrates was analyzed by the atomic force microscope Solver PRO (AFM, NT-MDT Spectrum Instruments, Moscow, Russia) and by the 3D stylus profilometer (Bruker Dektak XT, Billerica, MA, USA). The scanning area of AFM was 5 µm × 5 µm. The evaluation area of the 3D stylus profilometer was 1 mm^2, while the resolution in x, y and z (vertical) directions were 0.2 µm, 1 µm and around 5 nm, respectively.

The morphology and layer structure of the coatings were studied using Zeiss Supra 35 VP (Jena, Germany), JEOL JSM 6500F and Sirion 400 NC (FEI, Eindhoven, Netherlands) scanning electron microscopes. All three microscopes were field emission SEM with the ability of EDX analysis (Oxford INCA 350, Oxford Instruments, Abingdon, UK). Cross-sections for SEM investigations were also prepared by focused ion beam techniques using an FIB source (Keithley Instruments Inc., Solon, Ohio) integrated into the scanning electron microscope Helios NanoLab NL650 (FEI, Eindhoven, Netherlands).

Detailed microstructural analysis of NL-TiAlN/CrN coating was performed by Jeol JEM-2100 transmission electron microscope (TEM) and a probe corrected Titan 60-300 (Thermo-Fisher, Eindhoven, Netherland) scanning transmission electron microscope (STEM) equipped with a high-angle annular dark-field detector and EDX analysis (ChemiStem set-up, Thermo Fisher, FEI, Eindhoven, Netherlands). It was used to study the interface characteristics at the atomic scale. Some specimens for TEM cross-sectional analysis were prepared by the conventional technique [12], while those for STEM were prepared by using a focused ion beam (FIB) workstation (FEI NOVA 200 Nanolab field emission scanning electron microscope, FEI, Eindhoven, The Netherlands). Four different modes were used to obtain different information of layer interfaces: high-resolution (HRTEM), high-angle annular dark-field (HAADF-STEM) [22], dark-field (DF-STEM) and bright-field (BF-STEM).

3. Results

3.1. Microstructure Characterization and Periodicity Analysis

Figure 1 shows the SEM images of the fractured surface of the NL-TiAlN/CrN hard coatings prepared in the same batch by one-, two- and three-fold rotation. The bright layers correspond to TiAlN and the dark ones to CrN. The coatings prepared by single, double and triple rotation differ not only in the total thickness but also in the microstructure and periodicity of the thin layers. All these differences can be better seen in cross-sectional TEM images. Figure 2 shows bright-field (BF) TEM images of the NL-TiAlN/CrN coating taken close to the substrate surface. For samples prepared using the one-fold substrate rotation, the individual layers had similar thicknesses (approximately 90 nm). For samples prepared using the two-fold substrate rotation, the thickness of the individual layers is reduced by about half, in comparison with the one-fold rotation, and varies slightly in coating thickness. In the case of the sample prepared by the three-fold rotation, the layer thickness is between 10 nm and 25 nm.

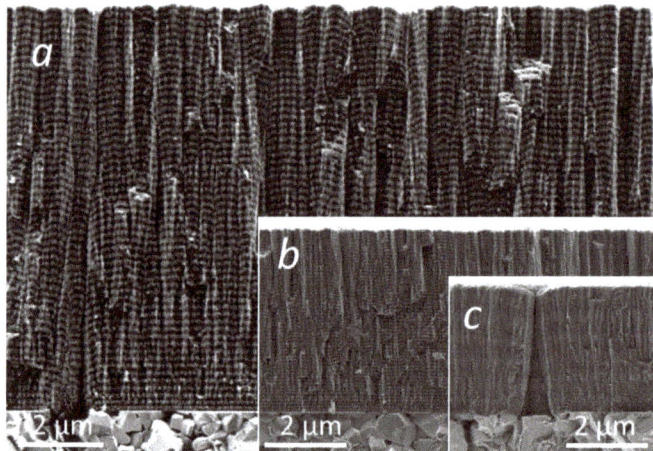

Figure 1. Fracture cross-sectional SEM images of NL-TiAlN/CrN hard coating sputter deposited on cemented carbide substrates. The samples were prepared in the same batch but rotated around a different number of axes: (**a**) one-fold, (**b**) two-fold and (**c**) three-fold rotation. The magnification scales are identical.

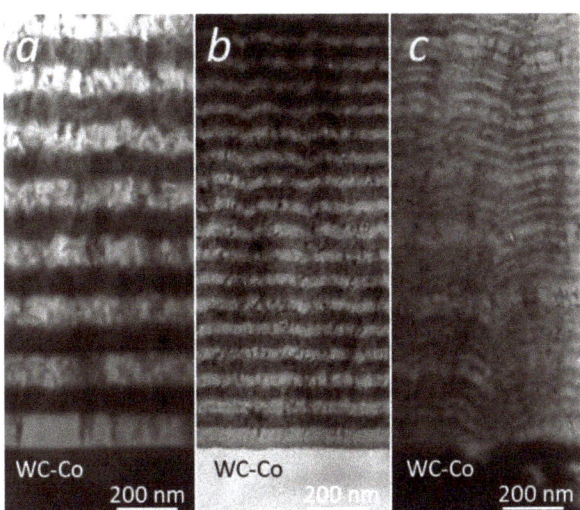

Figure 2. Cross-sectional TEM images of NL-TiAlN/CrN hard coating with a typical columnar grain growth and sputter deposited on cemented carbide substrates. The samples were prepared in the same batch but rotated around a different number of axes: (**a**) one-fold, (**b**) two-fold and (**c**) three-fold rotation. The magnification scales are identical.

In our previous articles, we showed that the periodicity of NL-TiAlN/CrN coatings strongly depends on the substrate rotation mode [20,21]. This topic has also been addressed by other authors [23]. In the case of the simplest one-fold rotation, the trajectory of the substrate is a circle. Therefore, the deposition rate and the layer structure are periodic. In the case of the two-fold rotation, the trajectory of the sample is more complex. The number of rotations required to return the sample to the same position and orientation depends on the gear ratio of the planetary substrate holder system. As a result, the layered structure is not fully periodic. This is clearly visible from the STEM image and corresponding EDX analysis (Figure 3). The EDX line scan performed continuously from the aluminum oxide inclusion in the D2 tool steel substrate through the first ten layers shows the relative changes in the content of each element that composes the NL-TiAlN/CrN coating. We should also note that the thickness of the first few layers changes due to an increasing cathode power at the beginning of the deposition. It can be also observed that all interfaces are well defined, without visible intermixing. The addition of a third rotation (in our case non-continuous) results in the formation of still more complex multilayer structures. In this case, the sample practically never returns to the same position and orientation, while the periodicity of the multilayer structure even depends on the initial position of the substrates. Therefore, the layer structure prepared by the three-fold rotation is aperiodic.

It can be also seen that the microstructure of the NL-TiAlN/CrN coating is considerably different for the samples prepared by different rotation modes. The TEM bright-field image reveals polycrystalline elongated columns, perpendicular to the substrate surface. The average width of columns with a curved top surface is around 600 nm, 280 nm and 110 nm, for one-fold, two-fold and three-fold rotation, respectively. In the case of a one-fold rotation, unlike in a two-fold and three-fold rotation, a lot of columnar grains extend from the substrate to the coating surface. The microstructure of the coatings prepared by the two-fold rotation is denser than that prepared by the one-fold rotation. The most fine-grained and compact are the coatings prepared by the three-fold rotation. Although the plasma conditions are identical for the different rotation modes, the trajectories and, particularly, the orientation of the substrate during rotation are very different. In the one-fold rotation, the substrates pass individual magnetron sources following the same trajectory (i.e., the distance and orientation changes are the same in every rotational cycle). However, this is not the case with two-fold and three-fold rotations. The substrate distance and orientation from the center of the turntable and from individual magnetron sources change significantly during the rotation. In certain positions, the sample may still be in the plasma but may face away from the target. In such a position, the coating surface is still exposed to the ion bombardment, while the coating does not grow. Additionally, the incidence angle of ions and the atom flux change constantly during the deposition. It is understandable that such inhomogeneous growth conditions strongly influence the microstructure and surface topography of the coating.

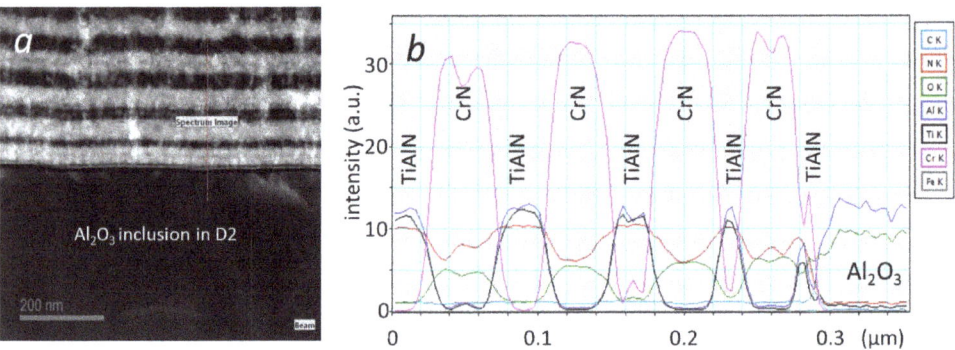

Figure 3. High-angle annular dark-field (HAADF) STEM image (Z-contrast image) of the NL-TiAlN/CrN hard coating sputter deposited on an Al_2O_3 inclusion in the D2 tool steel substrate using the two-fold rotation (**a**) and EDS line scan along the dashed red line indicated in the STEM image (**b**).

The NL-TiAlN/CrN coating is isostructural as it is composed of TiAlN and CrN individual layers that have the same (fcc—face-centered cubic) crystal structure. Because the lattice mismatch of both fcc crystal structures is small (<2%), epitaxial growth occurred, as can be seen in the high-resolution TEM image (Figure 4). We can see that the lattice fringes are continuous across neighbouring layers, which proves the layer coherency within the columnar grain (Figure 4c). The phenomenon of epitaxial growth in the NL-TiAlN/CrN layer was already discussed in more detail in our previous paper [12].

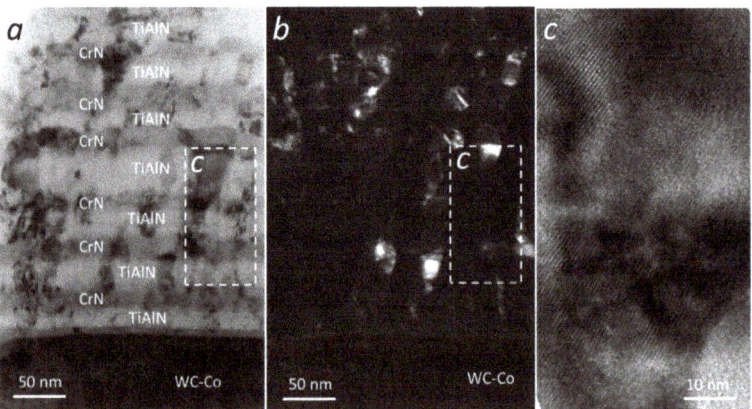

Figure 4. Bright- (**a**) and dark- (**b**) field TEM images of the same area of the NL-TiAlN/CrN hard coating deposited on cemented carbide substrate using the three-fold rotation; (**c**) high resolution TEM image of the area inside of the frame; the lattice fringes clearly show coherency between the layers.

3.2. Interlayer Roughness

Multilayer structures produced by PVD show layer interfaces that are not atomically sharp. Extended boundary regions due to, e.g., interdiffusion or faceting are often visible. The composition, volume and structure of the interlayer regions determine to a large extent the properties and performance of the coating. The coherency of interfaces depends on the crystal structure of materials in contact at the interface. It plays a key role in energy dissipation and stress relaxation. Therefore, the knowledge and control of the quality of interfaces in multilayer structures are important for many applications, especially in optics and microelectronics [24]. For example, interlayer roughness, non-parallel layer interfaces and non-uniformity of the optical constants can significantly affect the optical reflectivity and transmittance of a multilayer coating [25].

In this study, the interlayer roughness was measured on cross-sectional TEM (Figures 2 and 4), STEM (Figure 3) and SEM images (Figure 5) of the NL-TiAlN/CrN coating. Such images provide a more detailed insight into the microstructure and evolution of the interlayer roughness from the substrate–coating boundary to the top surface. Figure 5 shows a cross-sectional scanning electron micrograph of the multilayer obtained with a backscattered electron (BSE) detector. The contrast between the CrN and TiAlN layers is evident because CrN produces more back-scattered electrons, and it is therefore brighter. The thickness of the CrN and TiAlN layers is about 51 nm and 62 nm, respectively. The curvature in the CrN/TiAlN nanolayers reflects the growth front of the coating. The interlayer roughness, which is rather smooth near the substrate, gradually increases up to the coating surface. Cumulative accumulation of the interlayer roughness takes place with the increasing number of bi-layers. It increases more after the deposition of each TiAlN layer, while the upper CrN layer slightly smooths it. This phenomenon can be explained by the crystallographic features of the individual layer material. Namely, the crystal grains of the TiAlN layer are coarser in comparison with the CrN ones (see inset in Figure 5). The TiAlN grains occasionally grow out of the TiAlN layer and continue in the CrN layer (Figure 4). The formation of TiAlN grains larger than the average layer thickness increases the interlayer roughness.

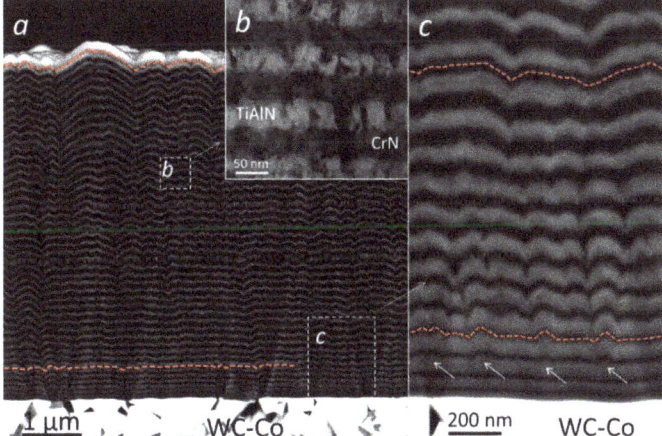

Figure 5. Bright-field cross-sectional SEM image of the NL-TiAlN/CrN coating deposited on cemented carbide substrate using the two-fold rotation (**a**). The interlayer roughness gradually increases up to the top of the coating. Crystal grains in both types of layers are clearly visible in the inset TEM image (**b**). The area marked with the dashed frame is shown at higher magnification in bright-field cross-sectional STEM image (**c**). The first increase in interlayer roughness occurs in the TiAlN layer (see sites marked with arrows).

Only a few studies concerning the interlayer roughness in the nanolayer hard coatings can be found in the literature. Recently, Beltrami et al. studied the development of interlayer roughness in nanolayer coatings composed of alternated stoichiometric CrN and understochiometric WN layers [26]. The thickness of individual layers was between 10 nm and 100 nm. All coatings were nanocrystalline with grain sizes in the order of 7 to 20 nm with a face-centered cubic (fcc) crystal structure. They found that the surface roughness of nanolayer coatings is lower in comparison to monolayer CrN coatings of comparable total thickness. The difference is enhanced by reducing the thickness of individual constituent layers, i.e., with a larger number of individual layers. They analyzed the relative contribution of CrN and WN layers to the overall roughness from the SEM image of FIB cross-sections, and they found that the interlayer roughness builds up with the deposition of each additional layer. They also observed that WN layers increase the overall roughness, while CrN layers slightly smoothed the surface. According to their explanation, the alternated deposition of two different material blocks the growth process of crystal grains, resulting in a finer grain size. Therefore, the volume fraction of the amorphous phase at the layer interfaces increases. The presence of the amorphous phase at the interface constrains the growth of crystal grains. Consequently, asperities due to misoriented grains are effectively reduced. Despite the same crystal structure, coherent growth was probably impossible due to a too-large mismatch between the CrN and WN lattices. Their findings generally agree with ours, only the increase in the interlayer roughness with increasing coating thickness is less pronounced than in our case. We explain the difference by the fact that the lattice mismatch in the case of the CrN/WN nanolayer coating is greater than in the case of the NL-TiAlN/CrN coating. Coherent growth within the columnar grain is therefore disabled, and their size is consequently smaller.

Zimmer and Kaufuss showed that, in some cases, the surface of CrN/TiN multilayer coatings prepared by the cathodic arc deposition technique can be smoothed [27]. The individual layer thickness in their nanolayer structure was between 5 and 10 nm. In monolayer CrN or TiN coatings, the size of nodular defects increases with the coating thickness. However, if two different materials are deposited successively, an interface is defined, and a new crystalline structure can start to grow. The precondition is that the new material cannot grow in the same structure or orientation as the droplets, or other types of seed particles incorporated into the coating, during the deposition process. Under such conditions, the nodular defects formed at the site of seed particles are "buried" during further deposition, and after a few layers, a smoothing effect is visible. They also found that the next step to improve the homogeneity of the film structure is a reduction in the individual layer thickness. They explained this phenomenon by the growth of very small crystal grains with random orientations. This was confirmed by X-ray measurements. Namely, they found that in the X-ray spectrum of the CrN/TiN multilayer coating, the intensities of diffraction peaks were strongly reduced in comparison with CrN and TiN monolayers. We did not observe this phenomenon in the case of sputter-deposited NL-TiAlN/CrN coating. The reason could be that the conditions for coherent growth in the NL-TiAlN/CrN coating at the layer interface are more favorable due to a better matching of the lattice parameters than in the case of the CrN/TiN multilayer.

3.3. Surface Topography

Different analytical techniques can be used to characterize the surface morphology of the NL-TiAlN/CrN coating at the micrometer, submicrometer and atomic scale. For example, a 3D image of the coating surface over a large scan area (from a several hundred micrometers square to a several millimeters square) with all the micrometer-sized details can be obtained using a stylus profilometer (Figure 6a). Scanning electron microscopy can be used to look at individual features on the coating surface (Figure 6b), while at the nano level, the surface topography can be evaluated by atomic force microscopy (Figure 6c). In an AFM image, we can resolve the nano-scale sub-cells on the top of every column, which can be further investigated in detail using TEM [9]. The topographical features of the coating surface seen in an AFM image are related to the growth mode.

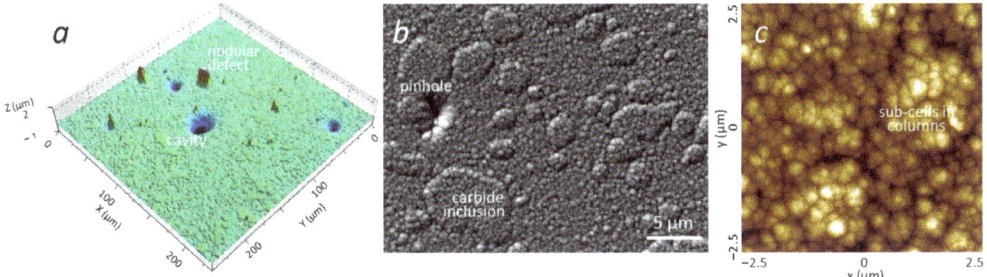

Figure 6. Surface topography appearance of the NL-TiAlN/CrN hard coating sputter deposited on the D2 tool steel substrate using three different surface analytical techniques: (**a**) 3D profile image of a scanning area of 250 µm × 250 µm, (**b**) plain-view SEM image of an area ten times smaller and (**c**) AFM image at a scanning area of 5 µm × 5 µm.

Figure 7 shows the SEM images of the same surface area of the steel substrates ASP30, M2, D2 and SS316L after polishing, MF and booster ion etching and deposition of the NL-TiAlN/CrN hard coating. All carbide and non-metallic inclusions on the polished surface of all four bare substrates were identified by EDX analysis. By comparing SEM images of the same surface of the substrate, we followed the topographical changes at the sites of these inclusions.

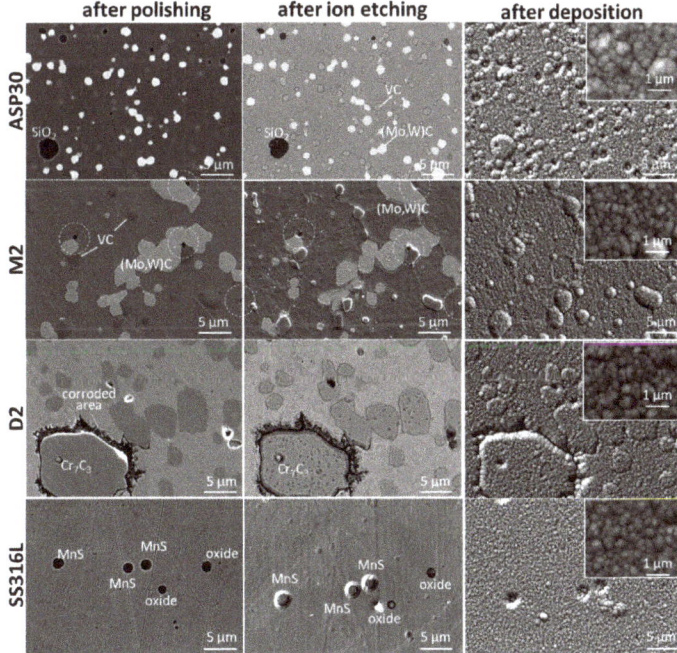

Figure 7. Top-view SEM images of the same surface area of ASP30, M2, D2 and SS316L steel substrates after polishing, ion etching (MF and booster) and deposition of the NL-TiAlN/CrN hard coatings. SEM images of the coating morphology at high magnification are shown in the insets.

During polishing, the removal rate of harder inclusions (e.g., carbides or oxides) is lower than the matrix; however, it is higher if inclusions are softer than the matrix (e.g., sulfides) [7,8]. Therefore, shallow protrusions and craters are formed at these sites. A similar effect arises during ion etching. In this case, the sputtering yield determines the etching rate of various types of inclusions and the matrix. After polishing and ion etching, the geometrical extension from the matrix level (typically up to a few hundred nanometers either in a positive or a negative direction) at the site of inclusions depends on the net removal rate [7,8]. The height of protrusions and depth of craters are much larger if a more intensive booster etching is used. If the net removal rate is higher than that of the matrix, the inclusion appears like a shallow crater, and after coating deposition, pinholes form at these sites. If it is lower than that of the matrix, the inclusion appears like a shallow protrusion, where a nodular defect forms after deposition. It is also possible that the net removal rates of the selected inclusion and matrix are similar. In this case, no geometric effect can be expected.

During the deposition process, all topographical irregularities on the substrate surface formed during polishing and ion etching are transferred onto the coating surface and are often magnified due to the geometrical shadowing effect. However, after the deposition of the coating, its surface roughness increases significantly due to the formation of growth defects. At a high magnification of the area between growth defects and inclusions (see insets in Figure 7), we did not notice any significant difference in the surface topography of coatings deposited on the four different types of substrates (ASP, M2, D2 and SS316L). Columns with dome-shaped tops of comparable size can be observed, regardless of the type of the substrate. This is probably because the roughness of the matrix of all four types of substrates after ion etching is comparable, while the intensity of ion bombardment during the deposition process is identical. However, the surface topography of the NL-TiAlN/CrN coating is considerably different for samples prepared by different rotation modes (Figure 8). The topography of the coating surface prepared by different rotation modes reflects their microstructural differences, as described in Section 3.1.

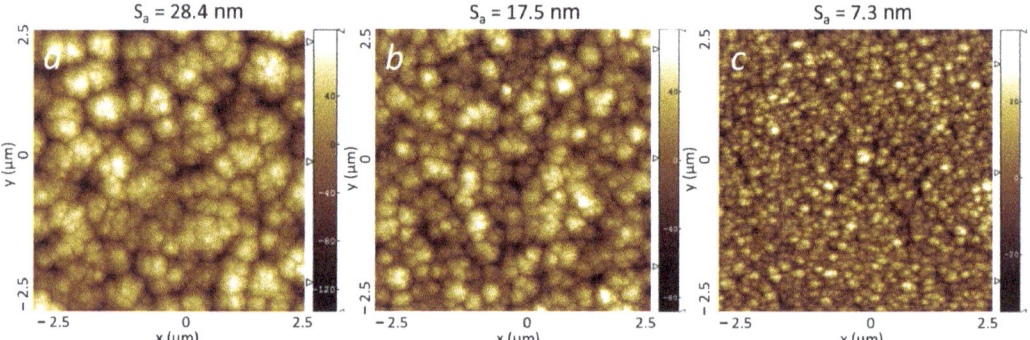

Figure 8. AFM images (scan area 5 μm × 5 μm) show the surface morphology of the NL-TiAlN/CrN hard coating sputter deposited on D2 tool steel substrates. The samples were prepared in the same batch but rotated around a different number of axes: (**a**) one-fold, (**b**) two-fold and (**c**) three-fold rotation.

Figure 9a shows the AFM surface roughness measured on the substrate after polishing, MF ion etching and deposition. Figure 9b shows similar measurements where MF and booster etching technique were applied. The roughness measurements were performed on an area of 20 μm × 20 μm where no growth defects were present. We can see that after MF ion etching, the roughness increases only a little. However, it increases significantly after the deposition of the coating. On samples etched by the MF and booster technique, the roughness increases significantly after both ion etching and deposition.

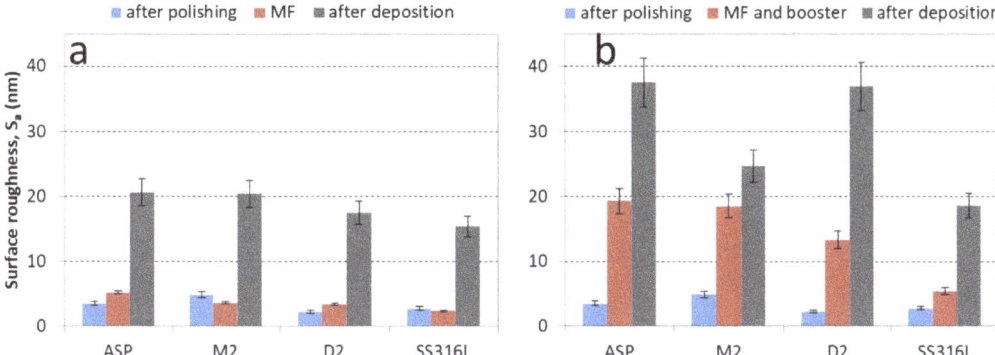

Figure 9. AFM surface roughness of four different steel substrates (ASP, M2, D2, SS 316L) after polishing, ion etching and deposition. One set of substrates was ion etched by the MF technique (**a**) while the other one by MF and booster (**b**). The scan area was 20 µm × 20 µm.

We also checked whether different etching methods affect the surface topography in the coating area without growth defects. For this test, one set of the substrates was cleaned in a separate batch using standard mid-frequency (MF) ion etching, while another set of substrates was cleaned by the more intensive booster etching. After that, both sets of substrates were loaded again in the deposition system where they were once more cleaned by MF ion etching for 10 min to remove any possible impurities that formed during the previous analysis. This was followed by the deposition of the NL-TiAlN/CrN coating. Figure 10 shows the AFM images of the surface topography of the coatings deposited on four different substrate materials, which were previously cleaned by two different ion etching procedures. No significant difference in surface topography can be observed. The larger clusters in Figure 10e,g have grown on the sites of carbide inclusions in the ASP and D2 substrates.

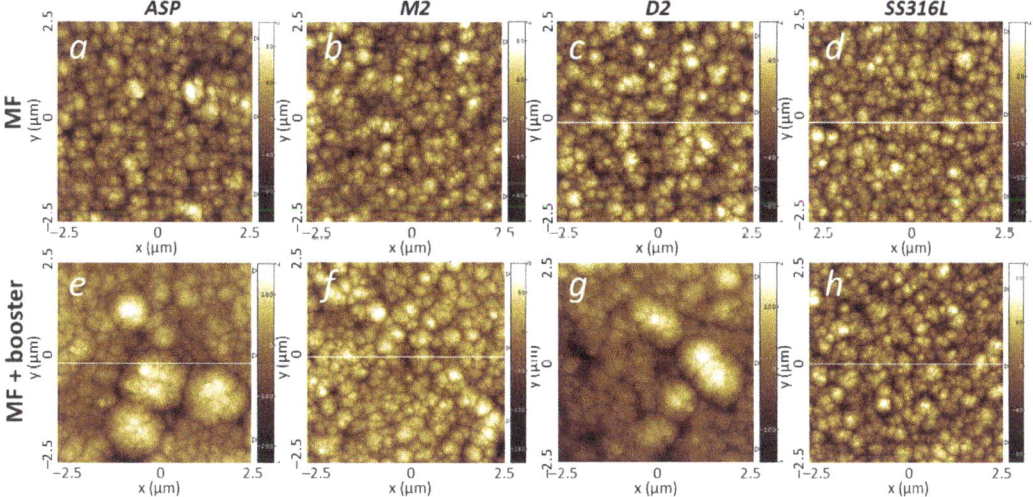

Figure 10. AFM images of surface morphology of the NL-TiAlN/CrN coating deposited in the same batch on four different steel substrates (ASP30, M2, D2 and SS316L) ion etched by MF (**top** row) and MF and booster techniques (**bottom** row).

3.4. Formation of Growth Defects

The largest increase in coating surface roughness is caused by growth defects formed on different substrate surface irregularities. One origin of irregularities is a result of substrate pretreatment (grinding, polishing and ion etching). As was explained in the previous section, topographical irregularities in the inhomogeneous substrate material, such as tool steel, are formed at the sites of inclusions (e.g., carbide and non-metallic inclusions, see Figure 7). While the carbides are an essential component of tool steels, the non-metallic inclusions are undesired products formed during the steel production. All steels contain non-metallic inclusions to some extent, as they precipitate during the cooling and solidification of the steel [28]. They can vary widely in size, shape and composition. The typical dimensions of inclusions are in the range of 0.1–100 µm. Non-metallic inclusions that form separate phases are the chemical compounds of metals (e.g., iron, manganese, aluminum, silicon and calcium) with oxygen, sulfur, carbon, hydrogen or nitrogen. The majority of the inclusions in steels are oxides and sulfides [7]. The type of non-metallic inclusions depends on the steel grade, steel-making process, secondary metallurgy treatments and casting of steel. Though the concentration of non-metallic inclusions is low (less than 0.01%), they have a significant effect on steel properties. Namely, a high density of coarse inclusions can cause the formation of cracks and, thus, initiates fracture and decreases the toughness of steels [29]. Surface defects caused by them deteriorate mechanical properties such as corrosion resistance, deformability, brittle fracture and fatigue strength. They also affect the weldability, polishability and machinability of the steel.

In principle, the coating surface is a conformal replication of the substrate surface. Therefore, all protrusions and cavities formed during substrate pretreatment are transferred to the coating surface (Figures 7, 11 and 12). Even if the inclusions have a similar net removal rate as the matrix, they can affect the growth of the coating because they are chemically and structurally different from the matrix.

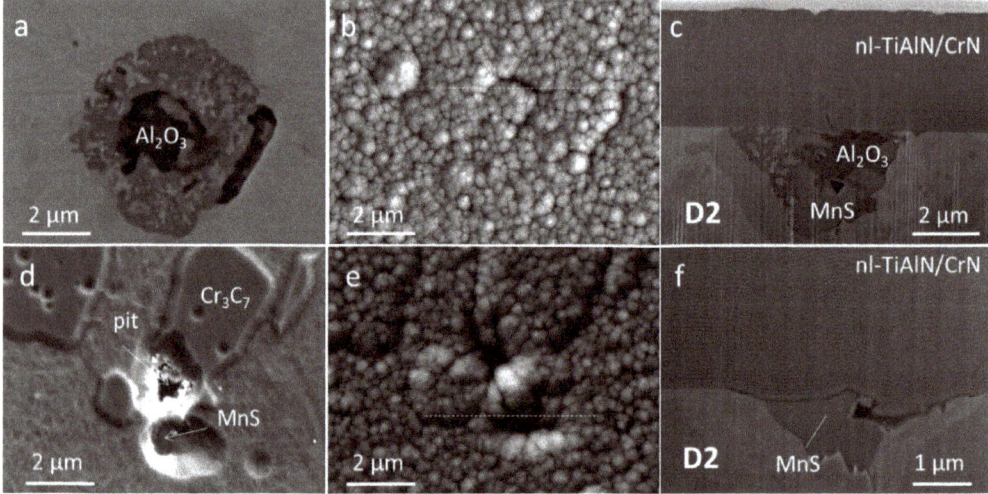

Figure 11. Plain-view SEM images of the same surface area at the site of a complex multicomponent non-metallic inclusion in a D2 tool steel substrate after MF ion etching (**a**), deposition of the NL-TiAlN/CrN coating (**b**) and FIB cross-sections of the coating (**c**). Similar SEM images were taken at the site of a MnS inclusion in the D2 tool steel substrate after MF and booster ion etching (**d**), deposition of the NL-TiAlN/CrN coating (**e**) and FIB cross-sections of the coating (**f**).

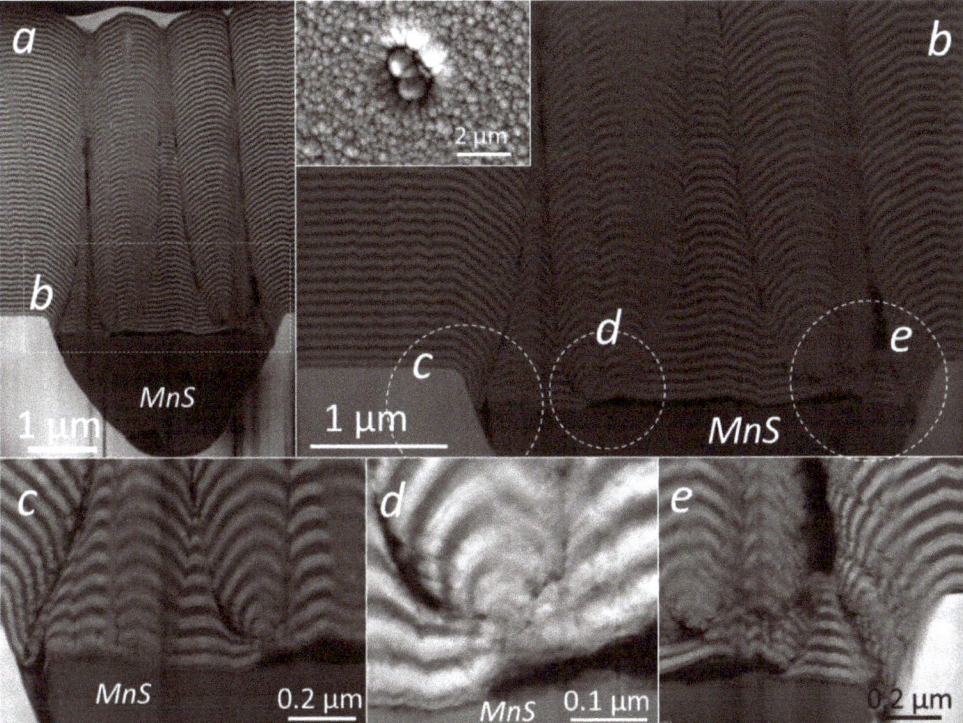

Figure 12. STEM images of the FIB cross-section of the NL-TiAlN/CrN hard coating at the site of a MnS inclusion in the stainless steel SS316L substrate (a–e). A shallow crater (a), which is formed at this site after booster ion etching, is the starting point for a pinhole formation at the boundary between the inclusion and the matrix. An additional effect is poor contact of the coating to the inclusion surface. We can also see nodular defects that have formed on smaller seed particles located at the bottom of the crater (c–e).

The largest nodular defects formed on foreign particles and droplets, which arrived on the substrate surface before or during the coating growth. Due to the geometrical shadowing effect, characteristic of the PVD deposition processes, the initial small seed particle is growing during the deposition of the coating (Figures 13 and 14). While a smaller portion of defects formed on seed particles embedded on the layer during the deposition process (Figure 14), a larger portion of defects formed on seed particles that were on the surface of the substrate, even before the start of the deposition (Figure 13). This is evident from a low-angle cross-section of the NL-TiAlN/CrN coating, prepared by the ball cratering technique, where the layer contours reveal the positions of defects (Figure 15). In our recently published paper, we showed that the distribution of nodular defects on the ground section of the NL-TiAlN/CrN coating is uniform, which means that most nodular defects started to grow at the substrate–coating interface [30]. The complex shapes of contours in the SEM image of the nodular defects' ground section reveal their internal structure (Figure 15e–g), which results from the growth of nodular defects on seeds with irregular geometry. Namely, the contours of individual layers in the multilayer coating follow the shape of the seed. The nodular defects are composed of several columns that start to grow on different parts of the seed particle independently of each other (Figure 13c,d).

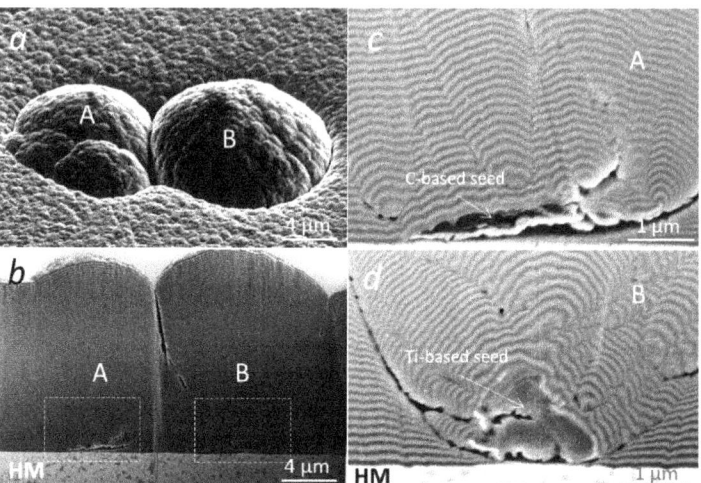

Figure 13. Top view SEM image (**a**) and SEM images of FIB cross-sections (**b**) of the nodular defects in the NL-TiAlN/CrN coating. The left nodular defect formed at the site of a carbon-based particle (**c**), while the right one was at a titanium-based particle (**d**). A small step on the substrate surface beneath the titanium-based particle proves that it arrived on the substrate surface during the ion etching process.

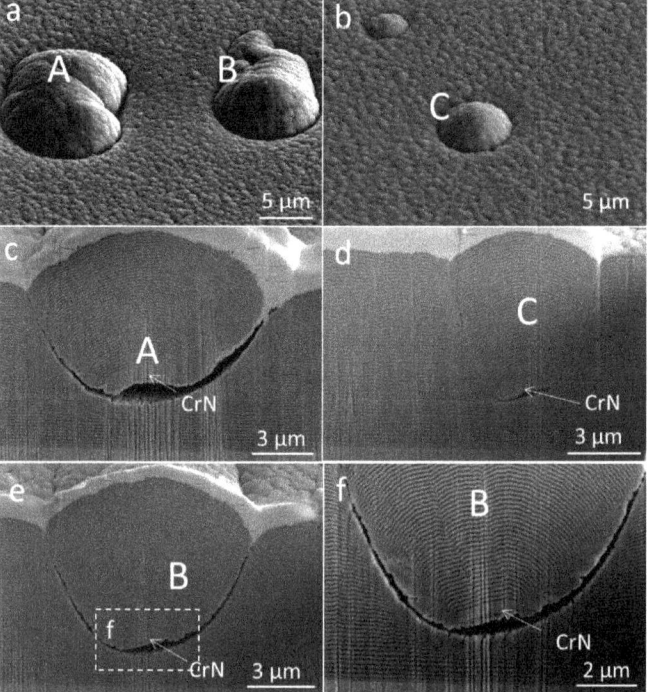

Figure 14. Top-view SEM images (**a**,**b**) and SEM images of the FIB cross-sections (**c**–**f**) of three nodular defects (A–C) in the NL-TiAlN/CrN coating. All three defects formed at the site of Cr-based flakes, which were incorporated in the coating at the same depth. All three flakes probably came from the same source. We believe that they originate from arcs occurring at the perimeter of the chromium target.

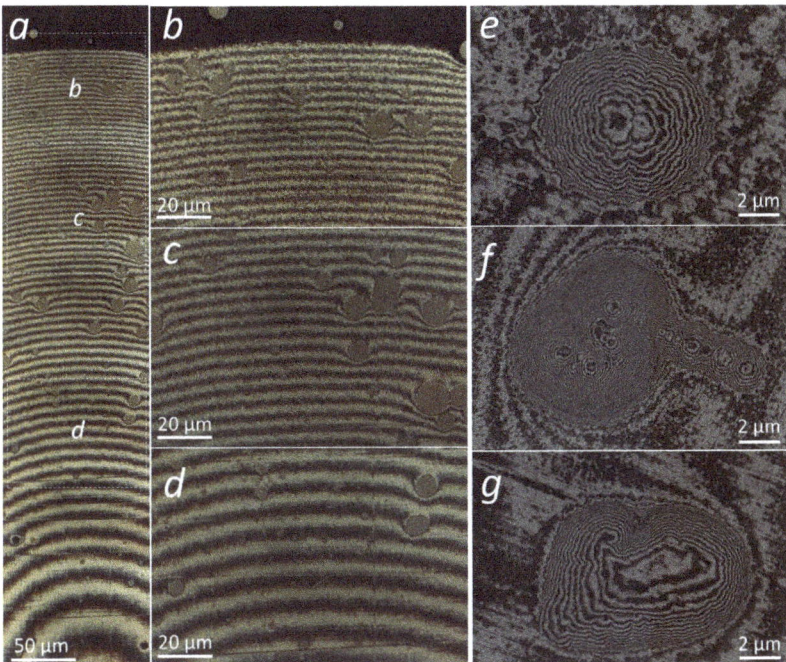

Figure 15. Optical microscopy images at lower (**a**) and higher (**b**–**d**) magnifications of a low-angle cross-section of the NL-TiAlN/CrN coating, prepared by the ball cratering technique. SEM images of the ground section of selected nodular defects (**e**–**g**) reveal their internal structure due to the complex geometry of seed particles.

4. Conclusions

The goal of this study was to better understand the growth and morphology of the NL-TiAlN/CrN coating sputter deposited on four different types of steels and cemented carbide substrates. We studied the microstructure, surface topography, layer periodicity, interlayer roughness and formation of growth defects at the sites of non-metallic inclusions and foreign seeds. Furthermore, these properties were analyzed with dependence on the substrate rotation mode, the type of the substrate material and method of ion etching. The results of our investigation can be summarized as follows:

- The microstructure, topography and periodicity of the NL-TiAlN/CrN hard coating strongly depend on the rotation mode. The coatings prepared by one-fold rotation have a periodic structure and a pronounced columnar microstructure that extends from the substrate to the coating surface. On the other hand, coatings prepared by three-fold rotation have an aperiodic, less columnar (with much smaller average column diameters) and a fine-grained microstructure.
- A coherent growth of TiAlN and CrN layers inside the columnar grains was observed.
- The conformity and uniformity of multilayer coating were analyzed from SEM images of the FIB cross-sections. The curvature of individual layers in the multilayer coating reflects the growth front of the coating. Although the initial roughness at the interface between the first layer and the substrate was rather small, it accumulated during deposition and gradually increased towards the top of the coating.
- Shallow craters or protrusions formed at the sites of non-metallic inclusions in the steel substrates, depending on whether the net removal rate after polishing and ion etching was higher or lower compared to the matrix. Even if the net removal rate of both materials is similar, the non-metallic inclusions can affect the growth of the coating,

as they are chemically and structurally different from the matrix. The influence of non-metallic inclusions on the microstructure and topography of the layer is also not negligible because they can cause a local loss of adhesion, pitting corrosion and other destructive effects.
- The largest increase in the coating roughness is due to the presence of nodular defects formed on seed particles that arrive on the substrate surface during its pretreatment or deposition. We showed that the majority of nodular defects start to grow at the substrate-coating interface, i.e., on particles that arrive on the substrate before the start of the deposition. We also demonstrated that the contours of the individual layers in the multilayered coating reveal the internal structure of the nodular defect. The irregular shapes of layer contours are the result of the complex geometry of the seed particles and the geometrical shadowing effect during the deposition.

Author Contributions: Design of experiments, 3D profilometry, interpretation of experimental results, manuscript writing and project administration, P.P.; design of experiment, preparation of specimen for TEM characterization, SEM analysis and manuscript review, P.G.; SEM and FIB analysis and manuscript review, T.B. and F.Z.; AFM, SEM and FIB analysis and manuscript review, A.D.; preparation of specimen for TEM characterization, manuscript review and project administration, M.P.; STEM investigations and data analysis, M.A.; manuscript review, M.Č. All authors have read and agreed to the published version of the manuscript.

Funding: This work was supported by the Slovenian Research Agency (program P2-0082, project J2-2509). We also acknowledge funding from the European Regional Development Funds (CENN Nanocenter, OP13.1.1.2.02.006) and the European Union Seventh Framework Programme under Grant Agreement 312483-ESTEEM2 (Integrated Infrastructure Initiative-I3).

Institutional Review Board Statement: Not applicable.

Informed Consent Statement: Not applicable.

Data Availability Statement: Not applicable.

Acknowledgments: The authors would also like to thank Martina Dienstleder (Graz Centre for Electron Microscopy, Graz, Austria) for the preparation of specimens for STEM analysis using the focused ion beam (FIB) workstation and Jožko Fišer (Jožef Stefan Institute, Ljubljana, Slovenia) for technical assistance.

Conflicts of Interest: The authors declare no conflict of interest.

References

1. Harlin, P.; Carlsson, P.; Bexell, U.; Olsson, M. Influence of surface roughness of PVD coatings on tribological performance in sliding contacts. *Surf. Coat. Technol.* **2006**, *201*, 4253–4259. [CrossRef]
2. Olofsson, J.; Gerth, J.; Nyberg, H.; Wiklund, U.; Jacobson, S. On the Influence from Micro Topography of PVD Coatings on Friction Behaviour, Material Transfer and Tribofilm Formation. *Wear* **2011**, *271*, 2046–2057. [CrossRef]
3. Saketi, S.; Östby, J.; Olsson, M. Influence of tool surface topography on the material transfer tendency and tool wear in the turning of 316L stainless s steel. *Wear* **2016**, *368–369*, 239–252. [CrossRef]
4. Panjan, P.; Drnovšek, A.; Kovač, J. Tribological Aspects Related to the Morphology of PVD Hard Coatings. *Surf. Coat. Technol.* **2018**, *343*, 138–147. [CrossRef]
5. Luo, Q. Origin of Friction in Running-in Sliding Wear of Nitride Coatings. *Tribo. Lett.* **2010**, *37*, 529–539. [CrossRef]
6. Panjan, P.; Drnovšek, A.; Terek, P.; Miletić, A.; Čekada, M.; Panjan, M. Comparative Study of Tribological Behavior of TiN Hard Coatings Deposited by Various PVD Deposition Techniques. *Coatings* **2022**, *12*, 294. [CrossRef]
7. Panjan, P.; Drnovšek, A.; Gselman, P.; Čekada, M.; Panjan, M. Review of Growth Defects in Thin Films Prepared by PVD Techniques. *Coatings* **2020**, *10*, 447. [CrossRef]
8. Panjan, P.; Drnovšek, A.; Mahne, N.; Čekada, M.; Panjan, M. Surface Topography of PVD Hard Coatings. *Coatings* **2021**, *11*, 1387. [CrossRef]
9. Panjan, P.; Drnovšek, A.; Dražič, G. Influence of Growth Defects on the Oxidation Resistance of Sputter-Deposited TiAlN Hard Coatings. *Coatings* **2021**, *11*, 123. [CrossRef]
10. Nordin, M.; Ericson, F. Growth characteristics of multilayered physical vapour deposited TiN/TaNx on high speed steel substrates. *Thin Solid Film.* **2001**, *385*, 174–181. [CrossRef]

11. Xu, Y.X.; Chen, L.; Fei, P.; Chang, K.K.; Yong, D. Effect of the modulation ratio on the interface structure of TiAlN/TiN and TiAlN/ZrN multilayers: First-principles and experimental investigations. *Acta Mater.* **2017**, *130*, 281–288. [CrossRef]
12. Panjan, M.; Šturm, S.; Panjan, P.; Čekada, M. TEM investigation of TiAlN/CrN multilayer coatings prepared by magnetron sputtering. *Surf. Coat. Technol.* **2007**, *202*, 815. [CrossRef]
13. Xu, Y.X.; Chen, L.; Pei, F.; Du, Y. Structure and thermal properties of TiAlN/CrN multilayered coatings with various modulation ratio. *Surf. Coat. Technol.* **2016**, *304*, 512–518. [CrossRef]
14. Helmersson, U.; Todorova, S.; Barnett, S.A.; Sundgren, J.-E.; Markert, L.C.; Greene, J.E. Growth of single-crystal TiN/VN strained-layer superlattices with extremely high mechanical hardness. *J. Appl. Phys.* **1987**, *62*, 481. [CrossRef]
15. Barshilia, H.C.; Prakash, M.S.; Jain, A.; Rajam, K.S. Structure, hardness and thermal stability of TiAlN and nanolayered TiAlN/CrN multilayer films. *Vacuum* **2005**, *77*, 169–179. [CrossRef]
16. Park, J.K.; Park, H.J.; Ahn, J.H.; Baik, Y.J. Effect of Ti to Al ratio on the crystalline structure and hardening of a $Ti_{1-x}Al_xN$/CrN nanoscale multilayered coating. *Surf. Coat. Technol.* **2009**, *203*, 3099–3103. [CrossRef]
17. Waddsworth, I.; Smith, I.J.; Donohue, L.A.; Munz, W.D. Thermal stability and oxidation resistance of TiAlN/CrN multilayer coatings. *Surf. Coat. Technol.* **1997**, *94–95*, 315–321. [CrossRef]
18. Povstugar, I.; Pyuck-Pa, C.; Darius, T.; Jae-Pyeong, A.; Dierk, R. Interface-directed spinodal decomposition in TiAlN/CrN multilayer hard coatings studied by atom probe tomography. *Acta Mater.* **2013**, *61*, 7534–7542. [CrossRef]
19. Wei, Y.; Zong, X.; Jiang, Z.; Tian, X. Characterization and mechanical properties of TiN/TiAlN multilayer coatings with different modulation periods. *Int. J. Adv. Manuf. Technol.* **2018**, *96*, 1677–1683. [CrossRef]
20. Panjan, M.; Peterman, T.; Čekada, M.; Panjan, P. Simulation of a multilayer structure in coatings prepared by magnetron sputtering. *Surf. Coat. Technol.* **2009**, *204*, 850–853. [CrossRef]
21. Panjan, M. Influence of substrate rotation and target arrangement on the periodicity and uniformity of layered coatings. *Surf. Coat. Technol.* **2013**, *235*, 32–44. [CrossRef]
22. Pennycook, S.J.; Jesson, D.E. High-resolution Z-contrast imaging of crystals. *Ultramicroscopy* **1991**, *37*, 14–38. [CrossRef]
23. Rother, B.; Jehn, H.A.; Gabriel, H.M. Multilayer hard coatings by coordinated substrate rotation modes in industrial PVD deposition systems. *Surf. Coat. Technol.* **1996**, *86–87*, 207–211. [CrossRef]
24. Esashi, Y.; Tanksalvala, M.; Zhang, Z.; Jenkins, N.W.; Kapteyn, H.C.; Murnane, M.M. Influence of surface and interface roughness on X-ray and extreme ultraviolet reflectance: A comparative numerical study. *OSA Contin.* **2021**, *4*, 1497–1518. [CrossRef]
25. Kaiser, N. Review of the Fundamentals of Thin-Film Growth. *Appl. Opt.* **2002**, *41*, 3053–3060. [CrossRef]
26. Beltrami, M.; Zilio, S.D.; Kapun, G.; Ciubotaru, C.D.; Rigoni, F.; Lazzarino, M.; Sbaizero, O. Surface roughness control in nanolaminate coatings of chromium and tungsten nitrides. *Micro Nano Eng.* **2022**, *14*, 100107. [CrossRef]
27. Zimmer, O.; Kaulfuß, F. Hard Coatings with High Film Thickness Prepared by PVD. *Plasma Process. Polym.* **2009**, *6*, S152–S156. [CrossRef]
28. da Costa e Silva, A.L.V. Non-metallic inclusions in steels—Origin and control. *J. Mater. Res. Technol.* **2018**, *7*, 283–299. [CrossRef]
29. Murakami, Y. *Metal Fatigue: Effects of Small Defects and Nonmetallic Inclusions*; Elsevier Science Ltd.: Oxford, UK, 2002; pp. 75–122.
30. Panjan, P.; Drnovšek, A.; Čekada, M.; Panjan, M. Contamination of Substrate Coating Interface Caused by Ion Etching. *Coatings* **2022**, *12*, 846. [CrossRef]

Article

Contamination of Substrate-Coating Interface Caused by Ion Etching

Peter Panjan *, Aljaž Drnovšek, Miha Čekada and Matjaž Panjan

Department of Thin Films and Surfaces, Jožef Stefan Institute, Jamova 39, 1000 Ljubljana, Slovenia; aljaz.drnovsek@ijs.si (A.D.); miha.cekada@ijs.si (M.Č.); matjaz.panjan@ijs.si (M.P.)
* Correspondence: peter.panjan@ijs.si

Abstract: In–situ cleaning of the substrate surface by ion etching is an integral part of all physical vapor deposition (PVD) processes. However, in industrial deposition systems, some side effects occur during the ion etching process that can cause re-contamination. For example, in a magnetron sputtering system with several sputter sources and with a substrate holder located centered between them, the ion etching causes the contamination of the unshielded target surfaces with the batching material. In the initial stage of deposition, this material is redeposited back on the substrate surface. The identification of the contamination layer at the substrate–coating interface is difficult because it contains both substrate and coating elements. To avoid this problem, we prepared a TiAlN double coating in two separate production batches on the same substrate. In such a double-layer TiAlN hard coating, the contamination layer, formed during the ion etching before the second deposition, is readily identifiable, and analysis of its chemical composition is easy. Contamination of the batching material was observed also on seed particles that caused the formation of nodular defects. We explain the origin of these particles and the mechanism of their transfer from the target surface to the substrate surface. By comparison of the same coating surface area after deposition of the first and second TiAlN layers, the changes in coating topography were analyzed. We also found that after the deposition of the second TiAlN coating, the surface roughness slightly decreased, which we explain by the planarization effect.

Keywords: TiAlN hard coating; unbalanced magnetron sputtering; ion etching; surface topography; growth defects

1. Introduction

The adhesion of PVD hard coatings to substrates is the deciding factor determining their performance and success in industrial applications. In general, PVD coating adhesion is affected by both substrate properties (e.g., composition, microstructure, roughness, thermal expansion coefficient) as well as by the deposition parameters (e.g., temperature, bias voltage, internal stresses, thickness). The key condition for good coating adhesion is a clean surface of the substrate. The next requirement is a strong chemical bonding between the substrate atoms and the depositing atoms that can be formed only if a sufficient number of nucleation sites are available on the substrate surface. Two additional conditions must be also fulfilled: the interface between the substrate and the coating must have low porosity and it must not contain brittle intermetallic phases [1]. Several approaches are used for improving the coating adhesion:

- Primary cleaning process: external mechanical and chemical pretreatment of the substrate surface prior to the insertion in the deposition chamber in order to roughly remove contaminants (such as grease, oxides).
- In situ pre-treatment (heating and ion etching) of the substrate before the coating deposition process in order to remove the contamination that has formed since the primary cleaning process had been performed. Ion bombardment also creates a

- number of new nucleation sites for the coating atoms, which significantly improves chemical bonds between the substrate atoms and the condensing atoms. The ion etching also increases the roughness of the substrate surface and thus strengthens the coating–substrate interface.
- After substrate cleaning by ion etching, the coating adhesion strength can be further improved by deposition of a metallic interlayer, which is typically Cr or Ti. An interlayer minimizes internal stresses in the deposited coating and it can also dissolve oxides that remain on the substrate surface after ion etching [2].

A precondition for good PVD coating adhesion is therefore the cleanliness of the entire tool surface that should be free from oxides and other undesirable contaminants. Otherwise, if the contaminant layer is not removed completely, then there is a high likelihood of coating delamination.

The first report about the cleaning of substrate surfaces by ion etching is from 1955 when Farnsworth et al. reported the use of ion etching with Ar$^+$ to prepare ultra-clean surfaces for low-energy electron diffraction studies [3]. Later, cleaning by ion etching was used by D. M. Mattox as a part of the ion plating deposition technique, first introduced by the author [4]. Today, in-situ cleaning of the substrate surface by ion etching is routinely used in all deposition systems for the preparation of PVD coatings. The cleaning procedure is performed by low-energy bombardment with inert gas or metal ions, extracted from glow discharge plasma, cathodic arc, or high-power impulse magnetron sputtering (HIPIMS) discharge.

Cleaning by ion etching is most often performed in Ar (or Ar + Kr) glow discharge plasma, where the gas ions are accelerated to several hundred eV by a DC or RF bias voltage on the conductive or non-conductive substrates, respectively [5,6]. The disadvantages of ion etching by inert ions are its small etching rate of oxides due to the low plasma density and the accumulation of insoluble inert gas atoms at the substrate–coating interface [7]. Namely, during sputter cleaning, the argon ions can be incorporated (up to several atomic percent) into the near-surface region of the substrate. Implanted ions of inert gas can cause tensile stresses and/or form an amorphous zone in the substrate surface region. During coating deposition or under exploitation of coated components, the incorporated gas atoms diffuse and agglomerate into bubbles, which introduce porosity and weakening of the interface [8]. To prevent the incorporation of gas atoms, ion etching should be performed at an elevated temperature (>300 °C) or the substrate should be annealed before film deposition.

An alternative approach to argon ion etching is metal ion etching with a cathodic arc discharge serving as an ion source. Cathodic arc discharge produces a highly ionized metal flux without using a process gas. Bombardment with metal ions is known to provide not only a cleaner substrate surface but to also produce very thin (a few nm thick) implantation zones that can promote localized epitaxial growth of the coating. The disadvantage is the contamination of the substrate surface with the macroparticles (droplets) of the cathode material. Because their bonding with the substrate material is poor, they reduce the adhesion of the coating. There is also often a risk (especially for small tools) that too intensive etching leads to local substrate overheating and consequently softening effects [9,10].

Contamination of the substrate surface with droplets can be avoided by using a high-power impulse magnetron sputtering (HIPIMS) system [11–13]. Highly ionized HIPIMS plasma contains a mixture of single-, double-, and sometimes even higher charged metal and argon ions. The bombardment of the substrate surface with metal ions causes the formation of a shallow metal implantation zone, while the crystallinity of the substrate surface is not destroyed. This improves the bond between the substrate material and the subsequently deposited coating and thus enhanced adhesion. The HIPIMS etching of the coatings exhibits a defect-free smooth surface without any droplets. Therefore, PVD coatings deposited after HIPIMS pretreatment exhibit a superior adhesion in comparison to pretreatments performed in argon glow discharge and cathodic vacuum arc environments.

In the literature, most data refer to the etching of flat substrates in experimental deposition systems. Much more demanding is the ion etching of substrates (tools) in industrial deposition systems equipped with more sputtering sources and several substrate holders (towers). There are only a few studies that address the problem of etching tools in industrial systems for deposition of PVD hard coatings, where a typical job-coating batch is filled with different types of tools, which can vary considerably in size and geometry [14,15]. The electric field between the plasma and the tool attracts ions from the plasma. Since the electric fields are more localized at the sharp edges, rounded surfaces and edges are subjected to different ion flux densities as compared to a flat surface. Therefore, highly non-uniform ion etching, not only on different tools within the batch but also on different areas of the same tool, can be expected Therefore, it is almost impossible to predict the etching efficiency within a batch. Most often, information on the efficiency of etching in industrial systems is acquired empirically. Avoiding this problem requires a very careful design of the substrate table and holders. In addition, a well thought-out loading of tools by skilled operators is needed.

A big challenge in sputter cleaning of tools in a typical job-coating batch is how to achieve uniform ion current density over the entire tool surface. A more uniform etching can be achieved by using multiple rotations of tools. During ion etching and deposition processes, the smaller tools, such as drills, milling cutters, taps, and inserts are exposed to a triple planetary rotation [16]. They are mounted on several rotating satellites and depending on tool size, each satellite has one or several levels. We must also take into account that during ion etching, not only the tool surface, but also structural components (such as load fixtures) of the deposition system are sputtered and can cause additional contamination of tool surfaces.

In our recently published studies, we described the topographic changes of the substrate surface during ion etching [5,6]. We also pointed out that the etching efficiency depends on batching configuration and etching parameters. In such a system, there is a high probability that re-deposition and cross-contamination of the substrates will occur. Namely, in areas where the intensity of ion bombardment is low, contaminates may accumulate due to the re-deposition of sputtered material from areas where the intensity is high. The next problem is the material already deposited on the substrate holder, which is sputtered off and can be redeposited on the new batch of tools. All these phenomena, which can reduce the usefulness of cleaning by ion etching, must be properly taken into account to minimize their influence. In this paper, therefore, we focus on the cross-contamination and re-deposition problems in industrial deposition systems equipped with four rectangular unbalanced magnetron sources.

2. Materials and Methods

The industrial magnetron sputtering system (modified CC800/7, CemeCon, Würselen, Germany) was used for the deposition of the single layer TiAlN and multilayer TiAlN/CrN hard coatings. All experiments were performed in production batches. In a typical batch, different tools made of cemented carbide, high-speed steel (HSS), or cold work tool steel, were loaded on specially designed substrate holding fixtures. The TiAlN coating composition was Ti 23 at.%, Al 27 at.%, and N 50 at.%. The test substrates in the form of discs were made of powder metallurgical (PM) ASP30 tool steel (Uddeholm, Hagfos, Sweden), cold work tool steel AISI D2 (Ravne steel factory, Ravne, Slovenia), and cemented carbide (HM). All substrates were first ground and polished to a mean roughness of R_a = 10 nm. Prior to the coating process, they were cleaned in detergents (alkaline cleaning agents, pH $\approx$ 11) and ultrasound, rinsed in deionized water, and dried in hot air. In the vacuum chamber, they were first heated to about 450 °C and then in-situ cleaned by radio frequency (RF) ion etching in an argon atmosphere. The RF power and the etching time were 2000 W and 90 min, respectively. The details of the ion etching and deposition process are described in a recently published article [17]. The TiAlN coating thickness was around 4 μm as measured using a ball crater technique. The total operating pressure was maintained at 0.75 Pa, with the flow rates of argon, nitrogen and krypton being 100, 160, and 80 mL/min, respectively. A DC bias of −100 V was applied to the substrates. The duration of the deposition process was 135 min. After this time, the deposition process was interrupted for an intermediate ion etching (for 60 min under the same conditions as during substrate cleaning). This was followed by an additional deposition of a TiAlN coating (deposition time was 30 min). The intermediate etching creates new nucleation sites for the subsequently deposited nitride coating, resulting in a fine-grained and less porous microstructure of the top layer [18].

After the deposition, the samples were taken out of the vacuum chamber and the TiAlN coating surface was analyzed by 3D stylus profilometry and SEM microscopy to verify the presence of growth defects and other morphological features. The surface topography characterization of the coated and uncoated substrate was carried out using scanning electron microscopy (SEM, JEOL JSM-7600F, Tokyo, Japan) and 3D stylus profilometry (Bruker Dektak XT, Billerica, MA, USA). The evaluation area of profilometer was 1 mm^2 with a resolution of 0.2 μm in x and 1 μm in the y direction, while the vertical resolution was around 5 nm. After these investigations, the coated samples were ultrasonically cleaned and then put again into the deposition system and coated with the second TiAlN layer prepared in the same manner as the first one. This means that before the second deposition, the coated substrates were exposed to the intensive ion etching again. After the second deposition, the surface topography and the microstructure of TiAlN double layer were analyzed again.

Two Cr targets and two mosaic TiAl targets were used for the deposition of TiAlN/CrN multilayer coating, where the Cr targets were positioned on one side of the vacuum chamber and TiAl targets on the other. The samples were mounted in a one-fold rotation mode. The total thickness of the coating deposited on the cemented carbide substrate was about 12 μm, while the thicknesses of individual TiAlN and CrN layers were around 50 nm and 70 nm, respectively. The Zeiss Axio CSM 700 confocal optical microscope (Zeiss, Jena, Germany) was used to observe the ground section of the coating.

The microstructure and the coating morphology were studied using fracture cross-sections examined using a field emission scanning electron microscope (FEI Helios Nanolab 650i, Amsterdam, The Netherlands). Cross-sections were also prepared by the focused ion beam technique (FIB) using an FIB source integrated into the FEI SEM scanning electron microscope. SEM images were recorded using the ion beam and the electron beam. EDS mapping was carried out by using the Oxford Instruments system attached to the SEM.

Scanning transmission electron microscopy (STEM) was performed using a JEOL ARM 200 CF (Jeol Ltd., Tokyo, Japan) operated at 200 kV and a Jeol Centurio 100 mm^2 SDD EDXS system. All TEM images and analyses were obtained on cross-sections. The diameter of the electron beam was around 0.1 nm, but due to beam broadening and taking into account sample thickness, chemical composition, and density, the lateral resolution was around 10 nm. The FIB lift-out method was used for the preparation of TEM specimens.

3. Results and Discussion

3.1. Surface Topography of Single and Double Layer TiAlN Coatings

After deposition of the first TiAlN coating, the designated surface area was examined by 3D stylus profilometry and SEM microscopy. In SEM images at low magnification (Figures 1 and 2), growth defects of various shapes and sizes can be observed. Protrusions at sites of carbides in the steel substrate that were formed during ion etching of the substrate are visible. In SEM images at high magnification of the area between various protrusions, we can see that the coating surface exhibited a faceted domain-like morphology related to its columnar growth (Figure 1). Selected topographical features of the TiAlN coating were pinpointed. After these analyses, the coated samples were put again into the deposition system and coated with the second TiAlN coating prepared by the same procedure as the first one (heating, ion etching, deposition of TiAlN coating). The optical image of the ball crater through such double-layer TiAlN coating is shown in Figure 3a. The bright rings belong to the TiAlN layers that were formed after the intermediate ion etching of the individual coatings at about three-quarters of the deposition time [18]. Such ion etching causes re-sputtering of otherwise rounded column tops and the formation of new sites for nucleation of the subsequently deposited coating. These processes change the topography and microstructure of the layer that grows after the interruption. Figure 3b shows the SEM image of the FIB cross-section of TiAlN double-layer deposited on the PM ASP30 tool steel substrate. The interface between the first and the second TiAlN coatings is clearly visible. The EDS analysis showed that tungsten, iron, and chromium are present at the interface. The reasons for such contamination are explained later.

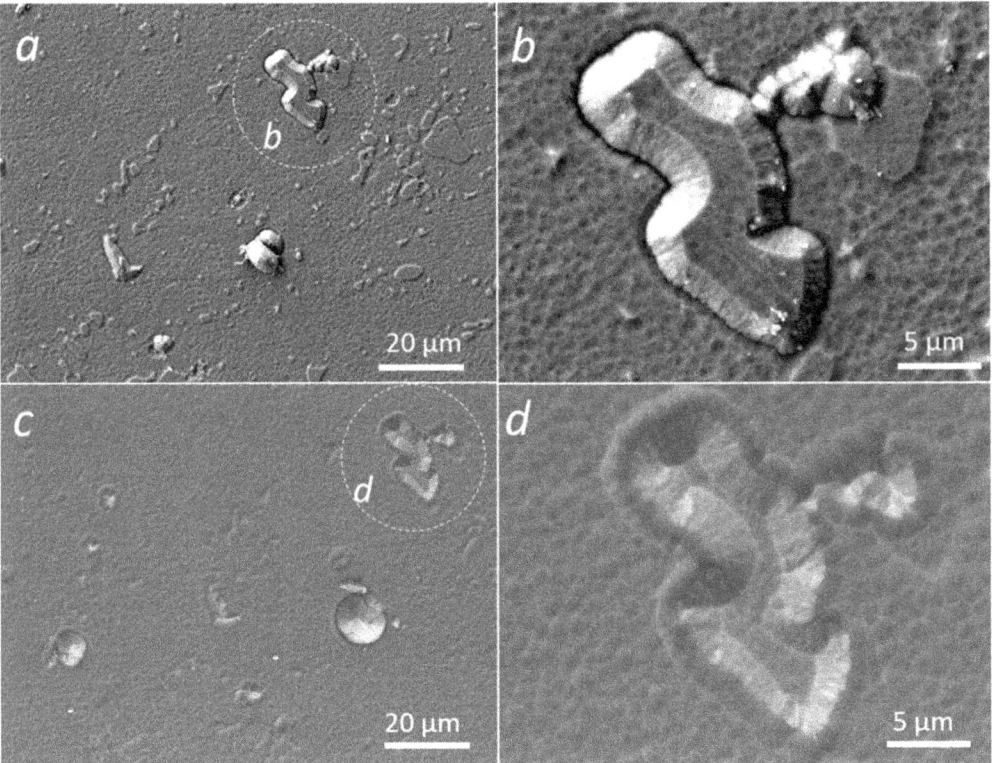

Figure 1. SEM micrographs at a lower (**a,c**) and higher magnification (**b,d**) of the same surface area of the single (**a,b**) and double layer TiAlN (**c,d**) coatings deposited on D2 substrate.

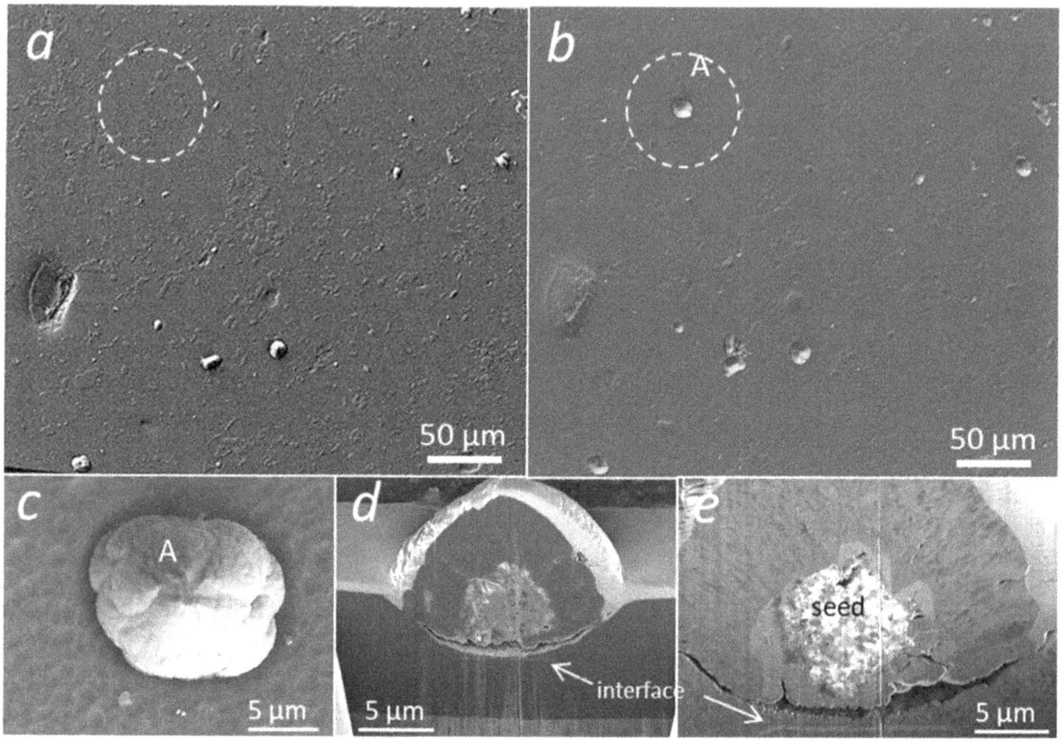

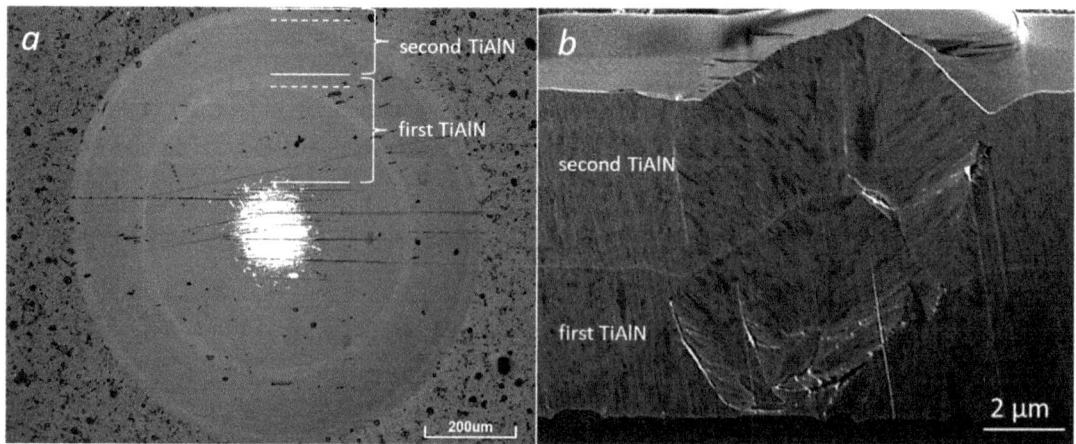

Figure 2. SEM micrographs at a low magnification of the same surface area of the single (**a**) and double layer (**b**) TiAlN coatings deposited on the D2 substrate. Plain-view SEM micrograph (**c**) and FIB secondary electron images (**d**,**e**) of cross-sections of a selected nodular defect (A) formed during deposition of the second TiAlN coating.

Figure 3. (**a**) Ball crater through a double layer TiAlN coating. The bright rings belong to the TiAlN layer formed after the intermediate ion etching of the coating. (**b**) SEM micrograph of an FIB cross-section of the TiAlN double layer deposited on the PM ASP30 tool steel substrate. The interface between the first and the second TiAlN coatings is visible as a continuous bright line.

The topography of the double-layer TiAlN coating surface was examined again by following the same surface area. SEM images in Figures 1 and 2 show the comparison of the size and shape of selected growth defects after the first and after the second deposition of TiAlN coating. We found that after the deposition of the second TiAlN coating, most nodular defects remained on the coating surface, some of them converted into craters, while some new nodular defects also appeared. The conversion of nodular defects into craters most likely occurs due to the high thermal load during ion etching. The resulting thermal stresses can cause the detachment of some weakly bonded nodular defects [19,20]. We can also observe that the surface of the double layer TiAlN coating is smoother in comparison with the single one. This planarization phenomenon is due to ion etching before the deposition of the second TiAlN coating. Namely, during ion etching, the side surface of all protrusions etches faster than the flat coating surface because the sputtering yield at the normal incidence of ions is much smaller than at a high incidence angle [6]. This effect leads to the shrinking of all protrusions and even elimination of the smaller ones [5]. This is the reason why the protrusions formed at carbide sites after the deposition of the first TiAlN coating on the ASP tool steel substrate almost disappear after the second deposition of the TiAlN coating (Figure 4).

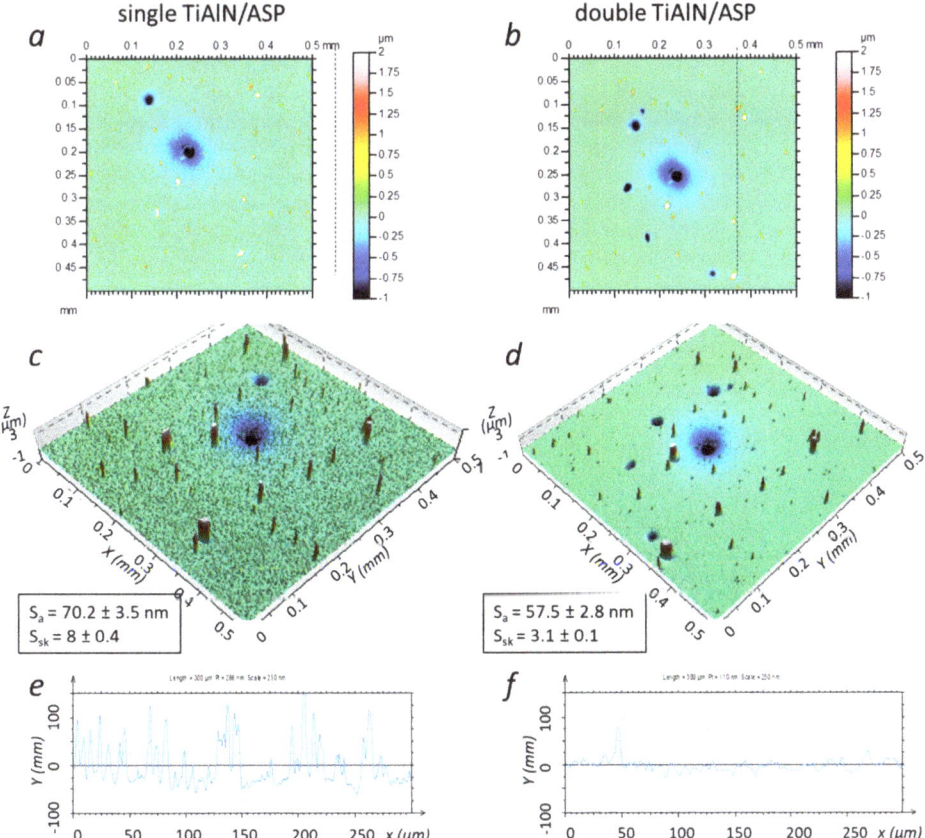

Figure 4. Top view (a,b) and 3D profilometer images (c,d) of the same surface area of single and double layer TiAlN coatings deposited on the ASP tool steel substrate. Roughness parameters (S_a, S_{sk}) of the substrate, after deposition of the first and the second deposition of TiAlN coatings, are given in the frames. The line profiles (e,f) were recorded on the defect-free areas of 3D profile images (c,d).

However, if the TiAlN coating is deposited on a D2 substrate, where the protrusions at carbide sites are much larger (Figure 5), they do not disappear completely. The effect of ion etching on protrusions at carbide sites is clearly seen from the roughness line profiles. They were recorded on the defect-free areas of the TiAlN coating (Figures 4 and 5). All peaks on these profiles belong to protrusions formed at carbide inclusions. It is evident that after ion etching before the deposition of the second TiAlN coating on the ASP tool steel substrate these peaks disappear, while they are not eliminated completely if D2 tool steel is used for the substrate. The planarization effect due to preferential ion etching is reflected in the reduction of surface roughness S_a after deposition of the second layer of TiAlN. The surface skewness S_{sk}, which is a measure of the asymmetry of the surface profile from the surface mean line, is also reduced. Due to carbide protrusions formed during the polishing of bare substrates, the value of the parameter S_{sk} was positive for both bare ASP and D2 tool steel substrates (Table 1). The S_{sk} roughness parameter for bare ASP substrate was much higher than that for bare D2 substrate. The reason for this phenomenon is that during mechanical pretreatment of D2 substrates, where large carbide inclusions are present, a part of protruded carbides is torn out due to large shear stress leaving a pit in the substrate. The formation of such pits causes the reduction of the S_{sk} roughness parameter. It remained positive but smaller after the deposition of the first TiAlN. Changes in the roughness parameters are caused both by ion etching of the bare steel substrate as well as by deposition of the TiAlN coating. Due to the inhomogeneity of the D2 and ASP tool materials and the consequent different etching rates of various phases and grains with different orientations, both shallow depressions and protrusions formed. This is reflected in both higher roughness S_a and lower skewness S_{sk} (Table 1). After deposition, the skewness parameter increased due to the formation of nodular defects. Shrinking of all protrusions during ion etching, performed before the deposition of the second TiAlN coating (Figures 1 and 2), is reflected in lower values of roughness parameters, both surface roughness S_a and surface skewness S_{sk}. For the reasons given above, the reduction in both roughness parameters is more pronounced in the layers applied to the ASP substrate.

Table 1. Roughness parameters (S_a, S_{sk}) of ASP and D2 substrates, after polishing, after deposition of the first TiAlN coating and after deposition of the second TiAlN coating. The scanning area was 0.5 mm × 0.5 mm.

Substrate	S_a (nm)	S_{sk}
bare ASP	11.8 ± 0.6	10 ± 0.5
ASP + TiAlN	70.2 ± 3.5	8 ± 0.4
ASP + TiAlN + TiAlN	57.5 ± 2.8	3.1 ± 0.1
bare D2	11.3 ± 0.6	2.9 ± 0.1
D2 + TiAlN	83.1 ± 4.1	7.5 ± 0.4
D2 + TiAlN + TiAlN	82.6 ± 4.1	3.2 ± 0.1

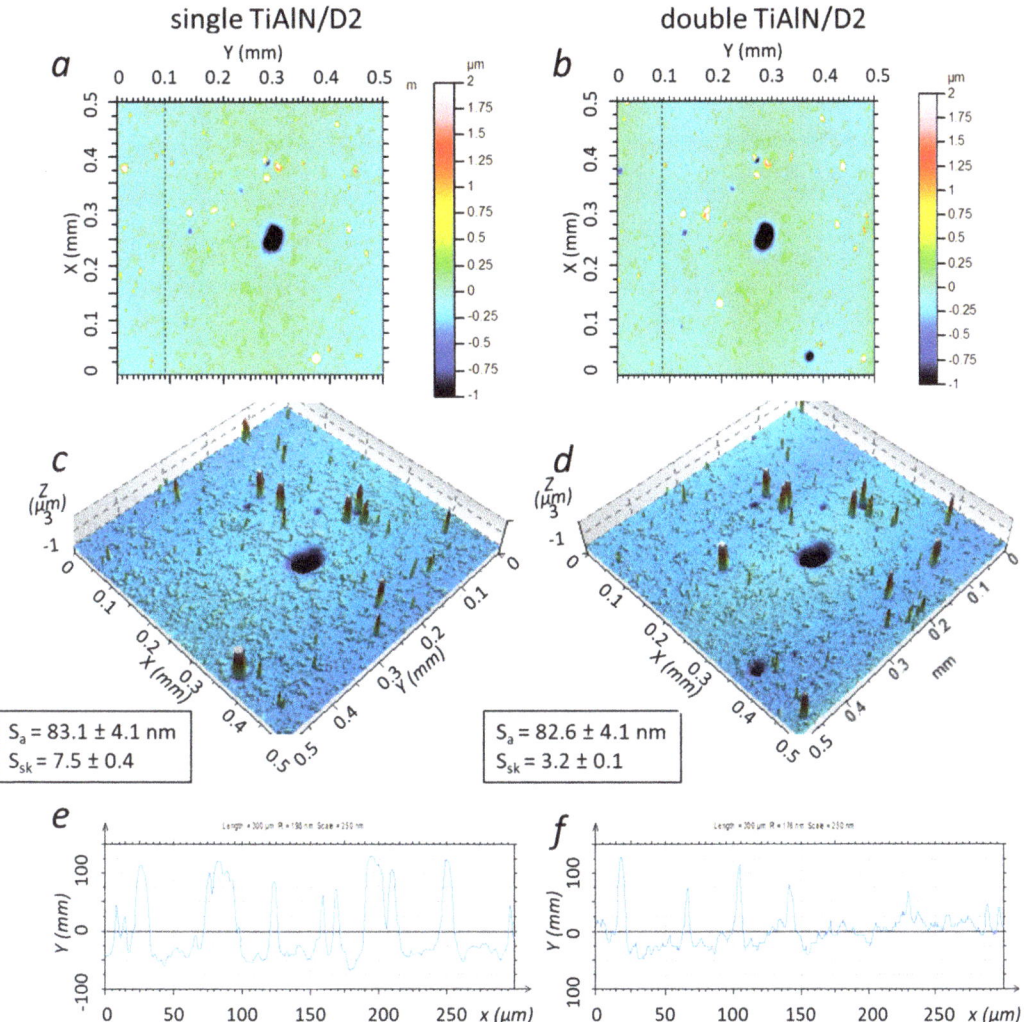

Figure 5. Top view (**a**,**b**) and 3D profilometer images (**c**,**d**) of the same surface area of single and double layer TiAlN coatings deposited on the D2 substrate. Roughness parameters (S_a, S_{sk}) of the substrate, after deposition of the first and the second deposition of TiAlN coatings, are given in the frames. The line profiles (**e**,**f**) were recorded on the defect-free areas of 3D profile images (**c**,**d**).

Additional phenomena are well-defined trenches or depressions occurring in the boundary region of sputtered nodular defects (Figure 6). Trenching is caused due to enhanced erosion near the sidewall of nodular defects (see Ref. [6]).

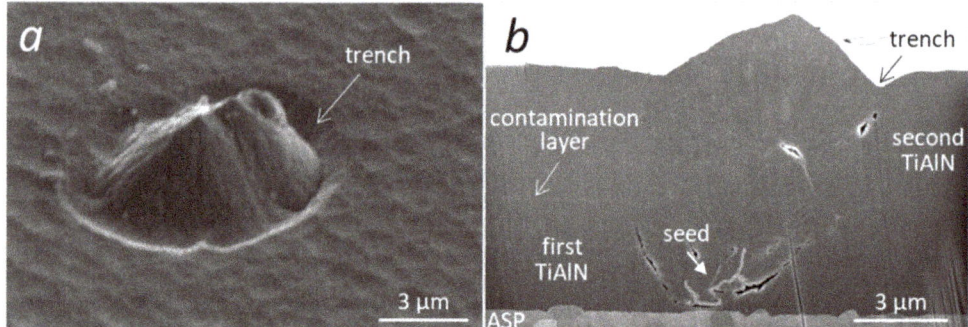

Figure 6. Plain-view SEM micrograph (**a**) and SEM micrograph of the FIB cross-section (**b**) of a nodule on the TiAlN coating. A trench around the nodular defect as well as a contamination layer on the seed particle is clearly visible.

3.2. Contamination of the Interfacial Region in Double-Layer TiAlN Coating

The microstructure and compositional depth profile of the resulting double-layer TiAlN coating were investigated by cross-sectional SEM (Figure 7) and STEM microscopy (Figure 8). High-resolution bright filed imaging indicated that the contamination interlayer formed with a well-defined thickness (around 5 nm) (Figure 8a). The interface between the substrate and the film is smooth and abrupt. The STEM-EDX line profiles show that it is composed mostly of iron, chromium, and tungsten (Figure 8b). The most likely origin of these metal elements is the batching material, which was in our case HSS and cemented carbide cutting tools. The detection of batching material in the interfacial region is probably due to the substrate ion etching before the coating growth. In an ion etching process, the sputtered atoms of the batching material are deposited on all surfaces in the vacuum chamber, including the target surfaces that are not protected by a shutter (Figure 9). Deposited material on the target surface subsequently re-deposited back on the substrates during the initial stages of the coating growth. In addition to the batching material, the interfacial region was also contaminated with the coating material, which was deposited on the substrate fixturing components in the previous batch.

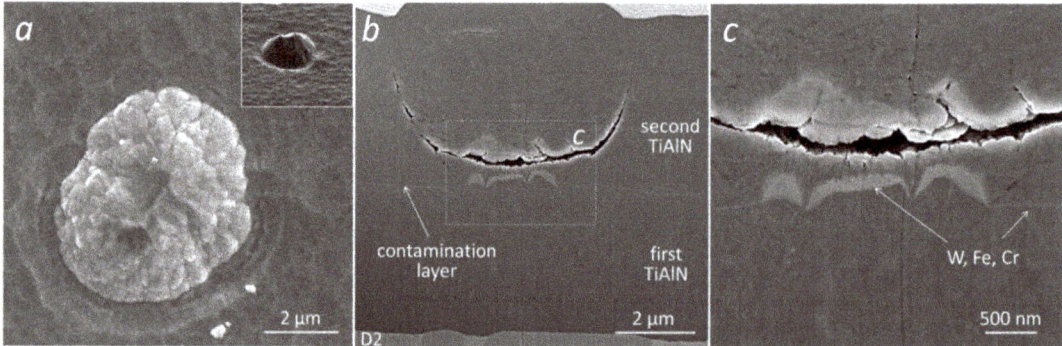

Figure 7. Plain-view SEM image (**a**) of a nodular defect formed in the initial stage of deposition of the second TiAl coating; (**b**,**c**) FIB cross-section of TiAlN double layer coating clearly shows the contamination layer between both TiAlN coatings and under the nodular defect.

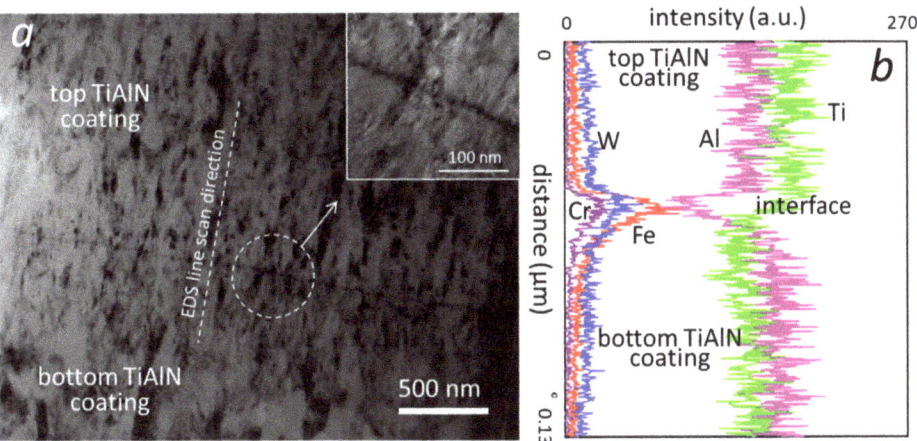

Figure 8. (a) Cross-sectional STEM micrograph and (b) STEM-EDX compositional profile of the interface region.

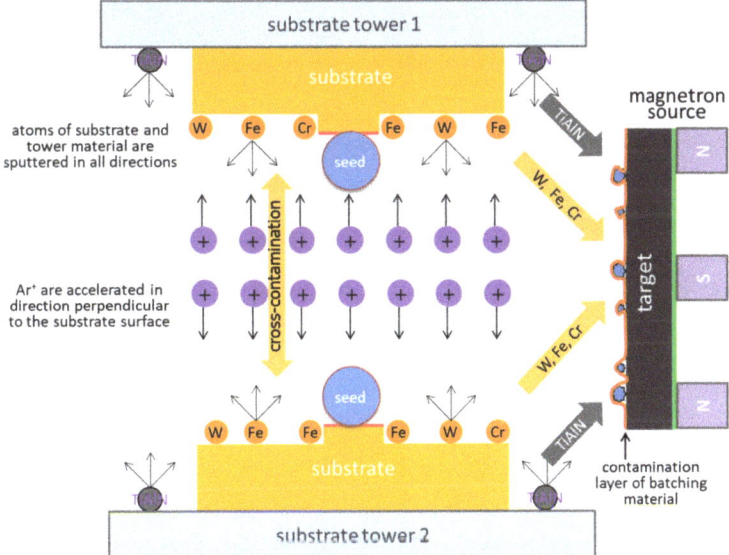

Figure 9. Schematic of the ion etching process and subsequent contamination of substrates and target surface with the batching material.

The interfacial region between the substrate and the first TiAlN coating is also contaminated in the same way. However, at this interface, it is very difficult to distinguish between substrate-related elements originating from contamination and those from the substrate itself [21,22]. The contamination of the target with the residual products from the etching process can be prevented by a movable shutter located close to the target, which collects the sputtered species and enables the pre-sputtering of targets. However, the majority of industrial deposition systems are usually not equipped with movable shutters because the complicated installation of such a system reduces the economy of the deposition process.

During the ion etching process, cross-contamination also occurs, i.e., when the sputtered atoms removed from one substrate tower contaminate the other one and vice versa.

An effective sputter cleaning is possible only if the mean number of sputtered atoms per unit surface area is higher than the number of re-sputtered atoms (and other molecules from the residual gas) hitting and sticking on the substrate surface. For the majority of industrial deposition systems, this requirement is fulfilled. However, contamination may occur on the substrate surfaces shaded beneath foreign particles (Figure 7b,c). These areas are not exposed to ion etching because the ions are directional perpendicular to the substrate surface. On the other side, the same surface is covered with atoms of the batching material, which can arrive from areas where the intensity is high. Therefore, the thickness of the contamination layer at sites of foreign particles is significantly larger (see Figure 7c).

3.3. SEM Images of Broken Nodular Defects

During or after the deposition, some nodular defects are broken off due to the high residual stress in the coating. The backscattered electron (BSE) images of the broken nodular defects are shown in Figure 10. In these images, we can see that the seed particles are surrounded by a thin contamination layer. EDS analysis shows that this layer has a similar composition as the contamination layer formed at the interface between both TiAlN coatings. Based on this fact, we can conclude that the origin of metal elements (W, Fe, Cr) is the same in both cases. However, the key question is the origin of these seed particles that form growth defects. In the PVD deposition systems with magnetron sputtering sources, the formation of such particles (flakes) can be caused by (a) flaking of cones formed in the target racetrack, (b) flaking of the redeposited nodules from the target surface, and (c) by arcing [6,23]. Due to the electrostatic self-repulsion effect, a part of the flakes reach the substrate surface, where they are built into the growing coating.

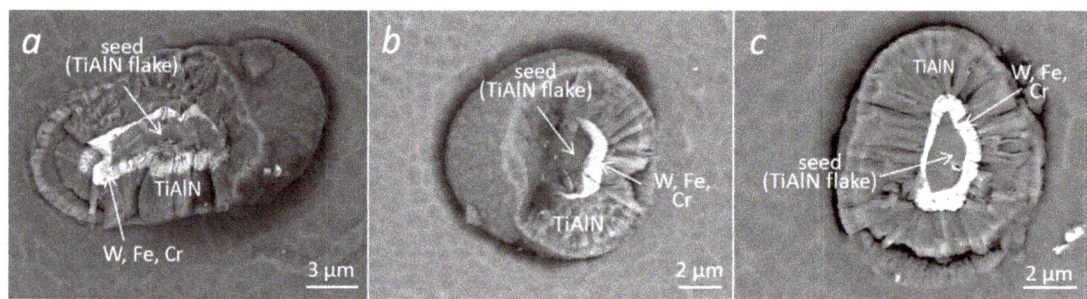

Figure 10. BSE images of broken nodular defects (a–c) show seed particles with an envelope that is composed of mostly of iron, chromium, and tungsten.

A still more intensive source of seed particles is in the target area outside of the racetrack. During the sputtering process, the target material is also re-deposited from the racetrack to the perimeter. Due to the target poisoning effect and low plasma density, sputtering does not take place in the perimeter region. High internal stress causes the re-deposited material to grow in the form of weakly bonded filaments [24,25]. Filaments formed near the racetrack eventually cross the high-density plasma region where they are heated. This causes their fracture into smaller fragments that are charged negatively accelerating them away from the target due to the electrostatic self-repulsion effect. A part of them reaches the substrate surface and during the deposition process, they cause the formation of nodules. In addition to the particles formed by the mechanism described above, there are also other particles on the target surface. These are, for example, flakes which have spalled off from the vacuum chamber components (e.g., shields, substrate fixture system) during the heating process or during the ion etching. There are also dust particles left on the surface of the target during the loading of the batching material since complete removal of these particles before deposition by blowing and wiping is not possible. During ion etching, these particles are also covered with a contaminant layer. At the beginning of the deposition process, an electric charge accumulates on these weakly bonded particles and electrostatic forces cause self-repulsion of them towards the substrate.

3.4. Depth Distribution of Nodular Defects

In the previous section, we found that many seed particles that caused the formation of nodular defects originated from the target surface, meaning that nodular defects start to form mostly at the beginning of the deposition process. However, the question is whether this mechanism of nodular defect formation is predominant. To answer this question, we tried to determine the depth distribution of nodular defects from the ground section prepared by the ball-cratering technique (Calotest). Because nodular defects on the ground section of a single TiAlN coating are not visible, we performed the analysis on a TiAlN/CrN multilayer structure deposited in the same system (Figure 11a).

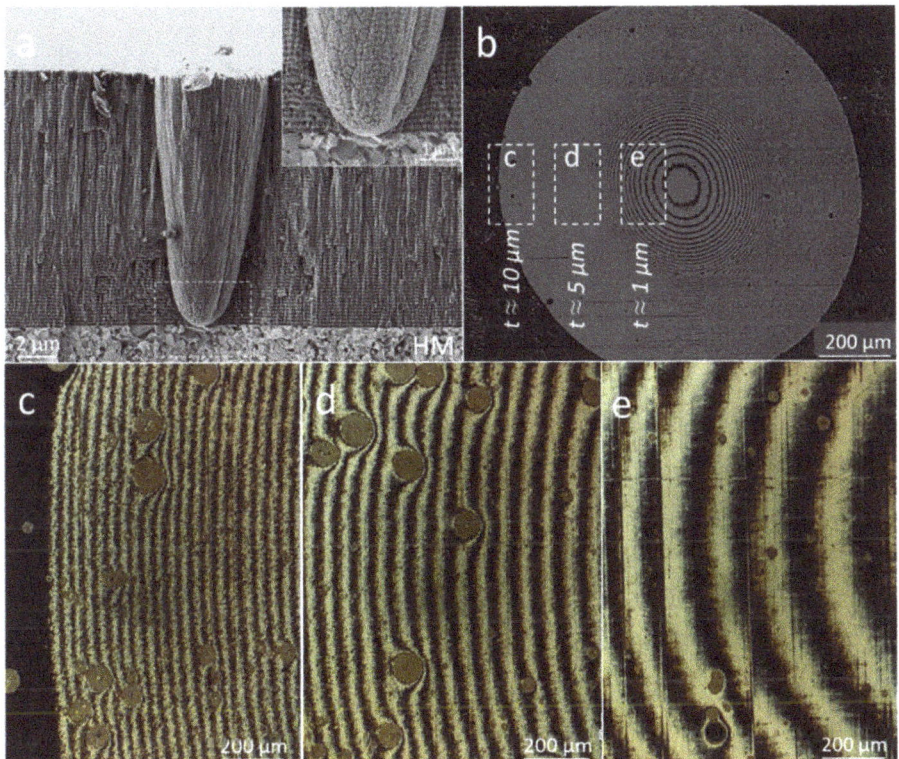

Figure 11. Fracture cross-sectional SEM micrograph of the TiAlN/CrN multilayer coating on HM substrate (**a**) and OM images of the ground section of TiAlN/CrN multilayer coating at low (**b**) and high magnifications (**c–e**). The density of nodular defects at different depths (frames (**c–e**)) is similar (20–30 defects/mm^2). Due to a parabolic shape, the nodule cross-section increases with coating thickness.

On a low-angle cross-section of such coating, the layer contours reveal the positions of defects (Figure 11c–e). Thus, we can count how many nodular defects are located at a certain coating depth. We found that the concentration of defects in the middle of the coating and close to the top surface is comparable with the concentration close to the substrate–coating interface. The more or less uniform distribution of nodular defects on the ground section of the TiAlN/CrN coating means that most nodules started to grow at the substrate–coating interface. Namely, we have to consider, that all nodular defects formed at the substrate–coating interface extend through the entire coating. Due to the parabolic shape, the cross-section of nodular defects close to the substrate is much smaller in comparison with that at the coating surface.

FIB cross-sectioning of selected nodular defects on the ground section showed that all of them were formed on the substrate surface (Figure 12). This could mean that the mechanism of defect formation described in the previous sections is most likely predominant. This is also confirmed by the fact that a thin layer of contamination from heavy elements is clearly visible on some seeds (Figure 12c–f).

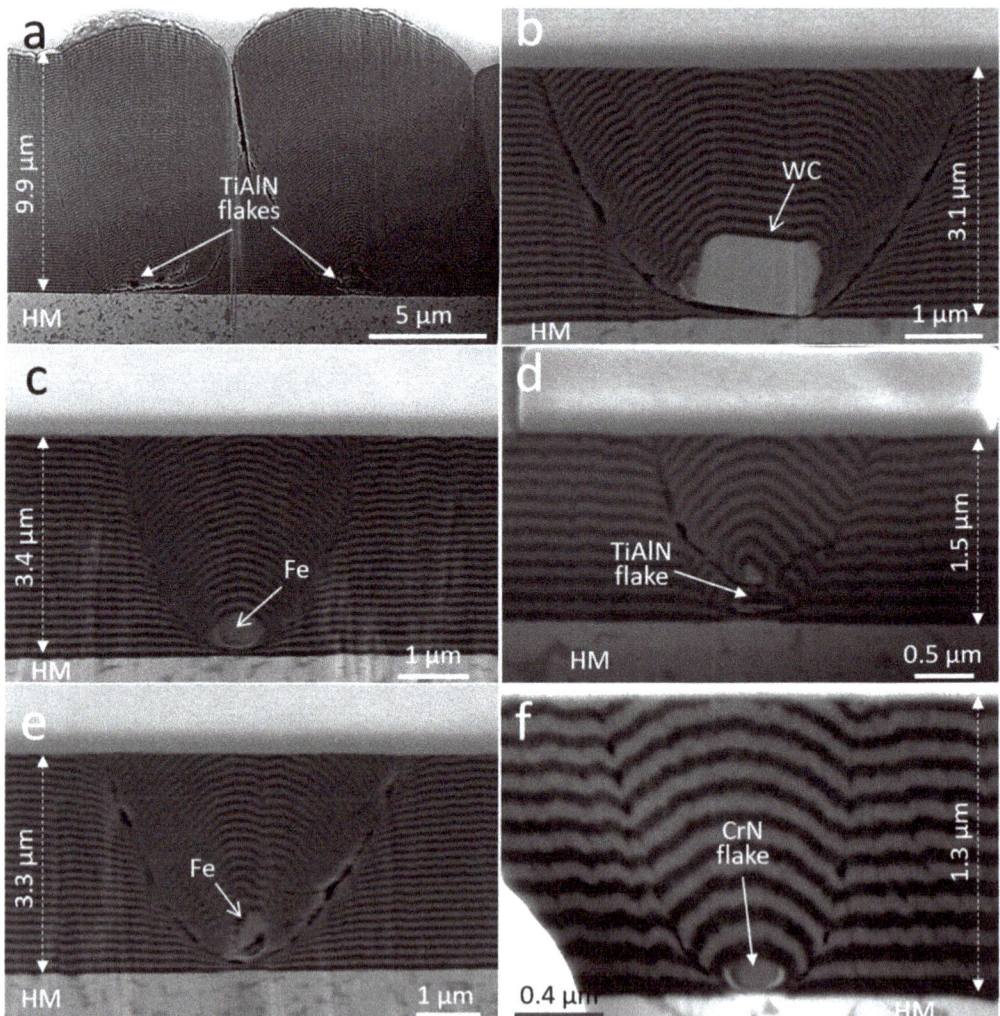

Figure 12. FIB cross-sectioning of selected nodular defects on the ground section of the TiAlN/CrN multilayer coating at different depths (**a**–**f**). SEM images were obtained with a BSE detector to emphasize the dissimilarity between TiAlN and CrN layers. Due to its high atomic number, the electron backscattering on CrN is more intense and is therefore displayed as layers with brighter contrast.

4. Conclusions

In this paper, we focused on the problem of target surface contamination in an industrial magnetron sputtering deposition system. Such contamination with the residual products from the etching process can be prevented by a movable shutter located close to the targets, but in order achieve a reasonable economics of the deposition process, complicated installations (including shielding and shuttering) are usually avoided. In this paper, all experiments were performed on samples prepared in the production batches. In a typical batch, several types of tools are loaded and different fixtures are used, which are designed for different tooling arrangements. The contamination layer at the interface between the substrate and the coating is difficult to identify because it contains both substrate elements (batching material) and coating elements (the coating material deposited on the substrate fixturing components in the previous batch) To avoid this problem, we deposited a TiAlN double coating in two separate production batches on the same substrate. Using this approach, the contamination layer between the two TiAlN coatings was easier to identify and determine its composition and thickness. We found that in such layer metal elements (W, Fe, and Cr) are present, which originate from the batching material (steel and cemented carbide tools).

We also found that many seed particles that cause the formation of nodules were covered with a similar contamination layer. We believe that these weakly bonded particles were formed on the target surface outside of the racetrack and that they were transferred to the substrate surface immediately after starting the deposition process by the self-repulsion effect. We observed such a contamination layer in the BSE plain-view images on the broken nodular defects as well as in SEM images of FIB cross-sections of nodular defects.

Another result of this study is that the surface roughness of the second TiAlN coating is smaller in comparison with the first one. This phenomenon was explained by the planarization effect.

Author Contributions: Design of experiments, 3D profilometry, interpretation of experimental results, manuscript writing, P.P.; preparation of specimen for TEM characterization, SEM and FIB analysis, manuscript review, A.D.; SEM analysis, manuscript review, M.Č. and M.P. All authors have read and agreed to the published version of the manuscript.

Funding: This work was supported by the Slovenian Research Agency (program P2-0082, project J2-2509). We also acknowledge funding from the European Regional Development Funds (CENN Nanocenter, OP13.1.1.2.02.006).

Institutional Review Board Statement: Not applicable.

Informed Consent Statement: Not applicable.

Data Availability Statement: Not applicable.

Acknowledgments: The authors would also like to thank Goran Dražić for TEM analysis and Jožko Fišer for technical assistance.

Conflicts of Interest: The authors declare no conflict of interest.

References

1. Van Stappen, M.; Malliet, B.; De Schepper, L.; Stals, L.M.; Celis, J.P.; Roos, J.R. Influence of Ti intermediate layer on properties of TiN coatings deposited on various substrates. *Surf. Eng.* **1990**, *6*, 220. [CrossRef]
2. Gerth, J.; Wiklund, U. The influence of metallic interlayers on the adhesion of PVD TiN coatings on high-speed steel. *Wear* **2008**, *264*, 885–892. [CrossRef]
3. Farnsworth, H.E.; Schlier, R.E.; George, T.H.; Burger, R.M. Ion Bombardment-Cleaning of Germanium and Titanium as Determined by Low-Energy Electron Diffraction. *J. Appl. Phys.* **1955**, *26*, 252. [CrossRef]
4. Mattox, D.M. Film deposition using acclerated ions. *Electrochem. Technol.* **1964**, *2*, 95.
5. Panjan, P.; Drnovšek, A.; Gselman, P.; Čekada, M.; Panjan, M. Review of Growth Defects in Thin Films Prepared by PVD Techniques. *Coatings* **2020**, *10*, 447. [CrossRef]
6. Panjan, P.; Drnovšek, A.; Mahne, N.; Čekada, M.; Panjan, M. Surface Topography of PVD Hard Coatings. *Coatings* **2021**, *11*, 1387. [CrossRef]

7. Schönjahn, C.; Lewis, D.B.; Münz, W.-D.; Petrov, I. Shortlisted substrate ion etching in combined steered cathodic arc–ubm deposition system: Effects on interface architecture, adhesion, and tool performance. *Surf. Eng.* **2000**, *16*, 176–180. [CrossRef]
8. Ehiasarian, A.P.; Wen, J.G.; Petrov, I. Interface microstructure engineering by high power impulse magnetron sputtering for the enhancement of adhesion. *J. Appl. Phys* **2007**, *101*, 054301. [CrossRef]
9. Mattox, M.D. A Short History of In Situ Cleaning in Vacuum for Physical Vapor Deposition (PVD). *SVC Bull. Fall* **2014**, 50–52.
10. Broitman, E.; Hultman, L. Adhesion improvement of carbon-based coatings through a high ionization deposition technique. *J. Phys. Conf. Ser.* **2012**, *370*, 012009. [CrossRef]
11. Lattemann, M.; Ehiasarian, A.P.; Bohlmark, J.; Persson, P.A.O.; Helmersson, U. Investigation of high power impulse magnetron sputtering pretreated interfaces for adhesion enhancement of hard coatings on steel. *Surf. Coat. Technol.* **2006**, *200*, 6495–6499. [CrossRef]
12. Hovsepian, P.E.; Ehiasarian, A.P. Six strategies to produce application tailored nanoscale multilayer structured PVD coatings by conventional and High Power Impulse Magnetron Sputtering (HIPIMS). *Thin Solid Films* **2019**, *688*, 137409. [CrossRef]
13. Ehiasarian, A.P.; Münz, W.D.; Hultman, L.; Helmersson, U.; Petrov, I. High power pulsed magnetron sputtered CrNx films. *Surf. Coat. Technol.* **2003**, *163–164*, 267–272. [CrossRef]
14. Gassner, M.; Schalk, N.; Sartory, B.; Pohler, M.; Czettl, C.; Mitterer, C. Influence of Ar ion etching on the surface topography of cemented carbide cutting inserts. *Int. J. Refract. Met. Hard Mater.* **2017**, *69*, 234–239. [CrossRef]
15. Tillmann, W.; Hagen, L.; Stangier, D.; Krabiell, M.; Schröder, P.; Tiller, J.; Krumm, C.; Sternemann, C.; Paulus, M.; Elbers, M. Influence of etching-pretreatment on nano-grained WC-Co surfaces and properties of PVD/HVOF duplex coatings. *Surf. Coat. Technol.* **2019**, *374*, 32–43. [CrossRef]
16. Panjan, M. Influence of substrate rotation and target arrangement on the periodicity and uniformity of layered coatings. *Surf. Coat. Technol.* **2013**, *235*, 32–44. [CrossRef]
17. Panjan, P.; Drnovšek, A.; Dražić, G. Influence of Growth Defects on the Oxidation Resistance of Sputter-Deposited TiAlN Hard Coatings. *Coatings* **2021**, *11*, 123. [CrossRef]
18. Jehn, H.A. Improvement of the corrosion resistance of PVD hard coating–substrate system. *Surf. Coat. Technol.* **2000**, *125*, 212–217. [CrossRef]
19. Panjan, P.; Čekada, M.; Panjan, M.; Kek-Merl, D.; Zupanič, F.; Čurković, L.; Paskvale, S. Surface density of growth defects in different PVD hard coatings prepared by sputtering. *Vacuum* **2012**, *86*, 794–798. [CrossRef]
20. Sebastiani, M.; Piccoli, M.; Bemporad, E. Effect of micro-droplets on the local residual stress field in CAE-PVD thin coatings. *Surf. Coat. Technol.* **2013**, *215*, 407–412. [CrossRef]
21. Håkansson, G.; Hultman, L.; Sundgren, J.E.; Greene, J.E.; Münz, W.D. Microstructures of TiN films grown by various physical vapour deposition techniques. *Surf. Coat. Technol.* **1991**, *48*, 51–67. [CrossRef]
22. Maura e Silva, C.W.; Alves, E.; Ramos, A.R.; Sandu, C.S.; Cavaleiro, A. Adhesion failures on hard coatings induced by interface anomalies. *Vacuum* **2009**, *83*, 1213–1217. [CrossRef]
23. Wehner, G.K.; Hajicek, D.J. Cone Formation on Metal Targets during Sputtering. *J. Appl. Phys.* **1971**, *42*, 1145. [CrossRef]
24. Heintze, M.; Luciu, I. Nodule formation on sputtering targets: Causes and their control by MF power supplies. *Surf. Coat. Technol.* **2018**, *336*, 80–83. [CrossRef]
25. Selwyn, G.S.; Weiss, C.A.; Sequedac, F.; Huang, C. Particle contamination formation in magnetron sputtering processes. *J. Vac. Sci. Technol.* **1997**, *15*, 2023–2028. [CrossRef]

Article

Comparative Study of Tribological Behavior of TiN Hard Coatings Deposited by Various PVD Deposition Techniques

Peter Panjan [1,*], Aljaž Drnovšek [1], Pal Terek [2], Aleksandar Miletić [2], Miha Čekada [1] and Matjaž Panjan [1]

1. Jozef Stefan Institute, Jamova 30, 1000 Ljubljana, Slovenia; aljaz.drnovsek@ijs.si (A.D.); miha.cekada@ijs.si (M.Č.); matjaz.panjan@ijs.si (M.P.)
2. Faculty of Technical Sciences, University of Novi Sad, 21000 Novi Sad, Serbia; palterek@uns.ac.rs (P.T.); miletic@uns.ac.rs (A.M.)
* Correspondence: peter.panjan@ijs.si

Abstract: In this paper, we present a comparative study of tribological properties of TiN coatings deposited by low-voltage electron beam evaporation, magnetron sputtering and cathodic arc deposition. The correlation of tribological behavior of these coatings with their intrinsic properties and friction condition was studied. The influence of surface topography and the surrounding atmosphere was analyzed in more detail. We limited ourselves to the investigation of tribological processes that take place in the initial phase of the sliding test (the first 1000 cycles). A significant difference in the initial phase of the sliding test of three types of TiN coatings was observed. We found that nodular defects on the coating surface have an important role in this stage of the sliding test. The tribological response of TiN coatings, prepared by cathodic arc deposition, is also affected by the metal droplets on the coating surface, as well as those incorporated in the coating itself. Namely, the soft metal droplets increase the adhesion component of friction. The wear rates increased with the surface roughness of TiN coatings, the most for coatings prepared by cathodic arc deposition. The influences of post-polishing of the coating and the surrounding atmosphere were also investigated. The sliding tests on different types of TiN coatings were conducted in ambient air, oxygen and nitrogen. While oxygen promotes tribo-chemical reactions at the contact surface of the coating, nitrogen suppresses them. We found that the wear rate measured in ambient air, compared with that in an oxygen atmosphere, was lower. The difference is probably due to the influence of humidity in the ambient air. On the other hand, wear rates measured in a nitrogen atmosphere were much lower in comparison with those measured in an oxygen or ambient air atmosphere.

Keywords: TiN hard coating; low-voltage electron beam evaporation; unbalanced magnetron sputtering; cathodic arc deposition; surface topography; tribology

Citation: Panjan, P.; Drnovšek, A.; Terek, P.; Miletić, A.; Čekada, M.; Panjan, M. Comparative Study of Tribological Behavior of TiN Hard Coatings Deposited by Various PVD Deposition Techniques. *Coatings* **2022**, *12*, 294. https://doi.org/10.3390/coatings12030294

Academic Editor: Sakari Ruppi

Received: 20 January 2022
Accepted: 18 February 2022
Published: 22 February 2022

Publisher's Note: MDPI stays neutral with regard to jurisdictional claims in published maps and institutional affiliations.

Copyright: © 2022 by the authors. Licensee MDPI, Basel, Switzerland. This article is an open access article distributed under the terms and conditions of the Creative Commons Attribution (CC BY) license (https://creativecommons.org/licenses/by/4.0/).

1. Introduction

Hard protective coatings started to be used on an industrial scale in 1969, when Sandvik Coromant and Krupp-Widia companies introduced chemical vapor deposition (CVD) technique for deposition of TiC coatings almost at the same time [1]. The CVD titanium nitride (TiN) coating was developed soon after TiC and reached the commercial exploitation stage in the early 1970s [2]. The high temperature (typically about 1000 °C) limited the application of such coatings mainly to cemented carbide tools, but they were not suitable for wear protection of high speed steel (HSS) tools. This shortcoming of CVD processes was overcome a decade later by the introduction of physical vapor deposition (PVD) processes [3]. PVD techniques enabled deposition of TiN hard coatings at temperatures of less than 500 °C, which is below the tempering temperature for most types of tool steels. The commercial exploitation of the first PVD TiN hard coatings started in the late 1970s, and their use underwent a remarkable boom especially in the field of metal machining. Experts in the field of metal machining believe that the protection of cutting tools with wear-resistant TiN hard coatings has been one of the most significant technological advances in the development of modern tools. In the early years of PVD hard coatings, titanium nitride was just about the only coating. Although many other PVD hard coatings were later introduced in the market, the TiN coating is still today one of the most important.

The first PVD TiN hard coatings were deposited by *low-voltage electron beam evaporation* (Balzers) [3]. This technology was commercially very successful, and TiN coatings prepared using this technique are still a quality standard in this field. Almost at the same time, the USA Multi-Arc company started producing TiN hard coatings by *cathodic arc deposition* [4]. Their technology was based on a license bought in the former Soviet Union [5]. In the review paper [6], Anders reported that much of the early work on cathodic arc technology was performed in the former Soviet Union with the first industrial systems in 1974 (*Bulat technology*).

Initially, the TiN coatings prepared by the conventional DC magnetron sputtering technique were significantly inferior to those prepared by other PVD techniques. This process did not provide a sufficient intensity of ion bombardment of the substrate surface during the deposition process. Therefore, sputter-deposited coatings were usually not fully dense. In the second half of the 1980s, this problem was overcome with the invention of *the closed-field unbalanced magnetron sputtering* technique by Window and Savvides [7], which provided a degree of substrate ion bombardment during the deposition process equivalent to that of competing PVD hard coating techniques. An even greater advance in this field was made about 20 years later with the introduction of *high-power impulse magnetron sputtering* (HIPIMS) deposition, which provides an extremely high ionization fraction of the sputtered species [8]. All of these improvements have enabled the preparation of high-quality coatings (higher density, improved adhesion, reduced roughness, and more uniform coating of high aspect ratio features) by magnetron sputtering.

The chemical bonds in TiN material are a combination of covalent, metallic and ionic bonds. This results in a unique combination of properties that are typical for metals (good electrical and thermal conductivity) as well as those that are typical for ceramics (high hardness, chemical inertness and high melting temperature). Due to such an unusual combination of properties, TiN coatings have found widespread application in industrial production.

TiN is a simple and cost-effective coating for *wear protection* of cutting, plastic molding and cold forming tools. It is suited for wear protection of HSS cutting tools working at low and medium cutting speeds. Typical applications include drilling, milling and turning of mild steels at cutting speeds below 100 m/min. Due to its gold color, TiN coating is also an excellent indicator for wear. The wear can be clearly seen on the worn areas of the tool or components after a certain time in service. However, the TiN coating is not suitable for the protection of tools used for machining difficult-to-cut workpiece materials (hard or "sticky") or for high-speed machining due to low oxidation resistance (below 600 °C). The surface oxidation of a TiN coating has a great impact on its wear resistance. The oxide layer formed on TiN coating easily spalls off due to high compressive stresses that develop because of the large difference in the molar mass ratio of oxide compared to nitride.

The unique gold color, abrasion resistance, and chemical inertness of TiN coatings have also encouraged their widespread applications for *decorative purposes* [9–11]. Watchmakers were the first group interested in the decorative aspect of TiN hard coatings. It is especially important that the preparation of PVD decorative coatings, unlike electrochemical ones, is not environmentally hazardous. The color of solid materials originates from electronic band structure or density of states. The golden yellow color of the pure and stoichiometric TiN coating is the consequence of the reflection edge located in the visible region with a characteristic reflectivity minimum of about 450 nm [12]. The location of the reflection edge can be affected by variations in composition, formation of lattice defects, or incorporation of impurities (e.g., oxygen, carbon). However, the color and spectral reflectivity of TiN films can also depend on surface roughness, arising from the columnar growth and formation of growth defects. Today, decorative applications of TiN coatings include sanitary hardware, household appliances, door handles, jewelry luxury items and architectural glass (to obtain special optical functions such as IR reflection). The use of decorative coatings often also allows the replacement of substrates made of expensive materials with cheaper ones.

TiN is non-toxic, and for this reason, it is used in a number of *biomedical applications* [13–15]. Because of its intrinsic biocompatibility, antibacterial properties, chemical inertness and good tribological properties, TiN coating is a suitable material for wear protection of orthopedic implants (hips, knees and other joints) and dental implants (screws, abutments) [14]. To protect the prosthetic joint, it is important that the coating material be hard and chemically inert to prevent wear, corrosion and the formation of debris that causes tissue inflammation. Due to hemocompatibility, TiN coating is used also in cardiology for ventricular assist devices for patients with heart failure and for pacemaker leads. For a long time, the medical industry has used PVD coatings to improve surgical tools (e.g., scalpels, blades, drills, reamers, orthopedic bone saw blades), where corrosion protection, sharpness and cutting edge retention are important (sharp cutting edges on medical instruments allow wounds to heal faster while the antimicrobial properties of TiN coatings reduce the possibility of infections) [13]. TiN coatings also improve surgical tool identification and reduce glare in the operating room.

Due to its relatively high thermal and structural stability combined with low electrical resistivity, TiN is employed in microelectronic and photovoltaic devices as an advanced metallization material or *diffusion barrier* [16]. In these devices, a broad variety of materials, ranging from metals and semiconductors to insulators, are in contact with each other. TiN diffusion barrier films prevent the intermixing and interdiffusion of these materials during the fabrication and operation of such devices and the resulting loss of their functionality. Its remarkable properties make titanium nitride film an attractive candidate also for other applications. These include gate electrodes in advanced field-effect transistors and emerging resistive switching memory technologies [17], cathodes for high energy density lithium-sulfur batteries [18] and electrodes in photovoltaic cells [19].

In each of these applications, with the exception of microelectronics, the good tribological properties of TiN coating are of crucial importance. In the past, a number of reports on the tribological performance of TiN coated surfaces have been published [20–22]. In the majority of these studies, researchers focused on the steady-state value of the friction coefficient, while the mechanism of friction and wear in the initial phase of the sliding test was largely ignored. Further, only a few papers have described the impact of coating surface roughness and surrounding atmosphere on the tribological properties of PVD hard coatings [23–25]. In this paper, we compared the tribological behavior of stoichiometric single-phase TiN coatings deposited by low-voltage electron beam evaporation, magnetron sputtering and cathodic arc deposition. All experiments were performed on samples prepared in production batches. We analyzed how the tribological properties of TiN coatings, prepared by these techniques, depend on the microstructure, texture and especially surface topography. The influences of post-polishing of the coating and the surrounding atmosphere were also investigated.

2. Materials and Methods

All TiN stoichiometric coatings were prepared in industrial batch-type deposition systems: a BAI730 low-voltage electron beam evaporation system (Blazers, Vaduz, Liechtenstein), CC800/7 modified unbalanced magnetron sputter deposition system (Cemecon, Würselen, Germany) and AIPocket cathodic arc deposition system (KCS Europe GmbH, Monschau, Germany). In the remainder of this paper, the abbreviations BAI, CC7 and AIP will be used for these three deposition techniques, as well as for their corresponding coatings. A more extensive description of all three deposition techniques can be found in a recently published paper [26], while the essential process parameters are given in Table 1. Different deposition techniques result in different physical and mechanical properties of TiN coatings.

Table 1. Deposition methods and process parameters used for the preparation of TiN hard coatings. The deposition rates and ion current density are averages since all deposition techniques exhibited both spatial and temporal variations during a given deposition run as well as from one run to another run (depending on batching material).

-	-	BAI	CC7	AIP
Preheating	heating method	electron bombardment	infrared heating	infrared heating
-	preheating temperature (°C)	450	450	450
Etching	etching mode	DC	RF	pulsed DC
-	working gas	Ar	Ar + Kr	Ar
-	negative substrate etching voltage (V)	200	200	300/400
-	etching time (min)	15–30	80	30
Deposition	deposition method	low voltage electron beam evaporation	unbalanced magnetron sputtering	cathodic arc evaporation
-	temperature (°C)	450	450	450
-	working gas	Ar + N_2	Ar + Kr + N_2	N_2
-	pressure of working gas (Pa)	0.2	0.7	4
-	deposition time (min)	80	125	45
-	negative substrate bias voltage (V)	125	120	70
-	average deposition rate * (nm/s)	0.85	0.36	1.56
-	thickness (µm)	4.1	2.7	4.2
-	average substrate current density (mA/cm^2)	3–5	~2	-

* for one-fold rotation of substrates.

Cold work tool steel AISI D2 (~58 HRC), produced by Ravne steel factory (Ravne na Koroškem, Slovenia), was used as a substrate material. Substrates were ground and polished to a mirror-like finish (R_a < 12 nm). Before the deposition, substrates were degreased and cleaned in ultrasonic baths with detergents (alkaline cleaning agents, pH ~ 11), rinsed in deionized water, and dried in clean hot air. Prior to the coating deposition, substrates were heated up to the deposition temperature (450 °C) and sputter-etched in order to remove the native oxide and other contamination. One-fold rotation of the steel substrates was applied.

The crystal structure of as-deposited coatings was examined by X-ray diffraction (XRD) with CuK$_\alpha$ radiation using a Bruker diffractometer (AXS Endeavor D4, Billerica, MA, USA) in Bragg/Brentano mode and equipped with a CuK$_\alpha$ X-ray source (0.15406 nm). The spectra were collected using a scan step size of 0.02° in a diffraction angle between 20° and 100°.

The surface topography characterization of the coated and uncoated substrate was carried out using 3D stylus profilometry (Bruker Dektak XT, Billerica, MA, USA), and scanning electron microscopy (SEM, JEOL JSM-7600F, Tokyo, Japan). The microstructure and the coating morphology were studied using fracture cross-sections examined in a field emission scanning electron microscope (FEI Helios Nanolab 650i, Amsterdam, The Netherlands). Cross-sections were also prepared by focused ion beam techniques (FIB) using an FIB source integrated into the FEI SEM scanning electron microscope. SEM images were recorded using the ion beam and the electron beam.

Tribological properties were evaluated by ball-on-disk tribometer (CSM, Neuchatel, Switzerland) using a linear reciprocating mode. Tests were conducted at room temperature in ambient air, nitrogen, and oxygen atmospheres. An alumina ball with a diameter of 6 mm was used as a counter-body. The normal load was 5 N, displacement amplitude 5 mm, number of cycles 1000. All coatings were tested both in the as-deposited and the post-polished surface conditions. The sliding tests were conducted at least three times on each sample, and the results were averaged. The coating wear track depth, the coating wear volume, and wear scar on the alumina ball were measured by a 3D stylus profilometer. Each of the measurements covered an area of 1 mm × 1 mm. The average cross-section area of a wear track was calculated from a series of profiles that composed the 3D profilometer image of the wear track. Wear rate was calculated as $W = V/(F \times L)$, where V is the wear volume (mm^3), F is the normal load (N), and L is the total sliding distance (m). Post-test characterization by scanning electron microscopy was also performed in order to investigate the mechanical and tribological response of the coatings. In order to detect chemical changes on worn surface, backscattered electrons (BSE) were recorded, since the atomic number contrast enables the identification of oxides.

The Vickers hardness of the coatings was measured with a nanoindenter (Fischerscope, Sindelfingen-Maichingen, Germany) using a load in the range of 25–1000 mN. The indents using the lower load were all located in flat areas free from visible defects, resulting in hardness values for the defect-free coating material.

2.1. Coating Microstructure

A comparison of SEM images from fracture-cross sections and FIB cross-sections of the BAI, CC7, and AIP TiN coatings deposited on D2 tool steel substrate are shown in Figure 1a–f, respectively. The SEM images demonstrate that all coatings, regardless of the deposition technique, exhibit columnar microstructure. Columns extend along the coating growth direction and are composed of grains oriented in the direction of the surface normal. The columnar morphology of the coating is mainly determined by the substrate surface roughness, preferential growth of crystal grains, and the surface mobility of the condensing atoms [27]. The columnar microstructure is a result of the growth competition between the adjacent grains. Due to a geometrical shadowing effect, columns growing faster prevent the growth of the slower ones. However, the microstructure is not homogeneous throughout the coating thickness, but changes due to the renucleation of new grains. Renucleation, which occurs during the growth at the sites of surface defects created by the impinging ions, disrupts the columnar microstructure. The newly formed grains are smaller and more equiaxed, although still elongated in the growth direction. This is particularly noticeable in the BAI coating, where the average column size is the smallest. The fracture cross-section SEM image (Figure 1a) of this coating shows that only a small fraction of the columns extends from the substrate to the top surface of the coating. A SEM image of the FIB cross-section recorded by ions (Figure 1d) shows that the BAI coating is composed of small grains (5–20 nm). On the other hand, the largest column size was observed in the AIP coating (with a diameter in the range of 0.5–1 µm), while the size of columns in CC7 coating were somewhere in between.

TiN coatings prepared by the three different deposition techniques, however, differ not only in the details of columnar microstructure but also in their coating porosity, degree of preferred orientation, and surface roughness. These differences in the properties of the coating are directly related to differences in the deposition processes and deposition parameters. While all the coatings were deposited on the same type of tool steel substrates (D2) and at comparable substrate temperatures (about 450 °C), the system geometries, coating growth rates, discharge conditions and the substrate pretreatment (ion etching) were quite different (see Table 1). The microstructure of TiN coatings primarily depends on the energy of the impinging atoms and ions (Ti^+, Ar^+, N^+), and the ratio (j_i/j_{Ti}) of the accelerated-ion flux j_i to the flux j_{Ti} of the deposited Ti atoms [28]. It is preferred that the ion flux reaching the substrate is high, while the ion energy is low (<20 eV) in order to avoid coating damage. A moderate ion bombardment during the coating growth increases adatom mobility and promotes the formation of nucleation sites and chemical reactivity. All of these processes, together with recoil implantation and redeposition of depositing atoms, can disrupt the columnar microstructure (i.e., transition from columnar to equiaxed growth) and increase the coating density. Considering all these facts, the different microstructures of the BAI, CC7 and AIP coatings can be explained by the difference in the bias voltage (it was highest for the BAI coating and lowest for the AIP one) and also by the different (j_i/j_{Ti}) ratio, which was largest during deposition in the BAI system (around 2.5).

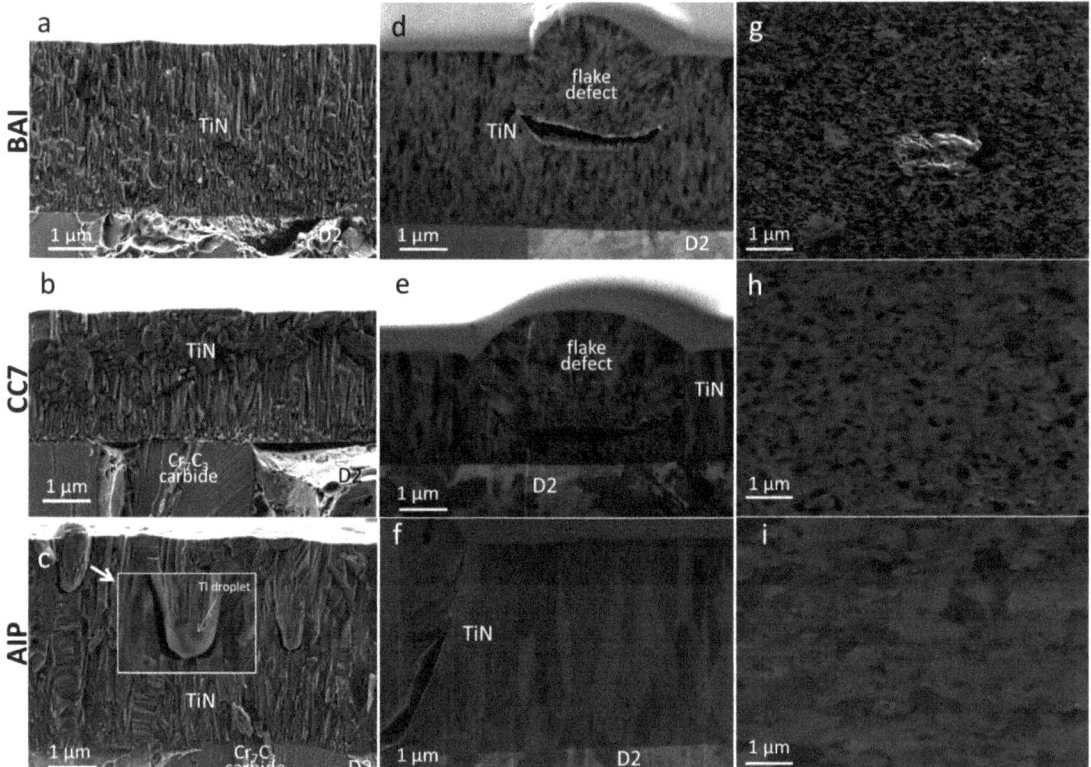

Figure 1. Fracture cross-sectional SEM images (**a**–**c**), FIB secondary electron images of cross-sections (**d**–**f**), and FIB secondary electron top view images (**g**–**i**) of TiN coatings deposited by low-voltage electron beam evaporation, magnetron sputtering and cathodic arc deposition techniques.

In order to obtain additional information about large area surface microstructural features, we used an innovative method based on top-view SEM images recorded using the ion beam. A pre-condition for performing an analysis is a very smooth coating surface, such as the one that formed in the wear track during a running-in period of a sliding test, where all protrusions were removed. The next step is a low-current ion etching with Ga$^+$ ions (no tilt applied on the sample stage) in the FIB module of the SEM microscope in order to remove the contamination layer from the coating surface. A clean and smooth coating surface is suitable for SEM imaging with an ion beam, in which the image contrast arises as a consequence of the ion channeling effect. When ions penetrate deeper into the crystal grain (the incident beam is parallel to a set of crystallographic planes inside a grain), the grain will appear darker due to a decrease in the number of secondary electrons that are emitted. However, if the grain has a no-channeling crystallographic orientation, it will appear bright. SEM imaging of coating surfaces with ions, therefore, provides information about grain preferential orientation. In order to produce a good quality image and to avoid any artifacts that may form as a result of ion exposure, the ion beam current density and ion dose need to be small. SEM images of polished BAI, CC7 and AIP TiN coating surfaces recorded using the ion beam are shown in Figure 1g–i. The differences in the size of crystal grains and their preferred orientation are evident.

2.2. Coating Topography

In our recently published study [26], we described in more detail the topography of PVD hard coatings, prepared by different PVD deposition techniques on various substrate materials. We showed that the surface topography can be distinguished on several size scales. It originates from the topography of the substrate surface, intrinsic coating microtopography and growth defects that form during the deposition process. The top-view SEM images of D2 tool steel substrates, after ion etching in the BAI, CC7 and AIP deposition systems, show that the etching efficiency of individual methods is different (Figure 2). The intensity of ion etching is reflected in the height of the step at the sites of protruding Cr_7C_3 carbides, which have a much smaller etching rate than the ferrous matrix. Outside the carbide area, the surface of substrate etched in the BAI system looks rather smooth, while it is rougher for samples etched in the CC7 and AIP systems. All topographic irregularities formed during ion etching are transferred to the coating surface and even magnified due to the geometrical shadowing effect. The protruding carbides of the substrate are, therefore, also visible in the SEM images of the TiN coating surfaces (Figure 3).

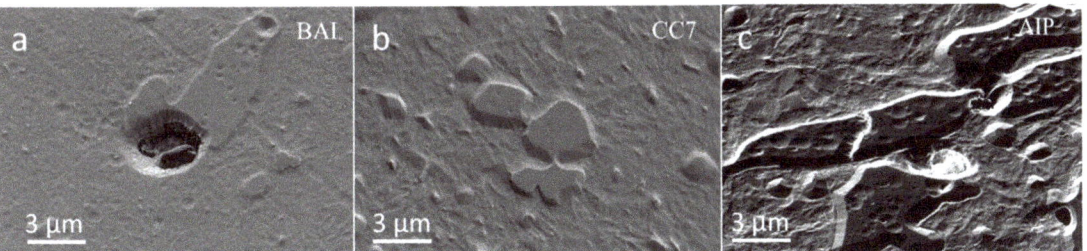

Figure 2. Top-view SEM images of D2 tool steel substrates after ion etching in BAI (**a**), CC7 (**b**) and AIP (**c**) deposition systems.

The topography of the coating surface is most affected by the growth defects (e.g., nodular defects, craters, droplets) formed during the deposition process [29]. The growth defects occur at sites of substrate topographical irregularities and at foreign particles that remain on the substrate surface after substrate pretreatment (cleaning, ion etching), and a part of them originates from the deposition process. The distribution of growth defects (nodular defects, droplets, pinholes, craters) on the coating surface and their shape and size were analyzed using a SEM microscope (Figure 3) and 3D stylus profilometer (Figure 4). The top-view SEM images of the BAI, CC7 and AIP TiN coating surfaces were taken at low (Figure 3a–c) and high magnifications (Figure 3d–f). In SEM images recorded at low magnification, we can see that growth defects are unevenly distributed over the coating surface. The lowest concentration of growth defects was observed on the BAI coating, while the highest one was on the AIP coating (Figures 3a–c and 4). High magnification SEM images (see insets in Figure 3d–f) of the area without growth defects reveals a dimpled surface topography for BAI coating, while the surfaces of the CC7 and AIP coatings are a little bit rougher, similar to substrate surfaces after ion etching.

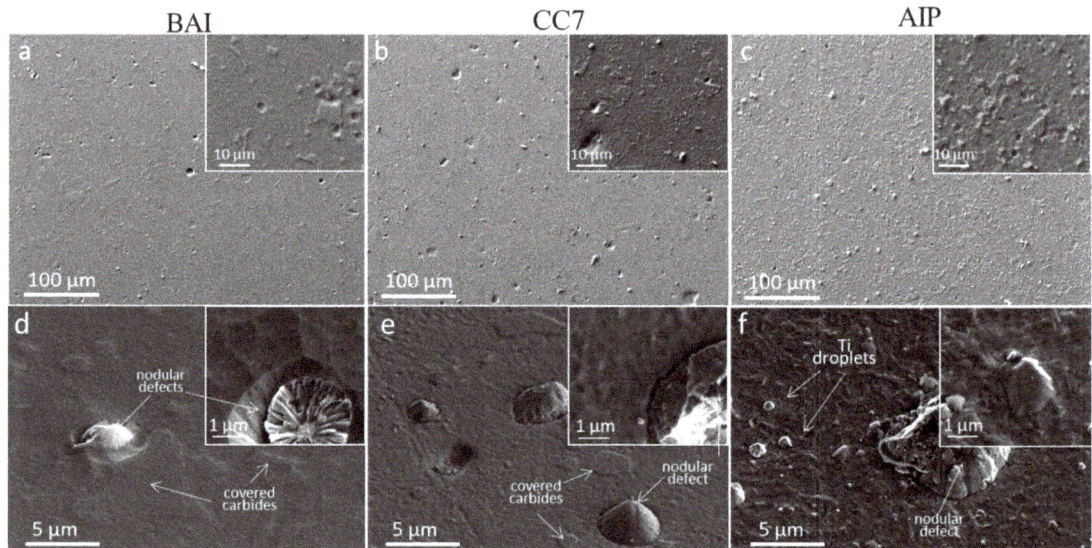

Figure 3. Top-view SEM images of TiN hard coatings deposited onto D2 tool steel substrate in BAI, CC7 and AIP deposition systems. The upper SEM images (**a–c**), including insets, were recorded at low magnifications, while the lower images (**d–f**), including insets, were taken at high magnification and by tilting the sample approximately 20°.

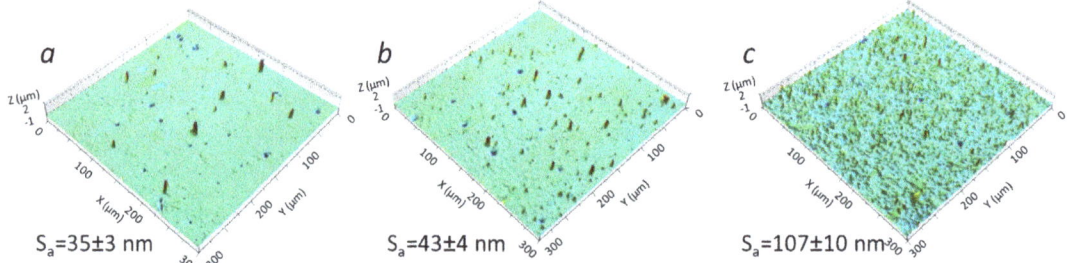

Figure 4. Three-dimensional profile images of TiN coatings deposited onto D2 tool steel substrate in BAI (**a**), CC7 (**b**) and AIP (**c**) deposition systems. The surface roughness values S_a are also added [26]. The nodular defects are the sharp peaks, while the blue dots are craters. The scan area was 300 μm × 300 μm, while the z-scale was 3 μm.

A 3D stylus profilometer was used to quantify the density of growth defects. Due to the relatively low surface density of growth defects and their uneven distribution, only large scanning areas provide a reliable result. In our case, the evaluation area was 1 mm² with a resolution of 2 μm in the x-direction and 0.2 μm in the y-direction, while the effective vertical resolution was around 5 nm. In order to distinguish the growth defects from the background, a threshold value has to be determined, and it should be larger than the step height of protrusions at carbide inclusions (which is typically about 200 nm). In our experience, a reasonable threshold value is 0.5 μm. The sharp peaks in Figure 4 are in most cases the nodular defects. The surface density of nodular defects, evaluated using these conditions, was about 180, 360 and 610 defects/mm² for the BAI, CC7 and AIP TiN coatings, respectively. On the AIP coatings, the growth defects originated mainly from the metal droplets, formed during the coating deposition process. Droplet-like defects (Figure 3f) were much smaller than the other protrusions (Figure 3d,e), but their density was much higher. Therefore, the roughness of the AIP coating was much higher than that of the other two. Parts of the droplets were torn out of the coating due to the high compressive stresses, leaving small craters (see inset in Figure 3c).

In general, the surface topography of a coating can be described with different surface roughness parameters [30]. The most commonly used surface roughness parameter is S_a, which is the absolute value of measured height deviations from the surface mean area. However, it does not give any information on the wavelength and is not sensitive to small changes in profile. Contrarily, the *surface skewness* S_{sk} is a measure of the asymmetry of the surface profile from the surface mean line. Zero skewness reflects symmetrical height distribution, while positive and negative values of S_{sk} indicate a prevalence of peaks and valleys, respectively. Due to carbide protrusions formed during polishing of bare D2 substrates, the value of the parameter S_{sk} was positive (Figure 5). It remained positive also after ion etching and deposition in all three deposition systems. However, after ion etching in BAI, CC7 and AIP systems, the parameter S_{sk} decreased from 2.86 for the polished substrate (where protruding carbides predominate) to 1.8, 1.3 and 0.53, respectively. This means that due to the inhomogeneity of the D2 tool material and the consequently different etching rates of various phases and grains with different orientations, both shallow depressions and protrusions were formed. This is reflected in both higher roughness S_a and lower skewness S_{sk}. After deposition, the skewness parameter significantly increased due to the formation of growth defects—the most for the BAI coating and the least for the AIP coating. This means that nodular defects predominated in the BAI coating, while there were also many craters in the CC7 and especially in the AIP coatings. Since there were many metal droplets on the surface of the AIP coating, the corresponding skewness was smaller for this coating. Useful information also can be obtained from the roughness parameter S_{ku} (*kurtosis*), which highlights whether the surface profile has steep or rounded peaks and valleys. For relatively flat surfaces, the S_{ku} value is smaller than 3; it is equal to 3 for perfectly Gaussian surfaces, while it is larger than 3 for surfaces with sharp peaks and valleys. In our case, the kurtosis was more than 3 for polished substrate surface, as well as for all substrates after ion etching and deposition in the BAI, CC7 and AIP systems. Kurtosis decreased after ion etching (the most in the CC7 system and the least in the BAI) but increased significantly after deposition (the most for BAI coating and the least for the AIP one). This means that the peaks formed during ion etching at sites of carbide and nonmetallic inclusions were less sharp, while during the deposition process, growth defects were formed, significantly increasing the S_{ku} roughness parameter.

2.3. X-ray Diffraction Analysis

Characterization of the crystallographic texture, grain size and crystal phases in the coatings was performed using XRD diffraction analysis (Figure 6). In addition to peaks originating from the TiN coating, substrate peaks (labeled »s« in Figure 6) were also present. The analysis showed that all three TiN coatings had a B1 NaCl crystal structure with varying degrees of preferred orientation. The positions of TiN diffraction peaks could be indexed to the face-centered cubic phase structure of TiN (JCPDS file No. 38-1420) with lattice constant a = 0.4241 nm [31]. Measured values of interplanar spacing d_{hkl} and the corresponding lattice parameters are given in Table 2. The spacings d_{hkl} for all reflecting planes (except (200)) parallel to the surface were 0.2–0.9% larger than the reference values obtained from a randomly oriented strain-free standard sample. This indicates the coatings were in a state of inhomogeneous compression stresses. The lattice parameters of the TiN coatings were in the range from 0.4219 to 0.4267 nm. Therefore, they slightly deviated from the bulk value of 0.424 nm. In the case of stoichiometric TiN coatings, this deviation can be explained by intrinsic stresses that can be caused by small grain size, high defect concentration, and/or interstitial incorporation of argon or nitrogen.

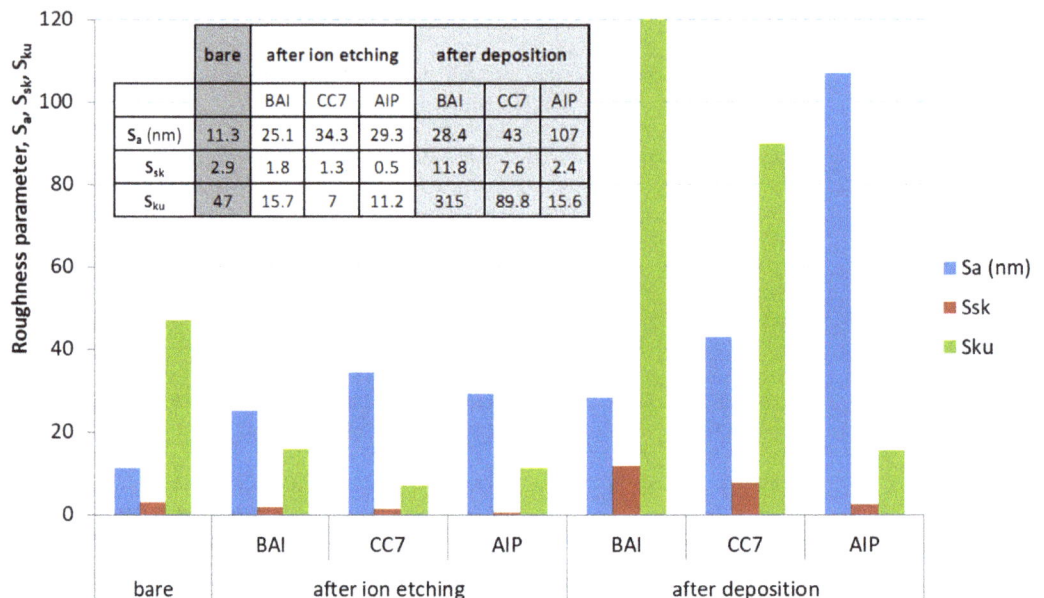

Figure 5. Roughness parameters (S_a, S_{sk} and S_{ku}) of D2 substrate, after polishing, after ion etching, and after deposition of TiN coatings in BAI, CC7 and AIP systems.

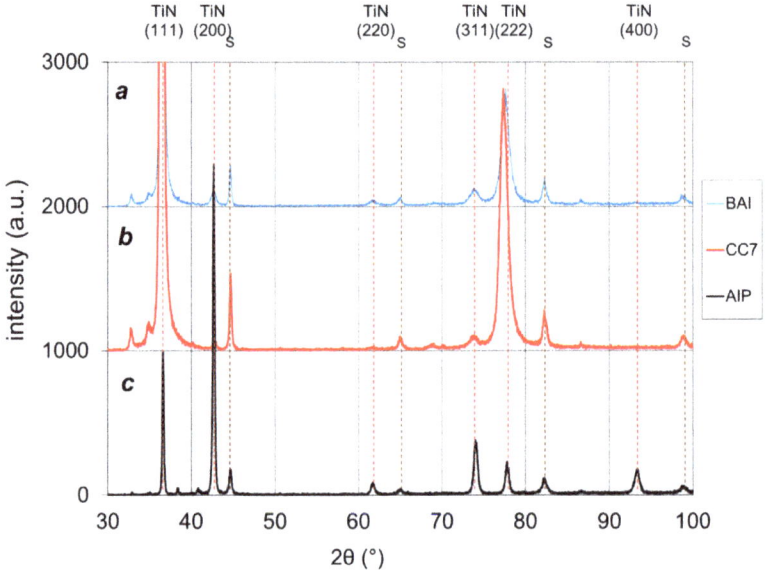

Figure 6. XRD spectra of TiN coatings deposited by low-voltage electron beam evaporation (**a**), magnetron sputtering (**b**), and cathodic arc deposition (**c**).

Table 2. Results of X-ray diffraction analysis of TiN coatings deposited by three different deposition methods: low-voltage electron beam evaporation (BAI), unbalanced magnetron sputtering (CC7) and cathodic arc deposition (AIP).

	Interplanar Distance, d_{hkl} (nm) Lattice Parameter, a_{hkl} (nm)				FWHM (°)			Relative Intensity (%)		
Deposition process	d_{111} a_{111}	d_{200} a_{200}	d_{220} a_{220}	d_{311} a_{311}	β_{111}	β_{200}	β_{220}	$I_{111}/\Sigma I_{hkl}$	$I_{200}/\Sigma I_{hkl}$	$I_{220}/\Sigma I_{hkl}$
Reference (6-0642) [31]	0.2449	0.212	0.1496	0.1277	-	-	-	26.9	35.8	19.7
BAI	0.2459 0.4259	0.2118 0.4236	0.1502 0.4248	0.1281 0.425	0.5	0.71	0.95	75	1.8	0.6
CC7	0.2462 0.4265	0.2110 0.4219	0.1502 0.4249	0.1282 0.4252	0.45	0.55	0.72	83	0.2	0.1
AIP	0.2451 0.4245	0.2117 0.4234	0.1502 0.4248	0.128 0.4244	0.23	0.27	0.51	24.8	55.6	1.8

The full width at half-maximum (FWHM) peak broadenings β_{hkl} are also given in Table 2. The measured values showed that peak broadening was the highest for BAI samples ($\beta_{111} = 0.5°$) and the lowest for AIP samples ($\beta_{111} = 0.23°$). A broader peak could be explained by smaller grain size and/or the presence of inhomogeneous microstrains in the crystallites. The high ratio of β_{222}/β_{111} indicates that the main contribution to peak broadening can be attributed to the inhomogeneous strain rather than small grain sizes. It is commonly known that the ion bombardment of growing coatings induces intrinsic compressive stresses through the entrapment of nitrogen and/or argon ions in non-equilibrium on interstitial (tetrahedral) lattice sites. The phenomenon is favored at low deposition rates and low deposition temperature.

The intensities of TiN reflections are also very different from those of a randomly oriented sample. There are, however, clear differences among the samples in the degree of preferred orientation. The preferred orientation of the coatings could be quantified by evaluating integrated intensities of four apparent peaks, namely, (111), (200), (220), and (311). Texture coefficients for all three types of coatings are included in Table 2. The degree of preferred orientation of the coatings is given by the texture coefficient γ_{hkl} ($\gamma_{hkl} = I_{hkl}/\Sigma I_{hkl}$, where I_{hkl} is the intensity of a specific peak and ΣI_{hkl} is the sum of the intensities of all detected TiN reflections). The BAI and CC7 samples had 75% and 83% degrees of (111) preferred orientation parallel to the substrate surface, respectively. All other diffraction reflections contributed very little to the total diffracted intensity. The (111) orientation became a little bit less dominant in the BAI sample. AIP samples, in contrast, had (200) preferred orientation ($\gamma_{200} = 55.6\%$) with a significant (111) component (25%) and much smaller (220) and (311) peaks (1.8% and 9%, respectively).

The preferred orientation of TiN coatings is determined by the minimization of the overall free energy per surface coating area, which resulted from the competition between the surface energy, the strain energy, and the stopping energy of different lattice planes [32,33]. For TiN coatings with a NaCl crystal structure, the (200) plane has the lowest surface energy, the (111) plane has the lowest strain energy, and the stopping energy is minimal in the (220) plane. The intensive ion bombardment of the growing coating increases the accumulation of strain energy in coatings and the (111) plane becomes the preferred orientation [34]. It should be stressed that the surface energy is independent of film thickness, whereas the strain energy increases linearly with the thickness of the film [35]. Therefore, at lower thickness, the surface energy term is significant and (100) orientation with minimum surface energy may be expected. At a larger film thickness, the strain energy becomes dominant and (111) preferred orientation is the result of relieving the strain energy. The stopping energy (defined as energy of ions deposited due to channeling effect) is minimal in the (220) plane, while <001> direction is the most open channeling direction, where the energy of impinging ions is distributed over larger volumes. Channeling energy is dominant (220 preferred orientation) only when the deposited ion energy is sufficiently high (bias voltage greater than −100 V). The texture of TiN coating can be changed either by increasing the ion energy (by increasing bias voltage) or by increasing the surface mobility of deposited species (by increasing the substrate temperature or ion-to-atom flux ratio) [36,37]. Increasing the adatom mobility through bias voltage may enable atoms to rearrange and thus form other textures to reduce competing energies. For example, Greene et al. found that a moderate ion bombardment of TiN films, prepared by conventional magnetron sputtering technique, leads to the (111) texture, while an intense ion bombardment leads to (001) texture [38].

X-ray structural analysis of TiN coatings showed that the dominant crystal plane parallel to the plane of the substrate was (111) for the BAI and CC7 samples and (200) for the AIP sample. These differences were most likely related to ion irradiation conditions during all three deposition processes [39]. Namely, the comparison of the three deposition techniques used for the preparation of the TiN coatings shows that the ion current densities on the substrates were about 2, 2.5 and 3 mA/cm^2 for CC7, AIP and BAI samples, respectively. On the other hand, the average deposition rate was 0.85, 0.36 and 1.56 nm/s for the CC7, BAI and AIP coatings, respectively. The bias voltage was the highest for the BAI sample (−125 V), while it was −100 and −90 V for the CC7 and AIP sample, respectively. Both ion energies and ion-to-atom flux ratio (j_i/j_{Ti}) were the highest in the case of the BAI deposition technique. That could be the reason why the BAI TiN coating had the most pronounced (111) preferential orientation.

The preferred crystallographic orientation has a strong influence on the resulting physical properties. Coating texture affects the mechanical behavior of the coatings. Thus, it has been reported that TiN coating with (111) preferred orientation possesses the highest hardness and superior resistance to abrasion and wear [40,41]. Ponnon et al., found that the (200) preferential orientation is more conductive than the (111) one [42]. They also found that the sample with a predominant (200) orientation had a considerably smoother surface than the one with (111) orientation.

2.4. Mechanical Properties of TiN Coatings

The tribological performance of hard coatings is strongly related to their hardness (H) and elastic properties (Young's modulus, E). In order to perform well in a tribological application, a coating should be both hard and ductile. Thus, high hardness and low modulus are desirable to enhance fracture toughness and elastic strain to failure (high yield strength). A high H/E ratio (*elasticity index*) is often a reliable indicator of good tribological behaviors of coatings [43,44]. The H/E ratio defines the amount of energy that a coating material can absorb without permanent damage. In general, it can be expected that the friction coefficient and wear rate of the coatings are improved as their Young's modulus is reduced. Another important parameter in the investigation of the relation between the mechanical and tribological behavior of coatings is the ratio H^3/E^2 (*plasticity index*). Namely, the coating resistance to plastic deformation is proportional to the ratio H^3/E^2. This means that the plastic deformation in coating materials with high hardness H and low modulus E is reduced because low modulus E allows the given load to be distributed over a wider area of contact.

Both the hardness and elastic modulus of TiN coatings prepared by BAI, CC7 and AIP deposition techniques were determined from nanoindentation measurements. The indents were all performed on flat areas free from visible growth defects. The measured values of hardness and elastic modulus, as well as the calculated H/E and H^3/E^2 ratios, are given in Table 3. It can be seen that the measured hardness was considerably lower for the AIP coating compared to the BAI and CC7 coatings. The main factor leading to the high hardness of both the BAI and CC7 TiN coatings was their fine-grained, compact and dense microstructure, while the AIP coating was characterized by a pronounced columnar structure. An additional reason for higher microhardness could be the (111) preferential growth of the BAI and CC7 coatings. On the other hand, fine-grained microstructure does not significantly affect elastic modulus. We found that all three coatings were characterized by similar elastic modulus values. Consequently, the elasticity index H/E and plasticity index H^3/E^2 for the AIP coating were lower in comparison with those of the BAI and CC7 TiN coatings. On the other hand, this meant less wear resistance of the AIP coating, which confirmed measurements of the wear coefficient (see Section 2.5). It can be expected that both the elasticity and plasticity index of coatings depend on the crystal structure, texture, internal stresses and microstructure. However, the wear resistance of a coating depends not only on the index of elasticity and plasticity, but, as we will see later, also on other factors (e.g., surface topography, tribo-oxidation processes).

Table 3. Hardness (HV0.05), elastic modulus (E), H/E ratio and H^3/E^2 ratio of TiN coatings deposited in BAI, CC7 and AIP deposition systems.

-	H (GPa)	E (GPa)	H/E	H^3/E^2 (GPa)
BAI	29 ± 2	376 ± 16	0.080 ± 0.008	0.17 ± 0.05
CC7	28 ± 2	377 ± 16	0.080 ± 0.008	0.15 ± 0.05
AIP	25 ± 2	378 ± 16	0.066 ± 0.007	0.11 ± 0.035

2.5. The Tribological Behavior of TiN Coatings

Although the surface topography of coatings and surrounding atmosphere strongly affect the tribological performance of PVD TiN coatings, literature that systematically addresses these subjects is scarce [23–25,45–48]. Therefore, in this study, we performed more systematic tribological tests of BAI, CC7 and AIP titanium nitride coatings in three different atmospheres. In particular, the dependence of the friction coefficient on the type of atmosphere in the initial phase of the sliding test (running-in period) was analyzed in more detail.

2.5.1. The Influence of Coating Topography on Friction and Wear

One of those properties of TiN coatings that has the greatest impact on its tribological behavior is surface topography. As discussed above (Section 2.2), the topographies of BAI, CC7 and AIP TiN coatings are rather different. Therefore, we would expect that these differences are reflected in their tribological behavior as well. In order to study the influence of topography during the tribological test, the formation of an oxide layer must be prevented. This can be achieved by performing the sliding test in a nitrogen atmosphere. We assumed that in the absence of such a layer, the tribological response of the TiN coating would be due mainly to its surface topography. Figure 7 shows representative friction curves obtained during the sliding test of an alumina ball on as-deposited and polished BAI, CC7 and AIP TiN coatings in the nitrogen atmosphere. We can see that different coatings showed different levels of friction development in the initial phase of the sliding test, as well as different levels of friction in the steady-state period. In the initial phase of the sliding test, the friction coefficient rapidly increased and reached a value slightly below 0.3 for the BAI coating after 15 cycles, slightly above 0.3 for the CC7 coating after 20 cycles, and 0.5 for the AIP coating after 50 cycles. Immediately afterwards, the coefficient of friction was reduced to 0.2 for the BAI and CC7 coatings, and to about 0.45 for the AIP coating. Afterwards, the coefficient of friction gradually increased with a small fluctuation to a steady value, which was 0.47, 0.44 and 0.57 for the BAI, CC7 and AIP coatings, respectively. A similar dependence of the coefficient of friction in the initial phase of the sliding test was reported by other authors, but for different types of hard coatings [47,49]. The initial rapid increase and then immediate decrease in the friction coefficients for the as-deposited coatings were due to decreased ploughing action of protrusions on the coating surfaces. This explanation was confirmed by measurements of friction coefficients on the post-polished coatings. Namely, on friction curves of the post-polished coatings (Figure 7), no initial rapid change could be observed. This was because polishing reduced both the height of protrusions and also their number. The initial value of the coefficient of friction was 0.2 for all three types of TiN post-polished coatings. This was also the value achieved on as-deposited BAI and CC7 coatings after the initial few tens of sliding cycles, while this was not the case for the AIP coating, which we will try to explain later. The running-in period of post-polishing BAI and CC7 coatings was significantly shorter as compared with the as-deposited ones, while there was no difference for AIP coating. The wear rate of the post-polished coating was much lower than that of the as-deposited one. In some cases, it decreased even more than 50%, depending on the degree of polishing. Namely, post-polishing of the as-deposited TiN coating significantly reduces the material pick-up tendency due to the generation of smoother, less abrasive, surfaces [46]. We have to mention another phenomenon that is due to post-polishing of coatings and that can affect their tribological behavior. Namely, due to high shearing stresses during post-polishing of a coating, a portion of the nodular defects and droplets are torn out, and craters are formed on the surface. During the sliding process, some newly formed wear particles are trapped in such craters (particle hiding); therefore, abrasive wear is reduced.

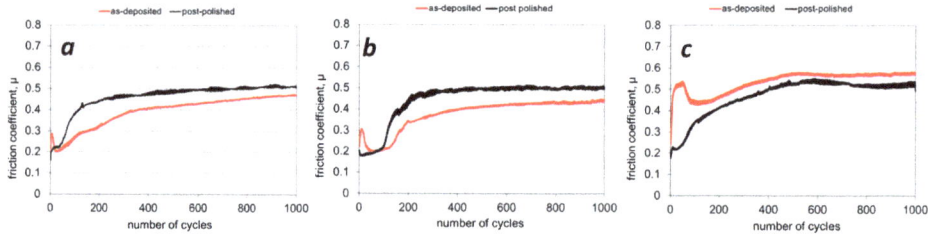

Figure 7. Friction curves of as-deposited and polished BAI (**a**), CC7 (**b**) and AIP (**c**) samples (TiN/D2).

In the initial phase of the sliding test, two mechanisms affected the friction and wear. We had to consider that the initial contact spots were at the coating protrusions (e.g., nodular defects), which cause friction due to ploughing action on the softer counter-body surface (alumina ball). A slight increase in the friction after the first sliding cycle was due to the increase in alumina ball contact area. The first contact between the flat coating surface and the alumina ball, which moved relative to each other, always took place at the highest peaks (nodular defects) of the surfaces, while the contact conditions were very complex. Protruding nodular defects cause abrasive wear of the counter material. Due to the ploughing and material pick-up effects, the friction coefficient increases. The pressure at these contact spots is very high due to the small actual contact area (about 10% of the surface area) [20]. However, the real area of contact depends not only on surface roughness but also on the applied normal load. Thus, for example, the contact area increases with the normal load if the contact is elastic. High pressure and shear stresses cause the collapse of the most protruding nodular defects into small fragments that are released into the wear track [46,50,51]. Therefore, the nodular defects represent the primary source of hard abrasive particles in the sliding contact during the running-in period. These particles have a strong effect on friction and wear because along with protrusions, they abrade the softer surface of the counter-body material.

The second reason for friction increase is the interlocking of asperities, on both the coating and counter-body surfaces. When two surfaces slide relative to each other, the friction force is not constant, because the motion changes periodically between the adhesion and sliding (i.e., the *stick-slip phenomenon*). After removal of the highest nodular defects at the beginning of sliding, contact is formed at the lower protrusions (e.g., sites of carbides and nonmetal inclusions) [51] (Figure 8). Although all of these asperities are much smaller, they cover a much larger surface area. They also offer additional anchoring points for counter-body material transfer. We found that a considerable amount of debris had already collected in front of such protrusions after a few cycles.

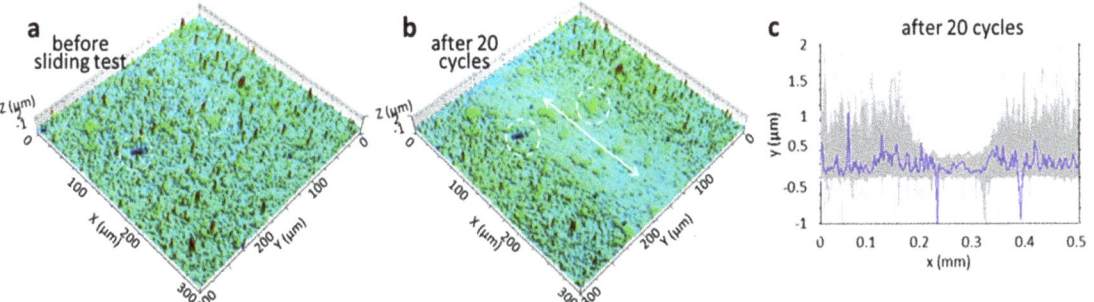

Figure 8. Three-dimensional profile images of the wear tracks on AIP TiN hard coating on D2 tool steel substrate before (**a**) and after 20 cycles of sliding against an alumina ball (**b**). The scanning area was 0.3 mm × 0.3 mm. Both 3D-profile images were taken from the same area on the substrate surface (the same sites before and after the test are labeled with the white circles). The sliding direction is indicated by arrows. The protrusions at sites of carbides are still visible in the wear track. The sharp peaks are the nodular defects, while the blue dots are craters (note the strong exaggeration in z-scale). (**c**) A series of a numerous line profiles were recorded perpendicular to the wear track and stacked together. The blue profile represents the level of coating, while grey peaks belong to protrusions.

During further sliding, all protrusions are gradually removed and a smooth surface is formed. As long as protrusions are present on the coating surface, the friction coefficient does not reach the steady-state. From the explanation described above, we can conclude that the rougher the coating is, the longer the time that is required to obtain the steady-state value of the friction. The period until a conformal sliding contact is formed is called the *running-in phase*. In this period we can expect a transition from *mechanical wear-dominated friction* to *adhesion-dominated friction*. The larger contact surface is reflected in a higher friction coefficient due to the higher adhesion component of the friction. It is also valid that as the contact area increases, the contact pressure decreases.

As already mentioned above, the initial value of the coefficient of friction of the as-deposited AIP coating was higher compared to that of the BAI and CC7 coatings. The reason for the different tribological behavior of the AIP coating in comparison to the BAI and CC7 coatings could be related to titanium droplets on the surface of the AIP coating. It can be expected that these relatively soft droplets can increase the adhesion component of friction. Because they are incorporated throughout the entire thickness of the coating, droplets also affect the friction coefficient after the top layer has been removed during the sliding test. This could be the reason why the steady-state value of the friction coefficient was highest for the as-deposited AIP coating (around 0.58), while it was around 0.45 and 0.47 for the CC7 and BAI coatings, respectively. The steady-state value was reached the fastest with the CC7 as-deposited coating (after about 300 cycles) and the slowest with the BAI coating (after about 500 cycles).

Still another aspect of the sliding contact should be mentioned. The characterization of microstructure (Section 2.1) showed that all three coatings had a more or less pronounced columnar microstructure. The columns grew normal to the coating substrate interface, while the individual columns were not in close contact. We could expect that the material transferred from the counter-body ball would be trapped in the open regions between the columns. In general, each column was characterized by a domed end-cap that also may have abraded the ball material as it cycled over the coating surface, resulting in the transfer of the plastically deformed material to it. Therefore, the more columnar and porous the coating is, the greater the contribution of the adhesive friction component that can be expected. The microstructures of our BAI, CC7 and AIP TiN coatings appeared quite compact; therefore, we do not believe that this mechanism played a significant role during the tribological test.

2.5.2. Friction and Wear in Different Surrounding Atmospheres

The tribological processes change drastically if the sliding test takes place in ambient air or in an oxygen atmosphere (Figures 9–11). In both cases, the tribological process was not affected only by the material transfer but also by the formation of the tribo-oxide layer. Oxides form in the sliding contact very quickly due to heating induced by friction. However, high compressive stresses appear in the oxide layer due to the large difference in the oxide layer molar volume in comparison to the molar volume of the nitride layer. These stresses, together with high shear forces, cause the formation of cracks in the oxide film and finally its spallation. Only small patches of oxides in the wear track can be observed. Delaminated oxide fragments can act as abrasive particles, which cause a significant increase in the coefficient of friction and wear rate. Repeated sliding results in the accumulation of oxide particles in the wear track (in the form of *roll-like debris*). Tribological tests performed in different atmospheres (Figures 9 and 11a) confirmed that all types of TiN coatings have the highest coefficient of friction in an oxygen atmosphere (about 1), while it is much smaller if the sliding test is performed in nitrogen (about 0.55) or in ambient air (about 0.4). Not only the friction coefficient but also the wear rate strongly depends on the type of atmosphere in which the test takes place (Figures 10 and 11). The wear coefficient is the highest in an oxygen atmosphere, lower in ambient air, and the lowest in nitrogen. The use of a nitrogen atmosphere suppresses the formation of an oxide layer, resulting in a strong reduction in the TiN coating wear rate. We believe that, besides the cooling effect, this mechanism is active in tool protection during cryogenic machining using liquid nitrogen.

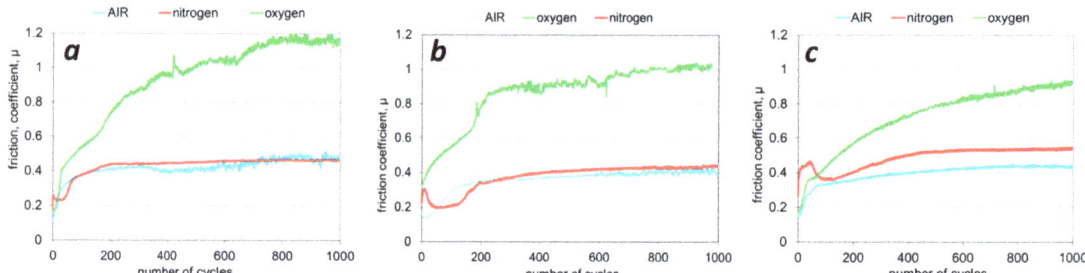

Figure 9. The friction curves of as-deposited BAI (**a**), CC7 (**b**), and AIP TiN coatings (**c**) for the tests performed at room temperature in different atmospheres (ambient air, nitrogen and oxygen) under a load of 5 N.

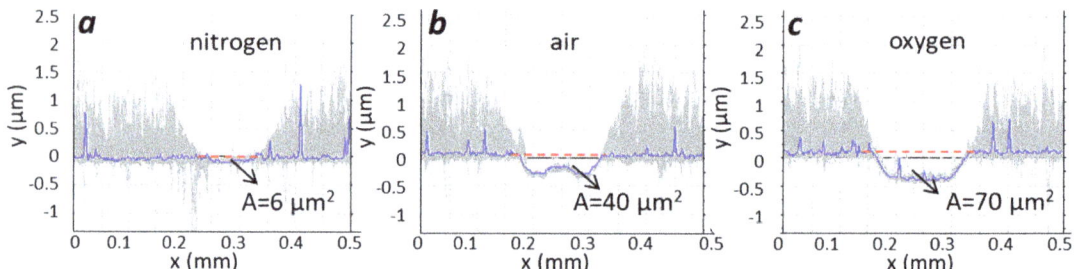

Figure 10. Series of 250 line profiles stacked together from the recorded profiles. The wear track of BAI TiN coating is shown for different surrounding atmospheres ((**a**)—nitrogen, (**b**)—air, (**c**)—oxygen). The scanning area was 0.5 mm × 0.5 mm. The blue profile represents the level of coating, while grey peaks represent protrusions. During the sliding test (1000 cycles), the protrusions were removed and the wear tracks with various depths were formed. Note the strong exaggeration in z-scale. "A" is the cross-section area of the wear track.

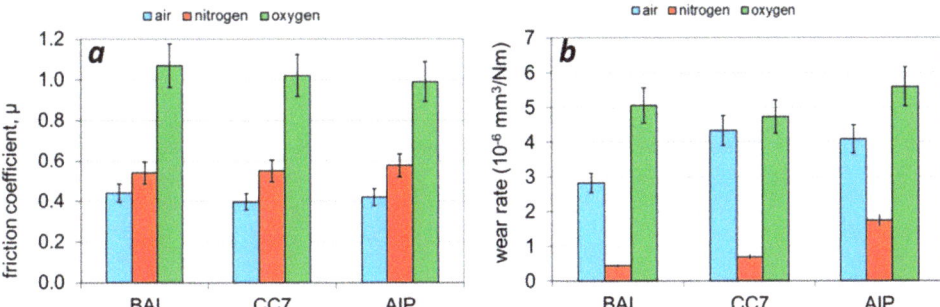

Figure 11. The friction coefficients (**a**) and the wear rates (**b**) in the ambient, oxygen and nitrogen atmospheres for three different coatings.

The influence of the surrounding atmosphere on the formation of the oxide layer was also analyzed by the following experiment. The sliding tests performed on the BAI TiN coating in the nitrogen, oxygen and ambient air atmospheres were interrupted after 50 cycles for 30 min (Figure 12a). After 30 min, the sliding tests were continued, and a different tribological response of TiN coating in the studied atmospheres was observed. It could be observed that such an interruption did not significantly affect the tribological process when the test took place in a nitrogen atmosphere. However, this was not the case in the tests performed in oxygen or ambient air atmosphere. In these cases, the coefficient of friction decreased to the initial value after each interruption and then, after a short running-in period, increased again to the value before the interruption. This indicates that a thin native oxide layer formed on the wear track during the interruption period.

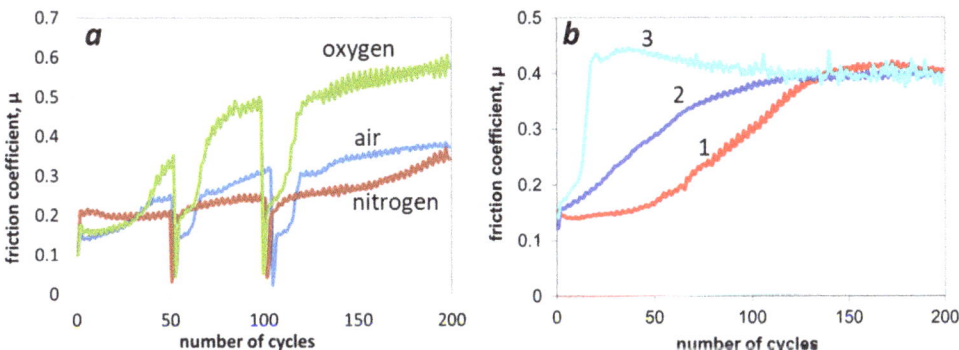

Figure 12. (**a**) The sliding test on sputter-deposited TiN coating was interrupted after 50 laps for 30 min and then continued using same conditions. After sliding test was continued, different friction development was observed in the three studied atmospheres. (**b**) The running-in period depends on the state of coating and ball surfaces: the sliding test was performed in ambient air on the fresh coating surface and a new spot on the alumina ball (1); on the worn ball surface and a new track on the coating surface (2); on a new spot on the alumina ball and the track formed on the coating surface during the previous test (3).

In the next experiment, we tried to separate the influence of the coating surface and alumina ball surface on the friction during the running-in period of the sliding test. The following three sliding tests in the ambient air atmosphere were performed: (i) in the first test, we selected a fresh as-deposited coating surface and a new spot on the alumina ball; (ii) in the second test we used the worn ball surface, but a new track on the coating surface; (iii) the third test was performed in a track formed during the previous test, while a new fresh surface on the alumina ball was selected. The results of friction coefficient measurements are shown in Figure 12b. In the case of a fresh coating surface and a new spot on an alumina ball, the duration of the running-in period was approximately 150 cycles. If we restarted the sliding test in the track formed during the previous test with fresh ball surface, then the running-in period was shortened to 20 cycles. In the case of worn alumina surface and fresh coating surface, the duration of the running-in period was about 115 cycles. Based on these results, we can conclude that the duration of the running-in period depends mainly on the coating surface conditions.

Wear rates given in Figure 11 are average values of measurements from a large number of samples. It should be noted, however, that all of these coatings differed in surface roughness. We have to consider that the surface roughness of the TiN coating does not depend only on the deposition method, but largely depends on the surface density of growth defects. The formation of such defects is a spatially localized and sporadic process [52]. Therefore, significant differences in surface roughness were observed not only for samples within the same batch, but even between samples from different batches. Figure 13 shows, for example, the peak (nodular defect) density distribution for BAI TiN hard coating. From a hundred measurements, we performed statistical analysis. The samples were divided into 14 classes with a width of 50 peaks/mm^2 ranging from 0 to 700 peaks/mm^2. For any series of samples, we constructed 14 classes (x, y), where x was the median number of defects of the class, and y was the number of samples falling into this class. In this way, we obtained the peak density distribution, which presents the samples per peak class versus the number of peaks. The data on the y-axis were normalized. For fitting of measurements, we used the Poisson distribution. In order to study the influence of surface roughness on the tribological response of TiN coatings prepared by one of three deposition methods, we selected the samples with different surface densities of growth defects. The surface density distribution of nodular defects for BAI TiN coatings showed a pronounced peak at 200 defects/mm^2. The surface density of defects was also reflected in the surface roughness, which increased linearly with the density of defects.

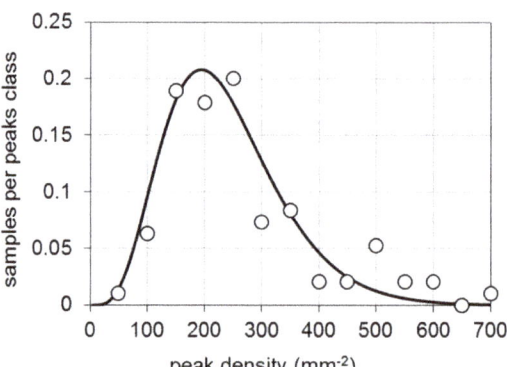

Figure 13. Peak (i.e., nodular defect) density distribution for TiN hard coatings prepared by low-voltage electron beam evaporation (BAI).

As can be seen from Figures 3 and 4, the surface densities of growth defects in the CC7 and AIP coatings were much higher than in the BAI coating, which was reflected in a greater surface roughness. In order to determine how the layer roughness affected its tribological properties, we performed tribological tests on a large number of samples prepared in different production batches. The wear rate measurements for BAI, CC7 and AIP TiN coatings with different levels of surface roughness were performed in nitrogen, oxygen and ambient air atmospheres. The results of the measurements are presented in Figure 14. The measurements clearly showed that the wear coefficient increased with the surface roughness of the coating and the most for the AIP coating. The wear rates in an oxygen atmosphere were much larger (by one order of magnitude), compared with the rates observed in a nitrogen atmosphere. The values of the wear rates for the coatings tested in the ambient air were also smaller than those in the oxygen atmosphere, but still much higher than those measured in the nitrogen atmosphere. The difference between the wear coefficients in the oxygen and ambient air atmospheres could be explained by the influence of humidity in the ambient air [53].

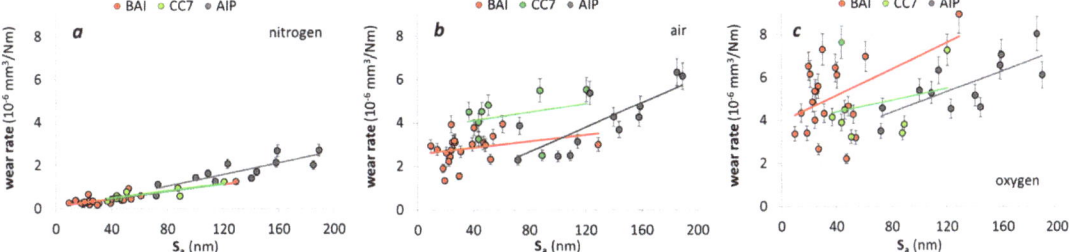

Figure 14. Wear rates of BAI, CC7 and AIP coatings as a function of surface roughness S_a. The wear tests were performed in nitrogen (**a**), ambient air (**b**) and oxygen (**c**) atmospheres.

The processes that take place in the wear track during the repeated sliding tests include: the breaking of nodular defects, coating surface smoothening, oxidation, formation of oxide patches, accumulation of wear debris at the track end and edges, and formation of roll-like debris (Figure 15). Previously, we explained that the formation of roll-like debris occurs due to plowing of the counter-body surface irregularities over the surface of the wear track [51]. Due to the large difference in the oxide layer molar volume in comparison with the molar volume of the nitride layer, the cracks formed in the oxide layer, which was mostly delaminated from the substrate. The dark regions on the backscattered-electron (BSE) image were oxide residues (Figure 15b). Oxides with lower atomic numbers than the coating appeared as dark areas, because they emit fewer back-scattered electrons in comparison to the coating material. We found that such oxide patches occurred preferentially at the sites of depressions and protrusions (at sites of carbide inclusions, nodular defects) and other surface irregularities (e.g., grooves, ridges).

Measurements on the alumina counterparts were also performed. 3D profile images of counterpart wear scars shown in Figure 16 reveal the existence of transfer layers on the ball sliding surface for all testing atmospheres. While the wear of the ball surface was similar for all three atmospheres, the amount of material transferred to the ball surface was the highest in the oxygen atmosphere and the lowest in nitrogen. As is evident from Figure 16, the lower wear of the coating in the nitrogen atmosphere was reflected in the lower transfer of material to the alumina ball. The tribofilm on the worn alumina surface was composed of the particles generated from both surfaces. Due to the high pressure and high temperature in the sliding contact, the wear particles were pressed together and sintered in the form of a tribofilm.

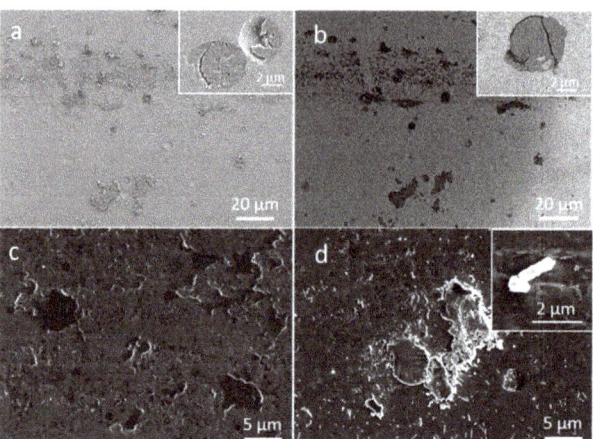

Figure 15. Secondary electron image (**a**) and backscattered electron image (**b**) of a wear track formed in BAI TiN coatings during a reciprocating sliding test performed in the ambient air; the anchoring points for wear debris and counter-body material transfer are depressions (see insets) as well as growth defects on the coating surface. SEM images of the wear track formed in AIP TiN coating during a reciprocating sliding test performed in the ambient air (**c**,**d**); the anchoring points for wear debris and counter-body material transfer are protrusions at the site of carbides (**c**), as well as broken nodular defects (**d**); roll-like debris are also visible in SEM image (**d**).

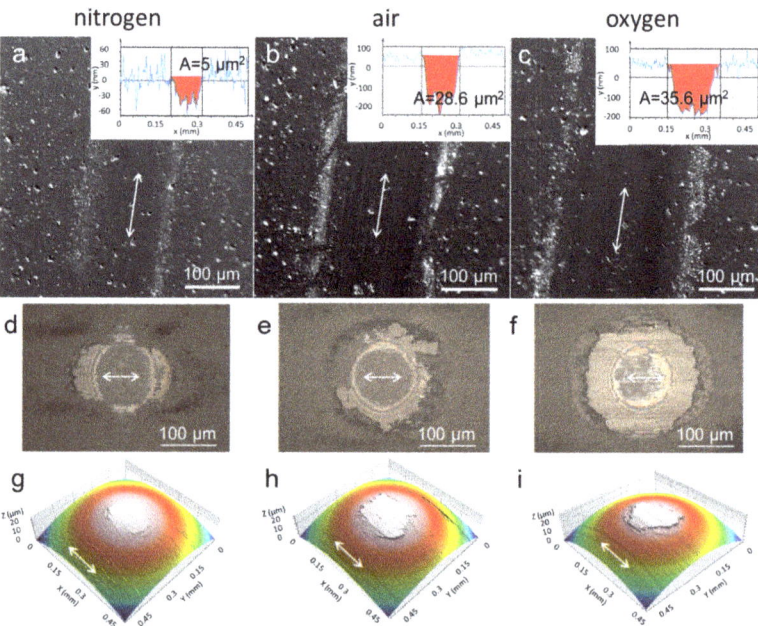

Figure 16. (**a–c**) SEM images of wear tracks formed in CC7 TiN coating during a reciprocating sliding test performed in nitrogen (**a**), ambient air (**b**), and oxygen (**c**) atmospheres. The cross-section areas (A) of corresponding wear tracks after 1000 cycles are shown in the insets. The areas of alumina ball wear scar are shown in the optical microscopy images (**d–f**) and 3D profilometer images (**g–i**). The images reveal the existence of transfer layers on the alumina ball for all testing atmospheres. Arrows indicate the sliding direction.

3. Conclusions

In this paper, we presented a comparative study of structural, microstructural, mechanical, topographical, and tribological properties of TiN coatings prepared by low-voltage electron beam evaporation, magnetron sputtering, and cathodic arc deposition. We tried to correlate the tribological behavior of these coatings with their intrinsic properties and friction conditions. We focused mainly on the investigation of tribological processes that take place in the initial phase of the sliding test. In this running-in period, the greatest impact was from different kinds of protrusions on the coating surfaces (e.g., nodular defects, droplets). Namely, all surface asperities caused abrasive wear to the softer counter-material surface and also the transfer of this material to the coating surface. Additionally, the protrusions (e.g., nodular defects) were crushed into small abrasive particles due to high pressure and shear forces during the sliding test. Therefore, we believe that in the running-in phase, nodular defects were the most intensive source of hard abrasive particles in the sliding contact. The longer running-in period and higher coefficient of friction of the coating prepared by the cathodic arc deposition were attributed to droplets on the surface of the as-deposited coating as well as those incorporated into the coating. Namely, relatively soft metal droplets increased the adhesion component of friction.

The negative impact of nodular defects on tribological performance can be reduced by post-polishing the as-deposited coating. Indeed, our tribological measurements showed that after post-polishing, the running-in period was shortened and the reduction in the coating wear rate was particularly enhanced.

In order to identify the influence of tribo-oxidation on friction and wear, the sliding tests on different types of TiN coatings were also conducted in different atmospheres (ambient air, nitrogen, oxygen). Oxygen promotes tribo-chemical reactions at the contact surface of the coating, while nitrogen suppresses them. This was particularly reflected in the coefficient of friction and wear, which were significantly higher in an oxygen atmosphere than in the nitrogen one. Oxides formed quickly in the sliding contact due to heating induced by friction. Due to the large difference in the oxide layer molar volume in comparison with the molar volume of the nitride layer, high compressive stresses appeared in the oxide layer. These stresses, together with high shear forces, caused the formation of cracks in the oxide film and finally its spallation. Only small patches of oxides in the wear track could be observed. Delaminated oxide fragments can act as abrasive particles, which cause an additional increase in the coefficient of friction and wear rate. In a test carried out in a nitrogen atmosphere, the formation of an oxide layer was avoided. Therefore, a more stable sliding contact between the two mating surfaces was formed. The more stable friction conditions were reflected in less fluctuation of the friction coefficient and, therefore, smoother friction curves. We found that the wear rate increased with the surface roughness of the coating. The wear rate in an oxygen atmosphere was much larger compared with that in a nitrogen atmosphere.

Author Contributions: Design of experiments, tribological and X-ray measurements, 3D profilometry, interpretation of experimental results, manuscript writing, P.P.; SEM and FIB analysis, manuscript review, A.D.; interpretation of tribological results, P.T. and A.M.; manuscript review, M.Č. and M.P. All authors have read and agreed to the published version of the manuscript.

Funding: This work was supported by Slovenian Research Agency (program P2-0082). We also acknowledge funding from the European Regional Development Funds (CENN Nanocenter, OP13.1.1.2.02.006).

Institutional Review Board Statement: Not applicable.

Informed Consent Statement: Not applicable.

Data Availability Statement: Not applicable.

Acknowledgments: The authors would also like to thank Jožko Fišer for performing some laboratory tests.

Conflicts of Interest: The authors declare no conflict of interest.

References

1. Sproul, W.D. Tribological Coatings: From Monolithic to Nanocomposite and Nanolayered Coatings. In Proceedings of the Annual Technical Conference-Society of Vacuum Coaters, Louiseville, KY, USA, 30 April 2007; pp. 46–49.
2. Feinberg, B. Longer Life from TiN Tools. *Mfg. Eng. Manag.* **1971**, *67*, 6–18.
3. Moll, E.; Daxinger, H. Method and Apparatus for Evaporating Materials in a Vacuum Coating Plant. U.S. Patent 4,197,175, 8 April 1980.
4. Sproul, W.D. *PVD Processes for Depositing Hard Tribological Coatings*; Donald, M.M., Vivienne, H.M., Eds.; Society of Vacuum Coaters: Albuquerque, NM, USA, 2009; pp. 36–40.
5. Vergason, G.; Hatto, P.; Tietema, R.; Anders, A. Early Years of Commercial Cathodic Arc Vapor Deposition: Hot Deals during the Cold War. In Proceedings of the 11th International Symposium Sputtering Plasma Processes, Kyoto, Japan, 8–11 November 2011.
6. Anders, A. A Review Comparing Cathodic Arcs and High Power Impulse Magnetron Sputtering (HiPIMS). *Surf. Coat. Technol.* **2014**, *257*, 308–325. [CrossRef]
7. Window, B.; Savvides, N. Charged Particle Fluxes from Planar Magnetron Sputtering Sources. *J. Vac. Sci. Technol.* **1986**, *4*, 196. [CrossRef]
8. Kouznetsov, V.; Macák, K.; Schneider, J.M.; Helmersson, U.; Petrov, I. A Novel Pulsed Magnetron Sputter Technique Utilizing Very High Target Power Densities. *Surf. Coat. Technol.* **1999**, *122*, 290. [CrossRef]
9. Zega, B. Hard Decorative Coatings by Reactive Physical Vapor Deposition: 12 Years of Development. *Surf. Coat. Technol.* **1989**, *39–40*, 507–520. [CrossRef]
10. Reiners, G.; Beck, U.; Jehn, H.A. Decorative Optical Coatings. *Thin Solid Film.* **1994**, *253*, 33–40. [CrossRef]
11. Constantin, R.; Steinmann, P.A.; Manasterski, C. Chapter 5 Decorative PVD Coatings. In *Nanomaterials and Surface Engineering*; Takadoum, J., Ed.; John and Wiley and Sons: Hoboken, NJ, USA, 2013.
12. Adachi, S.; Takahashi, M. Optical Properties of TiN Films Deposited by Direct Current Reactive Sputtering. *J. Appl. Phys.* **2002**, *87*, 1264–1269. [CrossRef]
13. Glocker, D.A.; Ranade, S.V. *Medical Coatings and Deposition Technologies*; Glocker, D.A., Ranade, S.V., Eds.; John and Wiley and Sons: Hoboken, NJ, USA, 2016.
14. van Hove, R.P.; Sierevelt, I.N.; van Royen, B.J.; Nolte, P.A. Titanium-Nitride Coating of Orthopaedic Implants: A Review of the Literature. *BioMed Res. Int.* **2015**, *2015*, 1–9. [CrossRef]
15. Damiati, L.; Eales, M.G.; Nobbs, A.H.; Su, B.; Tsimbouri, P.M.; Salmeron-Sanchez, M.; Dalby, M.J. Impact of Surface Topography and Coating on Osteogenesis and Bacterial Attachment on Titanium Implants. *J. Tissue Eng.* **2018**, *9*, 1–16. [CrossRef]
16. Mühlbacher, M.; Mendez-Martin, F.; Sartory, B.; Schalk, N.; Keckes, J.; Lu, J.; Hultman, L.; Mitterer, C. Copper Diffusion into Single-Crystalline TiN Studied by Transmission Electron Microscopy and Atom Probe Tomography. *Thin Solid Film.* **2015**, *574*, 103–109. [CrossRef]
17. Bradley, S.R.; McKenna, K.P.; Shluger, A.L. The Behaviour of Oxygen at Metal Electrodes in HfO_2 Based Resistive 3 Switching Devices. *Microelectron. Eng.* **2013**, *109*, 346. [CrossRef]
18. Mosavati, N.; Chitturi, V.R.; Salley, S.O.; Ng, K.S. Nanostructured Titanium Nitride as a Novel Cathode for High Performance Lithium/Dissolved Polysulfide Batteries. *J. Power Sources* **2016**, *321*, 87. [CrossRef]
19. Yang, X.; Liu, W.; De Bastiani, M.; Allen, T.; Kang, J.; Xu, H.; De Wolf, S. Dual-Function Electron-Conductive, Hole-Blocking Titanium Nitride Contacts for Efficient Silicon Solar Cells. *Joule* **2019**, *3*, 1314–1327. [CrossRef]
20. Holmberg, K.; Matthews, A. Coating Tribology. In *Tribology Series 56*, 2nd ed.; Elsevier: Amsterdam, The Netherlands, 2009; pp. 226–235.
21. Holmberg, K.; Ronkainen, H.; Laukkanen, A.; Wallin, K. Friction and Wear of Coated Surfaces—Scales, Modelling and Simulation of Tribomechanisms. *Surf. Coat. Technol.* **2007**, *202*, 1034–1049. [CrossRef]
22. Aiso, T.; Wiklund, U. Influence of Contact Parameters on Material Transfer from Steel to TiN Coated Tool–Optimisation of a Sliding Test for Simulation of Material Transfer in Milling, Tribology-Materials. *Surf. Interfaces* **2016**, *10*, 107–116. [CrossRef]
23. Heinrichs, J.; Gerth, J.; Bexell, U.; Larsson, M.; Wiklund, U. Influence from Surface Roughness on Steel Transfer to PVD Tool Coatings in Continuous and Intermittent Sliding Contacts. *Tribol. Int.* **2012**, *56*, 9–18. [CrossRef]
24. Drnovšek, A.; Panjan, P.; Panjan, M.; Paskvale, S.; Buh, J.; Čekada, M. The Influence of Surrounding Atmosphere on Tribological Properties of Hard Coating. *Surf. Coat. Technol.* **2015**, *267*, 15–20. [CrossRef]
25. Grant, A.J.; Gee, M.G.; Orkney, L.P. The Wear and Friction Behaviour of Engineering Coatings in Ambient Air and Dry Nitrogen. *Wear* **2011**, *271*, 2164–2175.
26. Panjan, P.; Drnovšek, A.; Mahne, N.; Čekada, M.; Panjan, M. Surface Topography of PVD Hard Coatings. *Coatings* **2021**, *11*, 1387. [CrossRef]
27. Mattox, D.M. *Handbook of Physical Vapor Deposition (PVD) Processing*; Elsevier: Amsterdam, The Netherlands, 2010.
28. Petrov, I.; Adibi, F.; Greene, J.E.; Hultman, L.; Sundgren, J.E. Average Energy Deposited Per Atom: A Universal Parameter for Describing Ion-Assisted Film Growth? *Appl. Phys. Lett.* **1993**, *63*, 36. [CrossRef]
29. Panjan, P.; Drnovšek, A.; Gselman, P.; Čekada, M.; Panjan, M. Review of Growth Defects in Thin Films Prepared by PVD Techniques. *Coatings* **2020**, *10*, 447. [CrossRef]

30. Sedlaček, M.; Podgornik, B.; Vižintin, J. Correlation between Standard Roughness Parameters Skewness and Kurtosis and Tribological Behaviour of Contact Surfaces. *Tribol. Int.* **2012**, *48*, 102–112. [CrossRef]
31. Smith, D.K. *Powder Diffraction*; JCPDS International Centre for Diffraction Data: Swarthmore, PA, USA, 1982.
32. Pelleg, J.; Zevin, L.Z.; Lungo, S.; Croitoru, N. Reactive Sputter Deposited TiN Films on Glass Substrates. *Thin Solid Film.* **1991**, *197*, 117. [CrossRef]
33. Zhang, Y.J.; Yan, P.X.; Wu, Z.G.; Zhang, W.W.; Zhang, G.A.; Liu, W.M.; Xue, Q.J. Effects of Substrate Bias and Argon Flux on the Structure of Titanium Nitride Films Deposited by Filtered Cathodic Arc Plasma. *Phys. Stat. Sol.* **2005**, *202*, 95–101. [CrossRef]
34. Saikia, P.; Joseph, A.; Rane, R.; Saikia, B.K.; Mukherjee, S. Role of Substrate and Deposition Conditions on the Texture Evolution of Titanium Nitride Thin Film on Bare and Plasma-Nitrided High-Speed Steel. *J. Theor. Appl. Phys.* **2013**, *7*, 66. [CrossRef]
35. Guruvenk, S.; Rao, G.M. Effect of Ion Bombardment and Substrate Orientation on Structure and Properties of Titanium Nitride Films Deposited by Unbalanced Magnetron Sputtering. *J. Vac. Sci. Technol.* **2002**, *20*, 678–682. [CrossRef]
36. Abadias, G. Stress and Preferred Orientation in Nitride-Based PVD Coatings. *Surf. Coat. Technol.* **2008**, *202*, 2223–2235. [CrossRef]
37. Mayrhofer, P.H.; Geier, M.; Locker, C.; Chen, L. Influence of Deposition Conditions on Texture Development and Mechanical Properties of TiN Coatings. *Int. J. Mater. Res.* **2009**, *100*, 8. [CrossRef]
38. Greene, J.E.; Sudgren, J.E.; Hultman, L.; Petrov, I.; Bergstrom, D.B. Development of Preferred Orientation in Polycrystalline TiN Layers Grown by Ultrahigh Vacuum Reactive Magnetron Sputtering. *Appl. Phys. Lett.* **1995**, *67*, 2928–2930. [CrossRef]
39. Håkansson, G.; Hultman, L.; Sundgren, J.E.; Greene, J.E.; Munz, W.D. Microstructures of TiN Films Grown by Various Physical Vapour Deposition Techniques. *Surf. Coat. Technol.* **1991**, *48*, 51–67. [CrossRef]
40. Martinez, G.; Shutthanandan, V.; Thevuthasan, S.; Chessa, J.F.; Ramana, C.V. Effect of Thickness on the Structure, Composition and Properties of Titanium Nitride Nano-Coatings. *Ceram. Int.* **2014**, *40*, 5757–5764. [CrossRef]
41. Wang, A.N.; Yu, G.P.; Huang, J.H. Fracture Toughness Measurement on TiN Hard Coatings Using Internal Energy Induced Cracking. *Surf. Coat. Technol.* **2014**, *239*, 20–27. [CrossRef]
42. Ponon, N.K.; Appleby, D.J.R.; Arac, E.; King, P.J.; Ganti, S.; Kwa, K.S.K.; O'Neill, A. Effect of Deposition Conditions and Post Deposition Anneal on Reactively Sputtered Titanium Nitride Thin Films. *Thin Solid Film.* **2015**, *578*, 31–37. [CrossRef]
43. Leyland, A.; Matthews, A. On the Significance of the H/E Ratio in Wear Control: A Nanocomposite Coating Approach to Optimized Tribological Behaviour. *Wear* **2000**, *246*, 1–11. [CrossRef]
44. Leyland, A.; Matthews, A. Optimization of Nanostructured Tribological Coatings. In *Nanostructured Coatings*; Cavaleiro, A., De Hosson, J.T.M., Eds.; Springer: New York, NY, USA, 2006; pp. 511–538.
45. Harlin, P.; Carlsson, P.; Bexell, U.; Olsson, M. Influence of Surface Roughness of PVD Coatings on Tribological Performance in Sliding Contacts. *Surf. Coat. Technol.* **2006**, *201*, 4253–4259. [CrossRef]
46. Harlin, P.; Bexell, U.; Olsson, M. Influence of Surface Topography of Arc-Deposited TiN and Sputter-Deposited WC/C Coatings on the Initial Material Transfer Tendency and Friction Characteristics under Dry Sliding Contact Conditions. *Surf. Coat. Technol.* **2009**, *203*, 1748–1755. [CrossRef]
47. Olofsson, J.; Gerth, J.; Nyberg, H.; Wiklund, U.; Jacobson, S. On the Influence from Micro Topography of PVD Coatings on Friction Behaviour, Material Transfer and Tribofilm Formation. *Wear* **2011**, *271*, 2046–2057. [CrossRef]
48. Podgursky, V.; Nisumaa, R.; Adoberg, E.; Surzhenkov, A.; Sivitski, A.; Kulu, P. Comparative Study of Surface Roughness and Tribological Behavior during Running-in Period of Hard Coatings Deposited by Lateral Rotating Cathode Arc. *Wear* **2010**, *268*, 751–755. [CrossRef]
49. Luo, Q. Origin of Friction in Running-in Sliding Wear of Nitride Coatings. *Tribol. Lett.* **2010**, *37*, 529–539. [CrossRef]
50. Drnovšek, A.; Panjan, P.; Panjan, M.; Čekada, M. The Influence of Growth Defects in Sputter-Deposited TiAlN Hard Coatings on Their Tribological Behavior. *Surf. Coat. Technol.* **2016**, *288*, 171–178. [CrossRef]
51. Panjan, P.; Drnovšek, A.; Kovač, J. Tribological Aspects Related to the Morphology of PVD Hard Coatings. *Surf. Coat. Technol.* **2018**, *343*, 138–147. [CrossRef]
52. Panjan, P.; Gselman, P.; Kek-Merl, D.; Čekada, M.; Panjan, M.; Dražić, G.; Bončina, T.; Zupanič, F. Growth Defect Density in PVD Hard Coatings Prepared by Different Deposition Techniques. *Surf. Coat. Technol.* **2013**, *237*, 349–356. [CrossRef]
53. Wilson, S.; Alpas, A.T. Tribo-Layer Formation during Sliding Wear of TiN Coatings. *Wear* **2000**, *245*, 223–229. [CrossRef]

Article

Influence of Growth Defects on the Oxidation Resistance of Sputter-Deposited TiAlN Hard Coatings

Peter Panjan [1,*], Aljaž Drnovšek [1] and Goran Dražić [2]

[1] Jožef Stefan Institute, Jamova 39, 1000 Ljubljana, Slovenia; aljaz.drnovsek@ijs.si
[2] National Institute of Chemistry, Hajdrihova 19, 1000 Ljubljana, Slovenia; goran.drazic@ki.si
* Correspondence: peter.panjan@ijs.si; Tel.: +386-1-477-3278

Abstract: This paper reports the results of an investigation of the oxidation of a sputter-deposited TiAlN hard coating in air at temperatures of 800 and 850 °C for times ranging from 15 min to 2 h. The study is focused on the role of growth defects in the oxidation process. The mechanism of oxidation at the site of the defect was studied on cross-sections made by the consecutive sectioning of oxidized coatings with the FIB technique. We found that in the early stage of oxidation, the locally intense oxidation always starts at such defects. Although the growth defects reduce the oxidation resistance of the coating locally, we believe that they do not have a decisive influence on the global oxidation resistance of the coating. There are several reasons for this. The first is that the surface area covered by growth defects is relatively low (less than 1%). Secondly, the coating is permeable only at those defects that extend through the entire coating thickness. Thirdly, the permeability at the rim of some defects strongly depends on the density of pores at the rim of defects and how open they are. The size and density of such pores depend on the shape and size of topographical irregularities on the substrate surface (e.g., seeds, pits), which are responsible for the formation of growth defects. We also found that oxidation of the TiAlN coating is accelerated by oxygen and titanium diffusion through the pores formed by crystal grain growth in the outer alumina overlayer. Such pores are formed due to the compressive stresses in the Ti-rich oxide layer, which are caused by the large difference in molar volumes between the oxide and nitride phases.

Keywords: TiAlN hard coating; magnetron sputtering; oxidation; growth defect; focused ion beam (FIB); scanning electron microscopy (SEM); scanning transmission electron microscopy (STEM)

1. Introduction

The increasing application of high-speed, dry cutting technologies and the machining of difficult-to-cut workpiece materials (hard or "sticky") demands that cutting tools be protected with high-performance PVD hard coatings. Due to the high friction between the tool surface and workpiece material during machining, the temperature at the tool's cutting edge may reach more than 800 °C. Therefore, the protective coating must maintain a high wear resistance (high hardness), thermal stability, and oxidation resistance at the elevated temperatures. The development of hard coatings that meet such severe requirements with subsequent improvement in the cost-effectiveness of the machining process is still a big challenge. The most common approach to meet all these requirements is the use of different multicomponent alloying hard coatings or hard coatings with a nanostructure architecture.

The first coating that largely met these requirements was TiAlN. Many conventional hard protective coatings on the market today are still based on it. The superior performance of TiAlN-coated tools in the machining of difficult-to-cut workpiece materials (e.g., stainless steels, tool steels, Ni- and Ti-alloys) is attributed to its high hot hardness [1], high oxidation resistance, and age-hardening effect [2]. TiAlN is a supersaturated pseudo-binary alloy, where Al atoms randomly substitute Ti atoms on the metal sublattice of the fcc-TiN lattice. Such substitution significantly improves the mechanical properties due to the solid solution hardening effect. However, this is not the only hardening mechanism present in TiAlN. It was recently found that under high thermal loading, such as during the cutting operation, the TiAlN coating decomposes by spinodal decomposition to nano-sized domains of cubic TiN and cubic AlN [3,4]. Such kinds of precipitate limits dislocation movement and generates an internal stress. Both processes, which start at a temperature around 700 °C and finish at around 900 °C, increase the hardness of the coating and enhance its wear resistance. However, the disadvantage of these post-treatments is the reduction of plastic properties and embrittlement of the interface between the coating and the substrate.

As mentioned above, the resistance of hard coating materials to high-temperature oxidation is one of the most important properties. The oxidation behavior of TiAlN coatings has previously been intensively examined and reported in the literature. McIntyre et al. [5] and Ihimura et al. [6] were the first to examine the oxidation mechanisms in TiAlN coatings exposed to air at high temperature (700–900 °C). They found that the remarkable oxidation resistance of TiAlN hard coatings relies on its ability to form a thermally stable oxide scale that comprises a continuous and protective Al-rich top-layer and a porous Ti-rich sub-layer below (adjacent to the nitride). They observed that the growth of the oxide over-layers initially followed a parabolic law and, after reaching a certain thickness (that depends on the oxidation temperature), the growth rate decreased. Using the inert-marker technique, they showed that the mobile species were mainly Al (outward) and oxygen (inward), while a limited mobility of Ti species was detected. The simultaneous diffusion of Al atoms toward the oxide/air interface and the diffusion of oxygen to the oxide/nitride interface, accompanied by nitrogen depletion, led to the formation of oxide scale near the surface [7]. Thus, the oxidation of the coating is controlled by diffusion of aluminum and oxygen through the alumina layer. However, such a layer can act as protection against further oxidation only if it is stable at high temperature, if the layer grows slowly with a high density and if its adhesion to the coating surface is good. The mobility of the aluminum and oxygen through the Ti-rich oxide layer is considerably higher than through the Al-rich oxide layer. Joshi et al. [8] found that the inward diffusion of oxygen controlled the oxidation process at temperatures smaller than 700 °C while both the inward diffusion of O and outward diffusion of Al controlled the oxidation at temperatures higher than 800 °C. At an oxidation temperature below 850 °C, the thicknesses of both oxide layers are approximately the same. This indicates that the mobility of Al and O species is similar. However, a further temperature increase up to 850 °C produces a fast growth of the rutile TiO_2 crystals. The reason for accelerated oxidation is the formation of micro-cracks in the oxide scale caused by the compressive stresses appearing due to the large difference in the oxide layer molar volume in comparison with the molar volume of the nitride layer. Cracks form during the heat treatment, also due to a difference in thermal expansion between the coating and the substrate. McIntyre et al. [5] and Lemke et al. [9] reported that rutile TiO_2 crystals not only appear along the crack boundaries but also at the local growth defects.

Many attempts have been made to enhance the oxidation resistance of TiAlN coatings in the last two decades [10–14]. Several ways have been proposed to increase the density of the oxide scale and to retard the diffusion through it. Thus, it was found that the oxidation resistance of the TiAlN coating strongly depends on the Ti/Al atomic ratio [2,15–19]. Increasing the content of aluminum within c-TiAlN promotes the formation of an Al_2O_3 layer, while the growth rate of the porous and highly permeable TiO_2 layer is reduced. However, if the content of Al in the $Ti_{1-x}Al_xN$ coating exceeds the maximum solubility limit (x_{max} = 0.7), the transition from the cubic structure to mixed cubic wurtzite structure takes place, while the oxidation resistance decreases [4]. A special case is that the Ti_2AlN coating is composed mainly of a nanolaminated MAX phase. Wang et al. found that such coating exhibited excellent oxidation resistance and thermal stability (up to 900 °C in the air) [15].

The oxidation resistance also depends on the microstructure and the morphology of the coating. Voided grain boundaries, inter-columnar porosity, and local growth defects in the TiAlN coating offer faster diffusion channels for the metal atoms outward and oxygen atoms inward during the oxidation, thereby lowering the oxidation resistance of a coating. A denser microstructure of coatings tends to suppress the diffusion rates of metallic and oxygen ions along the grain boundaries. The coating density and, with it, the coating oxidation resistance can be tuned to some extent by bias voltage used during deposition [5]. In general, increasing the energy of ions leads to the densification of porous column boundaries and promotes growth of smaller grains. A high coating density can also be achieved by high-power impulse magnetron sputtering (HIPIMS) [20]. This deposition technique provides conditions for effective ion bombardment, and thus increases ad-atom mobility on the surface of the growing coatings.

Another way to increase the oxidation resistance is based on the preparation of hard coatings in the form of nanolayers [19,21]. The layer interfaces act as a diffusion barrier for the diffusion of all atoms that participate in the oxidation process and thus improve the oxidation resistance of the coating.

The most common approach for tailoring the oxidation properties of TiAlN coatings is the addition of small concentrations (up to ~15 at.%) of different metallic (e.g., Cr, Y, Zr, Ta, Nb) or nonmetallic alloying elements (e.g., Si, B) [22]. Dopants such as yttrium, silicon, and boron improve the high-temperature oxidation resistance of TiAlN by plugging grain boundaries (Y) [23] or by forming very dense oxides (Si, B) [13,24]. It is shown that large and, therefore, relatively immobile yttrium atoms with a high affinity for oxygen segregate at grain boundaries, promoting grain refinements and blocking the diffusion pathways for oxygen and substrate elements along columns and grain boundaries [25,26]. If TiAlN is doped with silicon or boron, then an amorphous Si_3N_4 or BN phase is formed surrounding nanocrystalline TiAlN grains, forming nanocomposites and thus stimulating the further refinement of grains [24,27,28]. More grain boundaries prolong the diffusion paths of oxygen and thus improve the oxidation resistance of the coating. From the oxidation resistance point of view, doping the TiAlN coating with Cr is the most beneficial [23,29–31]. A better oxidation resistance was ascribed to: (a) the formation of mixed $(AlCr)_2O_3$ oxide, which crystallizes faster than Al_2O_3, and (b) the possibility of Cr retarding the transformation of TiO_2 from the anatase to rutile phase that occurs within the oxide scale at high temperature (>850 °C). Due to the reduction in molar volume, the phase transformation is accompanied by the formation of pores and cracks. Therefore, the oxidation rate increases. The phase transformation of anatase to rutile TiO_2 can also be delayed by Zr [32–34] or avoided by Ta alloying [14,35–37].

There are only few papers in the literature dealing with the role of growth defects during the oxidation process of hard coatings [5,38,39]. The influence of defects on the oxidation resistance of sputter-deposited TiAlN is briefly mentioned in an article by McIntyre et al. [5]. Their role in the oxidation process is discussed in more detail in the paper published by Lembe et al. [9]. They studied the influence of the growth defects on the localized oxidation behavior of TiAlN/CrN superlattice coatings deposited by arc bond sputtering (ABS) technology. The defect structure that surrounds a droplet leaves a gap between the growth defect and the coating. They observed localized oxidation at sites of such defects. Fernades et al. [39] also demonstrated that the oxidation of TiSiVN coatings is controlled by the formation of a silicon oxide diffusion barrier, which is affected by nodular defects in the as-deposited film. Recently, our research group published the results of the investigation of the high-temperature oxidation of nanolayered CrVN-based coatings [40,41]. We found that the first oxidation products appear around the nodular defects in the form of V_2O_5 patches.

In our recently published study [42], we discussed the influence of growth defects on the corrosion resistance of TiAlN coatings. We found that pitting corrosion occurring at growth defect sites has a very deleterious effect. In these sites, the corrosive medium dissolves the substrate material, which results in the collapse of the coating. The goal of present study is to examine in more detail if the growth defects also have an important influence during the high-temperature oxidation. Based on the experimental results, the oxidation mechanisms in the early stage of oxidation are discussed. A more detailed explanation of the growth defect phenomenon in PVD coatings can be found in our recently published review article [43].

2. Materials and Methods

The industrial magnetron sputtering system (modified CC800/7, CemeCon, Germany) was used for the deposition of the TiAlN hard coating. The coating composition was Ti 23 at.%, Al 27 at.%, and 50 at.% of N. The thickness of the coatings was around 4.1 µm, using a 2-fold planetary rotation. The samples analyzed in this study consisted of test plates made of cold work tool steel AISI D2 (~58 HRC) produced by Ravne steel factory (Slovenia). The D2 plates were first ground and polished up to the roughness of about 10 nm (R_a). Before deposition, they were cleaned in detergents (alkaline cleaning agents, pH ~ 11) and ultrasound, rinsed in deionized water, and dried in hot air. In the vacuum chamber, they were first heated to about 450 °C, then in-situ cleaned by radio frequency (RF) ion etching in argon atmosphere. The RF power and the etching time were 2000 W and 90 min, respectively. A standard TiAlN coating was deposited by DC-sputtering of four mosaic Ti-Al targets at 8 kW each. The total operating pressure was maintained at 0.75 Pa, with the flow rates of nitrogen, argon, and krypton being 100, 160, and 110 mL/min, respectively. A DC bias of −100 V was applied to the substrates. The deposition time was 135 min. After this time, the deposition process was interrupted for an intermediate ion etching (for 60 min at the same conditions as during substrate cleaning). This was followed by an additional deposition of a TiAlN coating (deposition time was 30 min). The intermediate etching creates new nucleation sites for the subsequently deposited nitride coating resulting in a fine-grained and less porous microstructure of the top layer [44,45].

Coated substrates were isothermally oxidized in an ambient atmosphere at temperatures of 800 °C and 850 °C for times ranging from 15 min to 2 h using a conventional tube furnace. Under such conditions, we were able to observe the early stage of the oxidation at the sites of the growth defect. The crystal structure of both as-deposited and oxidized coatings was examined by X-ray diffraction (XRD) with CuKα radiation using a Bruker diffractometer (AXS Endeavor D4) in Bragg/Brentano mode and equipped with a CuKα X-ray source (0.15406 nm). The microstructure and the coating morphology, before and after oxidation, were investigated by cross-sectional and plan-view scanning electron microscopy (SEM) using the Helios Nanolab 650i field emission scanning electron microscope. Cross-sections for detailed SEM investigation were prepared by the focused ion beam technique using the FIB source integrated into the SEM microscope. On this cross-section, SEM imaging using electrons and ions was performed. EDS element mapping was realized to determine the qualitative distribution of the elements on the FIB cross-section. EDS mapping was carried out by using the Oxford Instruments system attached to the SEM.

Scanning transmission electron microscopy (TEM) and energy-dispersive X-ray spectroscopy (EDS) analysis were performed, using a JEOL ARM 200 CF operated at 200 kV. All TEM images and analyses were obtained on cross-sections. The specimen for TEM characterization was prepared with a FEI Helios NanoLab 600i focus ion beam system using the standard lift-out technique.

3. Results

In order to understand the oxidation behavior, SEM, STEM, and EDS analyses were performed on the TiAlN hard coating before and after the oxidation test with the main emphasis focused on the role of the growth defects on the oxidation process.

3.1. Morphology and Microstructure of As-Deposited TiAlN Hard Coating

From the oxidation point of view, the morphology of a coating is important because the oxidation rate of a dense and fine-grained morphology is much smaller than that of the coating with a distinct columnar morphology [38]. The morphology of PVD coatings strongly depends on the deposition method and deposition parameters. Figure 1 shows the SEM plan-view image of the TiAlN hard coating sputter-deposited on a D2 tool steel substrate. The images were taken at low (Figure 1a,c) and high magnifications (Figure 1b,d). At a low magnification, a smooth coating surface interrupted by morphological defects (nodular defects, pinholes, craters) of various shapes and sizes can be observed. The growth defects are unevenly distributed. Sites of protruding carbides that were formed during ion etching of the substrate can also be observed. The growth defects are an unwanted phenomenon that significantly alters the morphology of PVD hard coatings [43]. Their origin is related to all topographical irregularities on the substrate surface formed during its pretreatment and ion etching, as well as by foreign particles introduced during the coating process. All these topographical irregularities on the substrate surface are transferred through the coating and even magnified due to the geometrical shadowing effect characteristic for the PVD deposition techniques [46]. Therefore, even relatively small imperfections of several tens of nanometers can grow into large micrometer-sized imperfections (growth defects) on the coating surface [43].

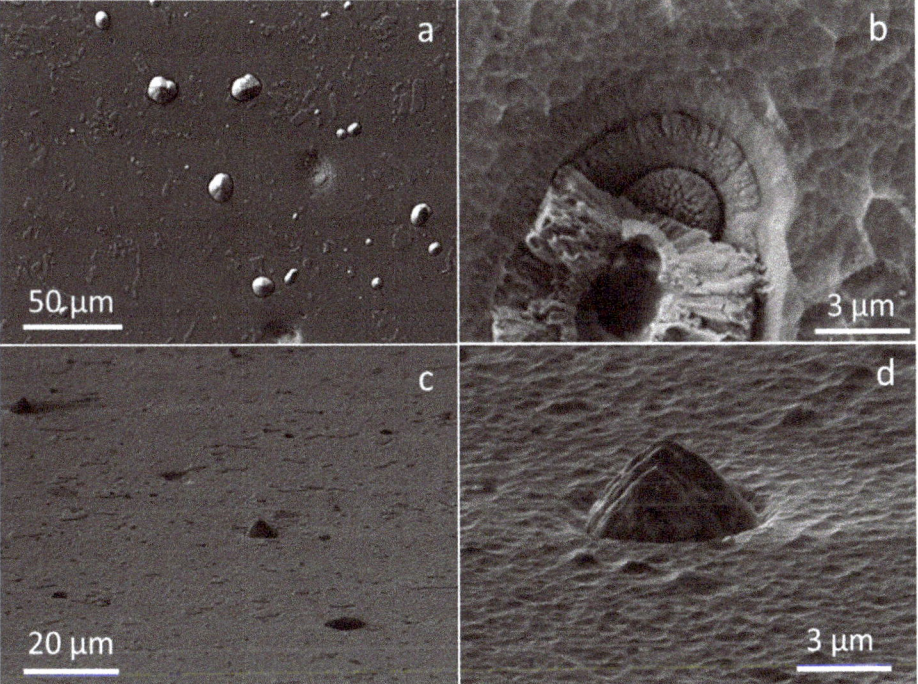

Figure 1. Plan-view SEM images of as-deposited TiAlN hard coating deposited onto D2 tool steel substrate. SEM images were recorded at low (**a,c**) and high (**b,d**) magnifications, while the images on the bottom (**c,d**) were taken by tilting the sample by approximately 20°.

At high magnification of the area between various protrusions, we can see that it exhibited a faceted domain-like morphology. The mean size of the domains is about 1 μm. Such topography is characteristic for coatings prepared using an intermediate etching. Interestingly, after such treatment, the roughness of the coating surface decreases. Namely, the sides of a nodular defect and other coating protrusions etch faster than the top as the etching rate at normal incidence is much smaller than at high incidence angles (~50°). This phenomenon leads to the shrinking of the nodular defects and even elimination of some smaller ones. Consequently, the surface topography also changes from the cap-shaped column tops to a dimpled surface after etching (Figure 1b,d).

Figure 2a presents the fracture cross-sectional SEM image of a TiAlN coating deposited on a D2 tool steel substrate. The image shows that the main part of the coating exhibits a pronounced coarse columnar structure, while the upper one is compact, fine-grained, and less porous. Such a microstructure of the top layer is the result of an intermediate ion etching of the coating after about three-quarters of the deposition time, which causes an interruption of the growing columns, re-sputtering of otherwise rounded column tops, and nucleation of new grains [44,45]. The coating surface looks rather smooth.

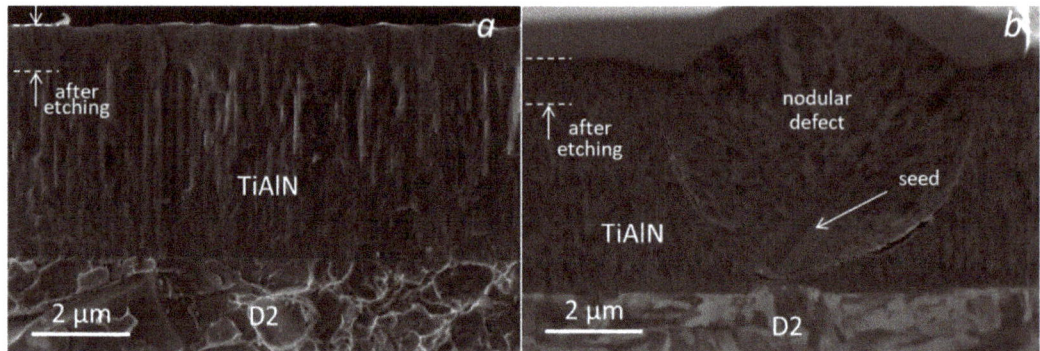

Figure 2. Fracture cross-sectional SEM image (**a**) and SEM image of FIB cross-section (**b**) of as-deposited TiAlN coating.

In the SEM image of the FIB cross-section taken using the ion beam (Figure 2b), the columnar structure of the coating cannot be noticed, while the crystal grains with different orientations are clearly visible. From a comparison of the fracture cross-sectional SEM image (Figure 2a) and SEM image of the FIB cross-section (Figure 2b), we can conclude that the individual column is not composed of a single grain but of several grains that are elongated in the growth direction. It can also be observed that the upper layer, which has grown after intermediate etching, has a more fine-grained microstructure. Beneath and around seed particles, a zone of lower density and large voids are formed due to the shadowing effect.

An even more detailed image of the microstructure and phase composition can be provided from the cross-sectional STEM image. The cross-sectional STEM bright-field micrographs of the as-deposited TiAlN coating reveal a columnar microstructure (Figure 3a,d). The diameter of the columns varies from 0.2 up to 0.4 µm. The columns stand tightly together but some voids are still visible in between (indicated by arrows in Figure 3a,d). STEM images clearly show that the individual columns are composed of several grains elongated in the growth direction. The grain diameter is less than 100 nm. Their growth is repeatedly interrupted by a re-nucleation process. Near the substrate, the microstructure consists of randomly oriented small grains. Some form V-shaped columns due to the crystal grains with a certain orientation, which grow faster and gradually overgrow the slower-growing ones. At the flat substrate surface, the crystal grains are elongated in the growth direction, while they have a feather-like appearance in the area above the seed particle (Figure 3d) or other protrusions (e.g., at site of carbide, Figure 3a). It is evident that at such substrate surface irregularities, the columns do not grow normal to the surface but grow toward the adatom source with a change in column shape.

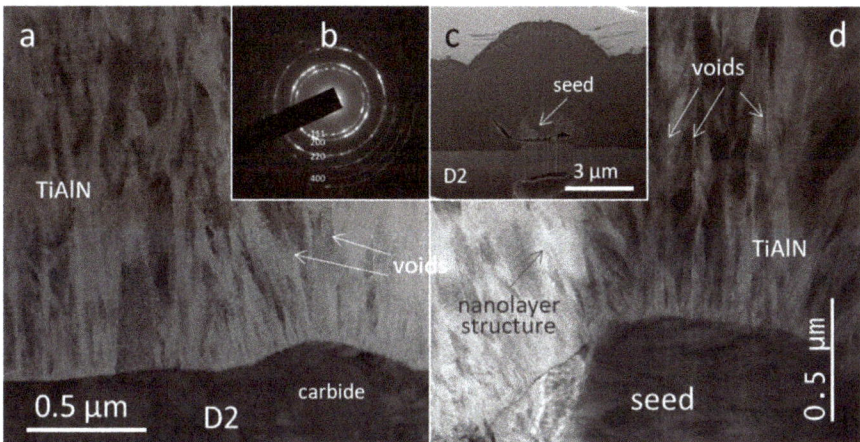

Figure 3. STEM image showing columnar morphology and nanolayer structure of sputter-deposited TiAlN coating outside (**a**) and inside the nodule defect (**d**). The corresponding electron diffraction patterns and SEM image of the FIB cross-section of the nodular defect are shown in insets (**b**) and (**c**), respectively. A step-like feature on the substrate surface under the seed particle (inset **c**) indicates that the seed existed on the substrate surface already before ion etching.

The STEM image of the as-deposited TiAlN coating also reveals that it has a nanolayered structure. The coating is composed of alternating Al-rich (bright) and Ti-rich (dark) layers with a modulation period of about 35 nm. Such a nanolayer structure is caused by the rotation of the sample that periodically approaches toward and moves away from the mosaic target. Both the angle-of-incidence of the depositing atom flux and deposition rates change constantly during the deposition process. A typical diffraction pattern recorded from a larger area of the TiAlN film is shown as inset b of Figure 3. The diffraction rings are continuous due to the small grain size and their different orientations. All of them can be attributed to the face-centered-cubic TiN-related structure. X-ray diffraction analyses also showed that the as-deposited coating has the fcc crystal structure and a (200) preferential orientation (Figure 4).

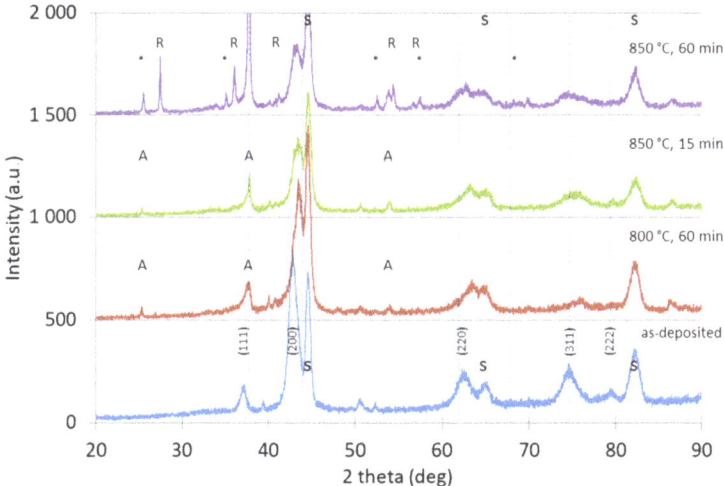

Figure 4. X-ray diffraction patterns of as-deposited and oxidized TiAlN hard coatings. The oxidation was performed for different temperatures and times (designation of peaks: s—substrate, A—anatase, R—rutile, *—α-aluminum oxide).

3.2. Coating Morphology and Microstructure after Oxidation at 800 °C

The first sample was oxidized at 800 °C for 60 min. The high-magnification SEM image shows that the whole surface is covered by very fine blade-like oxide crystals (Figure 5b,d and Figure 6a). EDS maps on the FIB cross-sections of the oxidized sample (Figures 6 and 7) show that the outermost region of the oxide scale is aluminum oxide. However, there is no indication of an alumina crystal structure by X-ray diffraction (Figure 4). Thus, it can be concluded that aluminum oxide has an amorphous or very-fine-grained microstructure. The XRD analysis of the oxidized sample revealed the presence of anatase TiO_2 in addition to nitride phases. As already mentioned in the introduction, the thin aluminum oxide layer that formed in the initial stage of oxidation on the top of the TiAlN coating strongly slows down the further oxidation process and thus provides an excellent protective effect in the area free of growth defects [5,38]. However, oxidation continues to take place intensively at the sites of certain growth defects. Around some of these defects, dark circular regions of various diameters (up to 40 µm in diameter) were formed (Figure 5a, inset in Figure 6a, inset in Figure 7a). The formation of such dark circles can be related to the diffusion of titanium on the oxide overlayer. The diffusion intensity depends on how open the pathways around the growth defect toward the substrate are. Therefore, such circular regions do not appear around all growth defects. Some defects where no such regions can be observed are marked with arrows in Figure 5a. Around those defects where they do occur, however, their diameter is very different, depending on oxidation intensity.

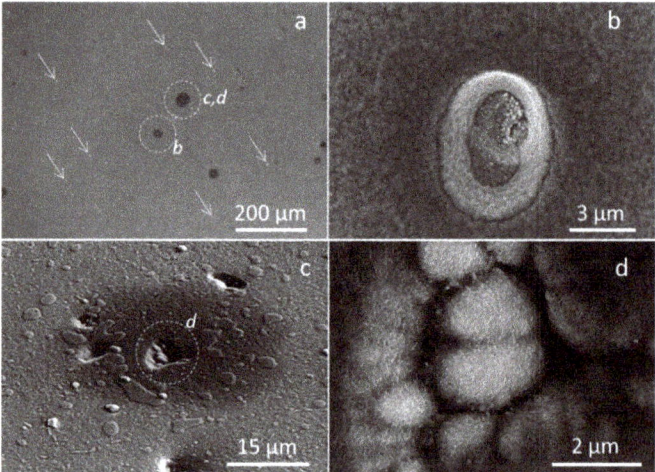

Figure 5. Plan-view SEM images of TiAlN hard coating oxidized at 800 °C for 60 min. SEM images were recorded at low (**a**,**c**) and high (**b**,**d**) magnification, while the image (**c**) was taken by tilting the sample by approximately 20°.

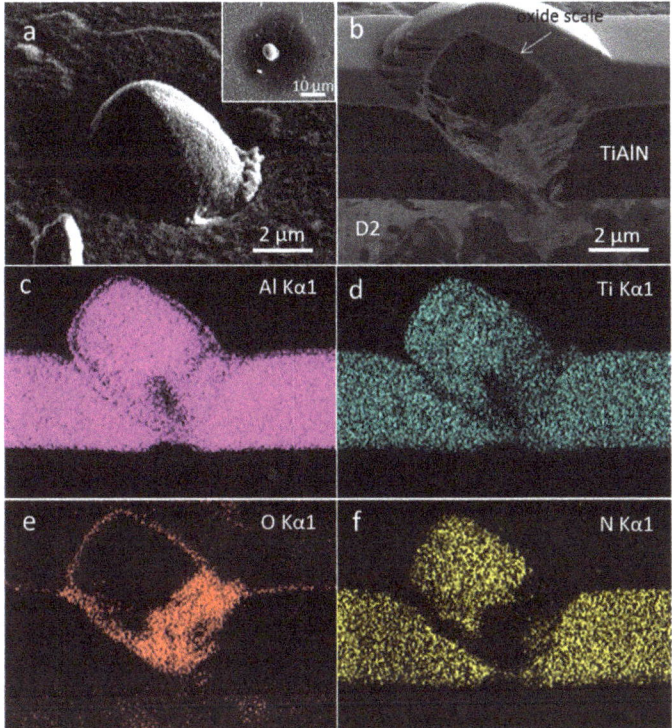

Figure 6. SEM images of nodular defect in sputter-deposited TiAlN hard coating oxidized at 800 °C for 60 min: (**a**) SEM images at high and low (inset) magnification, (**b**) SEM image of the FIB cross-section of the same defect and corresponding elemental maps (**c**–**f**).

Additional information about the oxidation process can be obtained from the SEM image of the FIB cross-section (Figures 6b and 7b). In such images, a thin two-layer oxide structure (total thickness ~200 nm) is visible. As expected, the EDS elemental maps revealed that the top layer is Al-rich, whereas the bottom one is Ti-rich. This is consistent with the results of the investigations by other researchers [5,38] who found that the layered oxide scale of TiAlN coatings is attributed to the simultaneous outward diffusion of aluminum (through the oxide/air interface) to grow the alumina layer and inward diffusion of oxygen (toward the oxide/nitride interface), to form a Ti-rich oxide layer. The continuous and homogeneous alumina layer prevents the further oxidation of the TiAlN coating. However, as mentioned before, such a continuous layer is interrupted at the sites of the growth defects. SEM images of the FIB cross-section confirm that the boundaries between the coating matrix and the defects are very porous (Figures 6b and 7b). Some of the pores are extended through the entire the coating thickness. Through such pores, oxygen can reach the substrate while elements from the substrate may penetrate into the coating surface. The EDS elemental maps shown in Figures 6 and 7 confirm an intense localized oxidation at sites of selected nodular defects and pinholes, respectively.

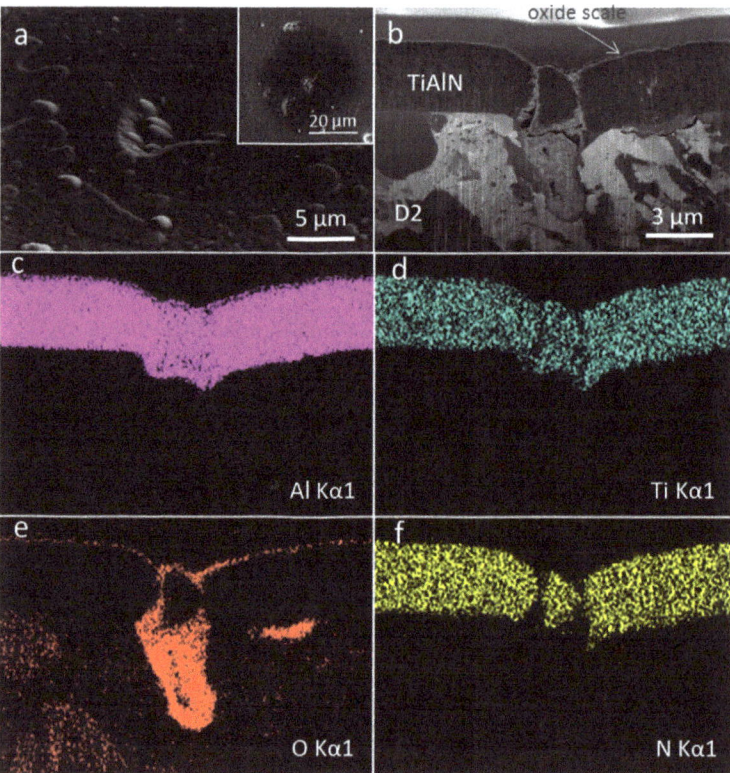

Figure 7. SEM images of pinhole in sputter-deposited TiAlN hard coating oxidized at 800 °C for 60 min: (**a**) SEM images at high and low (inset) magnification, (**b**) SEM image of FIB cross-section of the same defect, and (**c**–**f**) corresponding elemental maps [43].

3.3. Coating Morphology and Microstructure after Oxidation at 850 °C

The oxide double layer of almost the same thickness as in the previous case was formed on the sample that was oxidized at 850 °C for 15 min. The X-ray spectrum of this sample is similar to the sample that was oxidized at 800 °C (Figure 4). In addition to the expected peaks belonging to the nitride phase, peaks corresponding to the anatase TiO_2 phase can be identified, but there is no indication of crystalline alumina. From this, we can conclude that the alumina top layer is still amorphous. As in the case of the sample oxidized at 800 °C for 60 min, a dark circular region was noticed at some growth defects, but its diameter is about twice as large. However, in addition to these circles, a network of dark patches of irregular shape and different size (up to about 20 μm) was observed (Figure 8a, inset in Figure 9a). The occurrence of these patches cannot be associated with the diffusion processes along the sub-dense column boundaries, because the average distance between adjacent patches is much larger than the average inter-column distance. They also cannot be related to the growth defects, because the surface density of patches is much higher. As in the case of the dark circular region around the growth defects, the formation of these patches can be explained by the diffusion of titanium onto the surface of the oxide overlayer through the pores in the top alumina layer. How such pores can be formed will be explained later.

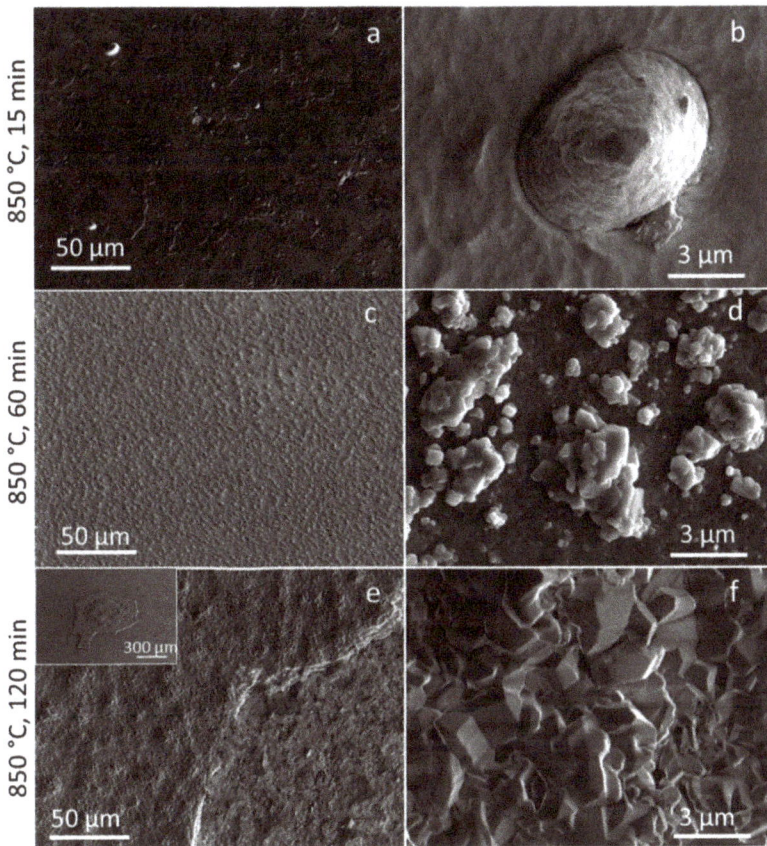

Figure 8. Plan-view SEM images of TiAlN hard coating oxidized at 850 °C for 15 (**a,b**), 60 (**c,d**), and 120 min (**e,f**). SEM images were recorded at low (**a,c,e**) and high (**b,d,f**) magnification.

The fracture cross-sectional SEM image of the TiAlN coating after short-term oxidation at 850 °C (Figure 10a) shows similar microstructure features as the as-deposited one, except a thin oxide scale (app. 200 nm thick) is present on the coating surface. The same finding applies for the SEM image of the FIB cross-section (Figure 10b).

It is evident from both types of images that the oxide overlayer is continuous and homogeneous except at sites of growth defects (Figures 9b and 10b). The EDS elemental maps on the FIB cross-section of the pinhole (Figure 9) shows that it is filled with oxidation products throughout its depth.

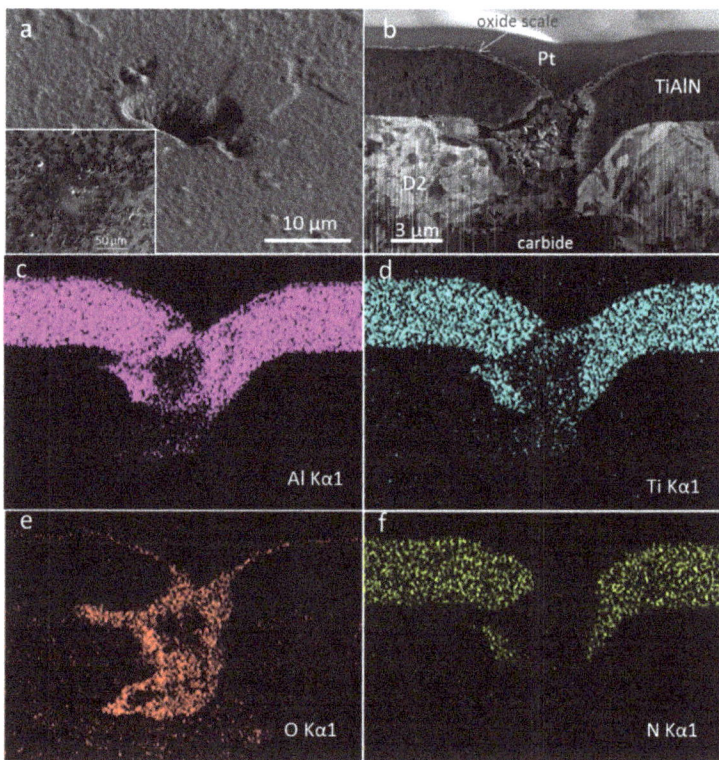

Figure 9. SEM images of pinhole in sputter-deposited TiAlN hard coating oxidized at 850 °C for 15 min: (**a**) SEM images at high and low (inset) magnification; (**b**) SEM image of the FIB cross-section of the same defect, and (**c–f**) corresponding elemental maps.

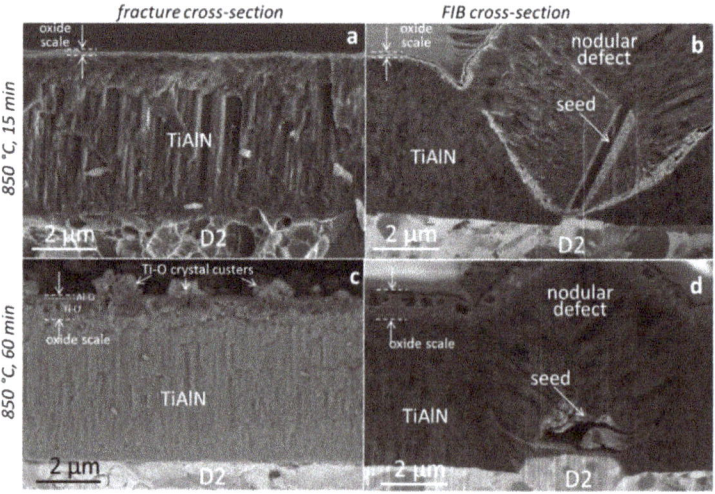

Figure 10. Fracture cross-sectional SEM images (**a,c**) and SEM images of FIB cross-sections (**b,d**) of TiAlN coating after oxidation at 850 °C for 15 (**a,b**) and 60 min (**c,d**).

As the oxidation time at 850 °C increased from 15 to 60 min, more pronounced changes occurred. Clusters of the large faceted TiO$_2$ crystallites with a rutile structure were observed to grow on the oxide overlayer (Figures 8c,d, 11a and 12a). X-ray analysis of the sample oxidized at 850 °C for 60 min confirms the formation of the rutile TiO$_2$ phase (Figure 4). However, the anatase phase, which is attributed to the titanium oxide layer at the interface with the nitride coating, is still present. Additionally, some peaks of the α-aluminum oxide phase were also detected. The EDS elemental maps, carried out at the oxidized surface, shows discontinuous and evenly distributed islands rich in the TiO$_2$ rutile phase on the top of an Al-rich oxide layer (Figure 11). It can be noticed that the size, shape, and surface density of these islands are comparable to the dark patches observed on the surface of samples oxidized at the same temperature for 15 min. The formation of similar TiO$_2$ crystallites in the microcracks of the alumina overlayer during oxidation of the TiAlN coating at 850 °C for 3 h was reported by McIntyre et al. [5]. They begin to appear after several hundred nanometers of oxide overlayer growth. The authors reported the appearance of crack networks upon oxidation of the Ti$_{0.5}$Al$_{0.5}$N hard coatings deposited at low bias voltages and their disappearance when the bias voltage was increased to −150 V. This dependency on the bias voltage was associated with an increase in residual stress with increasing bias voltage. The formation of cracks upon heating was explained by different thermal expansion coefficients of the substrate and coating.

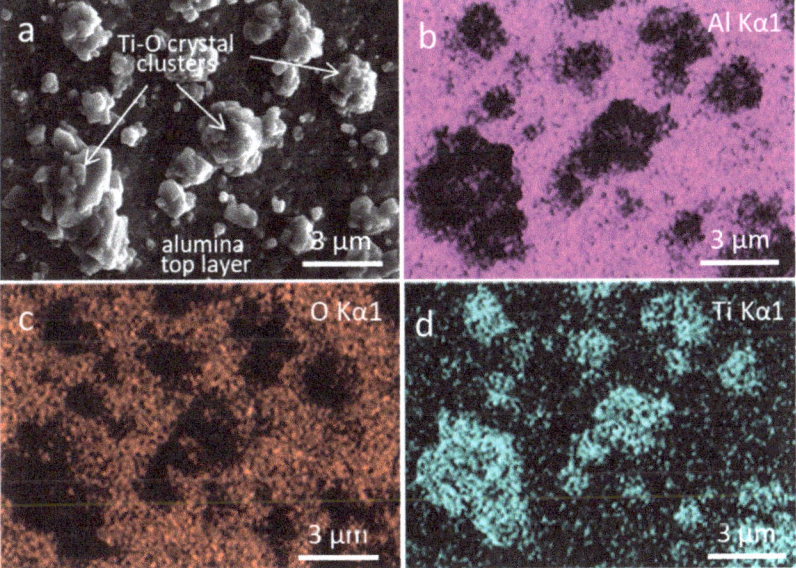

Figure 11. Top-view SEM image of the surface of the TiAlN coating oxidized at 850 °C for 60 min (**a**) and the corresponding elemental maps (**b–d**).

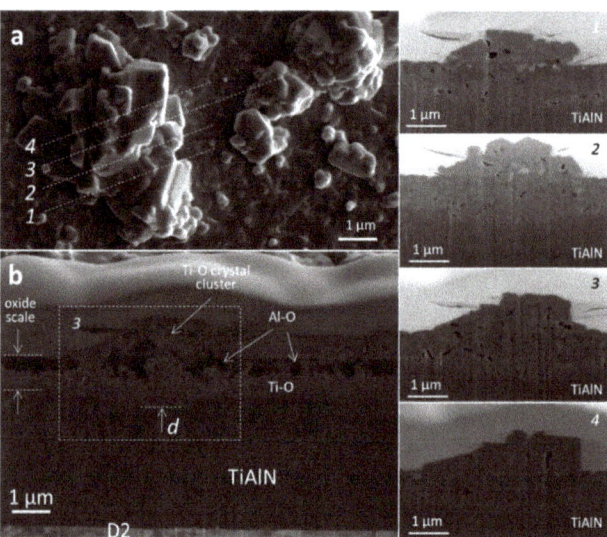

Figure 12. Top-view SEM image of the TiAlN coating surface after oxidation at 850 °C for 60 min (**a**); SEM images of FIB cross-sections through larger cluster of TiO$_2$ crystals in direction marked with dashed lines 1, 2, 3, and 4; (**b**) ion-generated SEM image of FIB cross-section in direction marked with dashed line 3.

We found that even if the oxidation time was extended to 60 min, no microcracks on the aluminum oxide surface were observed. In order to reveal the origin of TiO$_2$ crystallites on top of the Al-rich oxide layer, consecutive sectionings through the center of TiO$_2$ islands were made by the FIB technique. In Figure 12, four dashed lines marked with numbers 1 to 4 indicate the positions where the cross-sections were made. SEM images of consecutive cross-sections are shown in Figure 12 (images on the right). We were interested in whether there were any inhomogeneities, pores, or other defects in the nitride coating under the clusters of crystals. Such types of defects would allow faster diffusion of all species that participate in the oxidation process. The SEM images do not confirm that such defects are present in the nitride layer. It is obvious, however, that the oxidation front at the sites of the oxide crystallites reaches deeper into the nitride coating than outside of them (marked with "d" in Figure 12b). This means that the diffusion of oxygen at these sites is faster due to the higher permeability of the upper alumina layer. The ion-generated SEM image of the FIB cross-section (Figure 12b) reveals the reason behind the higher permeability of the A-rich oxide layer at these places. In this image, the aluminum and Ti-rich oxide layers are dark and bright areas, respectively. The boundary between the upper Al-rich oxide layer and the lower Ti-rich oxide layer is very uneven. We see that the titanium oxide has spread in the form of protrusions into the Al-rich oxide layer. The typical distance between the two protrusions is less than a micrometer. In some places, the titanium oxide phase also appears on the surface where clusters of large TiO$_2$ crystals are formed. We believe that the formation of such protrusions is related to compressive stresses in the Ti-rich layer that appeared due to the large difference in the titanium oxide molar volume in comparison with the molar volume of the nitride layer. Such stresses can separate the grain boundaries in the otherwise very-fine-grained aluminum oxide layer, and thus, open the way for the diffusion of titanium to the surface and oxygen to the oxidation front. Such a drastic change in oxidation of the TiAlN coating at 850 °C for 60 min could be related to the formation of crystalline aluminum oxide from the amorphous one. Such an interpretation is also supported by the findings of other authors. Namely, many years ago, some authors [6,47] reported that during high-temperature oxidation of the TiAlN coating, the oxygen can diffuse through the pores formed by the grain growth of the Al_2O_3 crystal in the outer alumina layer during high-temperature oxidation of TiAlN. Münz was the first who found that there is no indication of an Al_2O_3 crystal structure below 800 °C, while a drastic change in the oxidation behavior occurs above this temperature due to the formation of crystalline alumina from the amorphous one [47].

We must not forget, however, that after extending the oxidation time to 60 min, the oxidation taking place at sites of growth defects is even more intensive. Figure 13a shows that large oxide crystallites were formed at the rim of the nodular defect. However, as already mentioned, oxidation does not take place with the same intensity around all growth defects, which depends on the density of the pores at the rim of the defects and on how open they are.

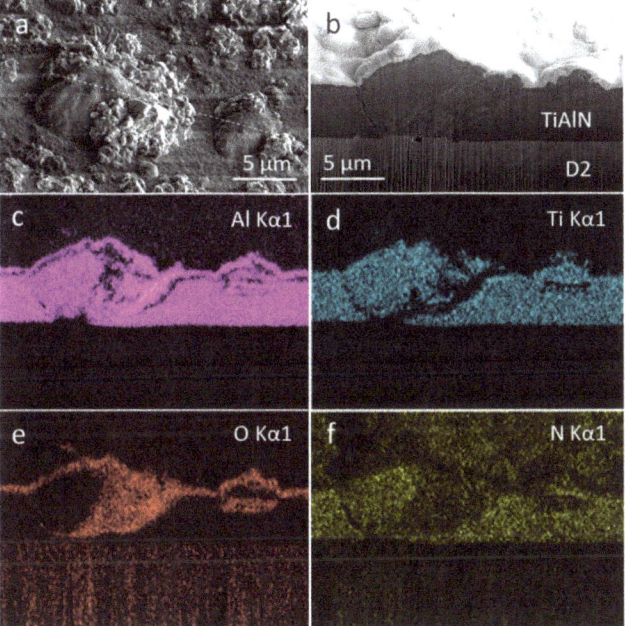

Figure 13. SEM images of nodular defect in sputter-deposited TiAlN hard coating oxidized at 850 °C for 60 min: (**a**) SEM image, (**b**) SEM image of FIB cross-section of the same defect, and (**c**–**f**) corresponding elemental maps.

The fracture cross-sectional SEM image of the oxidized film (Figure 10c) shows the presence of a thin two-layer oxide structure, with a porous inner Ti-rich oxide layer and an outer compact Al-rich oxide layer. The oxide bilayer thickness increases from 0.2 to 0.95 µm if the oxidation time increases from 15 to 60 min, respectively. The structure of the oxide scale can also clearly be seen in the SEM images of the FIB cross-section (Figures 10d, 12, 13b, and 14b). The EDS maps on FIB cross-sections revealed that the porous bottom layer and the denser top layer are Ti-rich and Al-rich oxides, respectively. Discrete and unevenly distributed TiO_2 crystals can be observed on the top alumina layer, while the rest of the nitride coating is still uninfluenced by the oxidation.

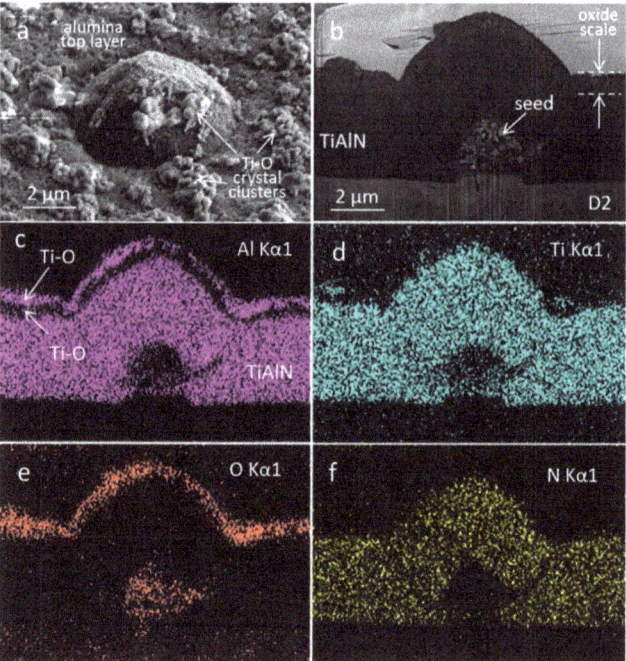

Figure 14. SEM images of nodular defect in sputter-deposited TiAlN hard coating oxidized at 850 °C for 60 min: (**a**) SEM image, (**b**) SEM image of the FIB cross-section of the same defect, and (**c**–**f**) corresponding elemental maps.

Plain-view SEM images (Figures 13a and 14a) show that a complex oxide structure was developed at sites of some growth defects. Due to the presence of under-dense regions and pores at the rim of the growth defect, a high amount of Ti ions is supplied in this zone. As a result, the localized oxidation and formation of large oxide crystals around many defects can be observed. This phenomenon is not equally pronounced around all growth defects, due to the different degree of porosity at the boundary between the coating matrix and the defect. Such porosity is influenced by the shape and size of the topographical irregularities (e.g., seeds, pits) on the substrate surface that are responsible for the formation of the growth defects [43,46]. If the seed has a smooth morphology, then the nodule looks like a cone, while a seed with complex morphology results in the growth of a nodular defect with irregular surface features. When an irregularly shaped seed is coated, the particle flux cannot reach the area underneath the seed, due to the shadowing of the particle flux, and thus causes formation of voids below the particles. The coating on such kinds of seeds results in a highly nonuniform coverage with a practically uncoated area underneath the seed. Figure 13 shows the top-view SEM image of the nodular defect where intensive oxidation has occurred. The SEM image of a FIB cross-section reveals open pathways at the rim of the defect. It is quite different in the case of a nodular defect in Figure 14, where almost no oxidation products at the edge of the defect are observed. From the SEM image of the FIB cross-section in Figure 14, we can see that the boundary between the nodular defect and coating matrix is much less porous than in the previous case. This is most likely due to the fact that the shape of the seed is more regular. In the case of cavities, the uniformity of coverage and, thus, formation of pores depend on the aspect ratio (depth/hole diameter), as well as from the angular distribution of the impinging vapor flux [43,46].

The microstructure of the oxide layers at the site of the nodular defect was analyzed in more detail with a STEM microscope. The FIB lift-out technique was applied for the preparation of the lamella from the nodular defect shown in Figure 14. Figures 15–17 show the bright-field STEM images and the STEM-EDS maps of the oxide scale at the nodular defect site and outside it. In these micrographs, a dense and fine-grained Al-rich oxide top layer can be seen, while the Ti-rich oxide underlayer has a coarser polycrystalline microstructure with a rough surface. The transformation of the amorphous alumina into the crystalline one was confirmed by both X-ray diffraction (Figure 4) as well as selected-area electron diffraction (SAED) analysis (Figure 17b). The thickness of the oxide double layer is app. 0.95 µm, while about 3.8 µm of the coating is not yet oxidized. Figure 15d,f and Figure 16d show the elemental maps of Al- and Ti-rich oxide areas within the oxide scale. It can be seen that the interface between Al-rich and Ti-rich oxide layers is not flat. The STEM-EDX maps in Figures 15 and 16 clearly show that titanium oxide crystallites penetrate deep into the alumina layer at some places. As already mentioned, the expansion of titanium oxide into the aluminum oxide layer can be caused by compressive stresses resulting due to a large molar volume difference of TiO_2 and TiAlN. Thus, the top Al-rich layer becomes porous and oxygen can more rapidly access the oxidation front. Hence, its passivation protection is significantly reduced. Based on this, it can be concluded that, in addition to localized oxidation at growth defect sites, this mechanism could also be responsible for the accelerated diffusion of titanium through the alumina layer. The result of both processes is large TiO_2 crystals forming on the oxidized TiAlN coating surface.

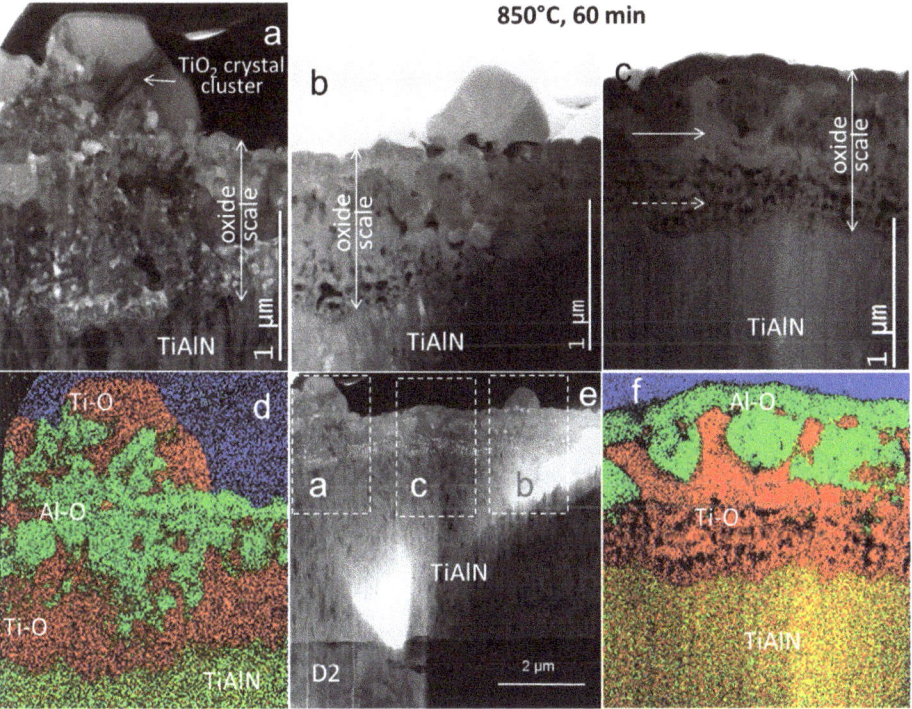

Figure 15. Bright-field STEM cross-sectional images of oxide scale (a–c,e) and corresponding STEM-EDS maps (d,f) of TiAlN coating oxidized at 850 °C for 60 min.

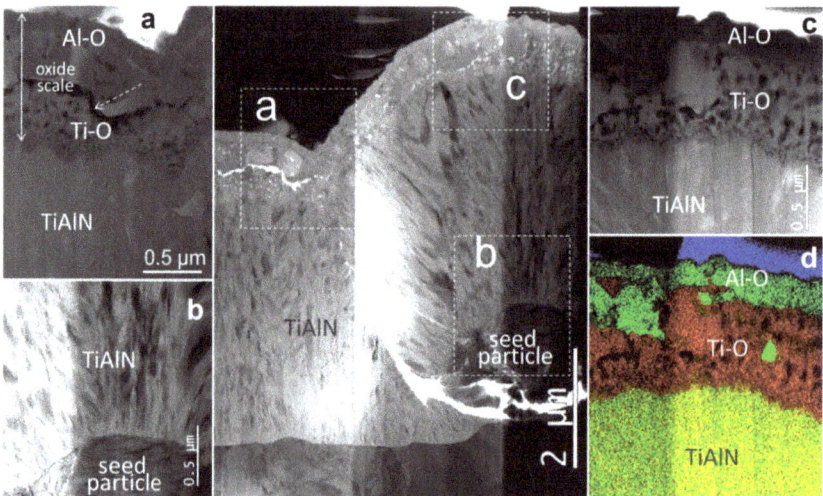

Figure 16. Bright-field STEM cross-sectional image of nodular defect (**a–c**) and corresponding STEM-EDS map (**d**) of TiAlN coating oxidized at 850 °C for 60 min.

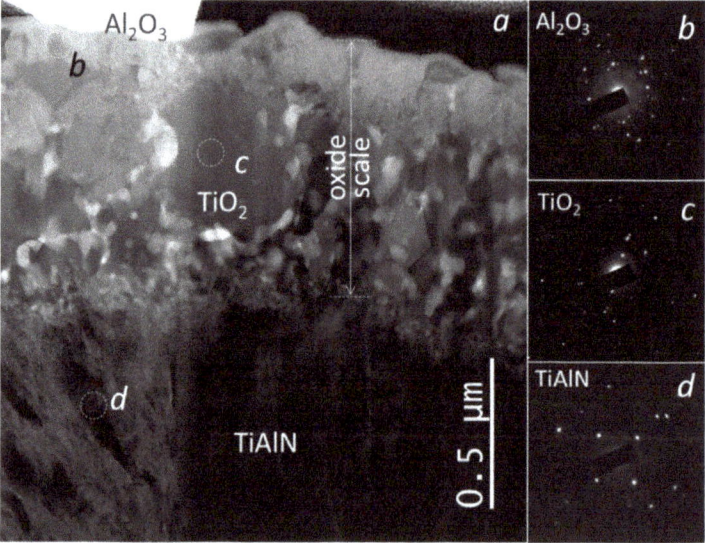

Figure 17. Bright-field STEM cross-sectional image of the oxide scale on TiAlN coating oxidized at 850 °C for 60 min (**a**) and SAED patterns of Al_2O_3 (**b**), TiO_2 (**c**), and TiAlN (**d**) crystallites at positions marked with dashed circles.

On the other hand, the outward diffusion of the Ti results in voids or pore formation at the nitride/oxide interface. The STEM images reveal that the Ti-rich oxide zone is composed of a two-layer oxide structure: An outer layer of well-defined crystals (marked by a solid arrow in Figure 15c) and a porous inner layer (adjacent to the rest of the nitride coating, marked by a dashed arrow in Figure 15c). The formation of pores at the nitride/oxide interfaces originates from different diffusion rates of the involved species. In the vicinity of the nodular defect and on its surface, some microcracks parallel to the surface can be observed (marked by a dashed arrow in Figure 16a). The formation of such cracks, located in the Ti-rich oxide layer, can be explained by the large compressive stresses that generated at the sites of the topographic irregularities. It can also be observed that during oxidation at 850 °C, the dense columnar microstructure of the as-deposited coating remains unchanged. The substrate/coating interface is also sharp and no changes can be observed.

A further oxidation test at 850 °C for 120 min causes the coating to completely oxidize and detach over a large area of the substrate (Figures 8e,f and 18). Only on some small patches is the partial integrity of the oxidized coating still preserved (see inset in Figure 8e). As it can be seen, the oxidized coating is composed of large faceted crystals while its surface is very rough. The titanium and iron diffused on the surface at sites of pores in the Al-rich oxide layer, while outside of this area, the iron diffused deep into the Ti-rich oxide layer. Cracks at the substrate/coating interface are probably due to a large difference in thermal expansion between the coating and substrate. It can be expected that the formation of such cracks accelerates the oxidation process.

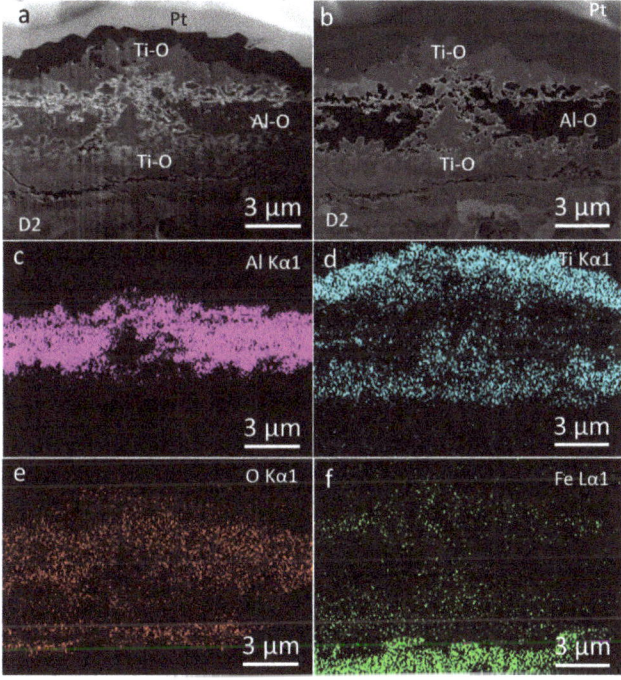

Figure 18. Conventional (**a**) and ion-generated (**b**) SEM images of FIB cross-section of sputter-deposited TiAlN hard coating oxidized at 850 °C for 120 min and corresponding elemental maps (**c**–**f**).

4. Conclusions

This study describes the influence of the growth defects on the oxidation process of sputter-deposited TiAlN coatings. The growth of oxide scale at different temperatures (800 and 850 °C) and different periods (15–120 min) was investigated. Simultaneously, the structure of the oxide scale and the interface integrity was also examined. It was found that in the early stage of oxidation, the aluminum oxide layer on the TiAlN grows steadily, maintains integrity, and thus retards the oxidation process. As oxidation continues, pores appear in the alumina layer due to the crystal grain growth in the outer alumina overlayer. The formation of pores is caused by compressive stresses that appear in the underlying Ti-rich oxide layer due to the large difference in molar volume between the oxide and nitride phase. Through such pores, oxygen can rapidly access the oxidation front and accelerate the oxidation process. Therefore, this phenomenon strongly affects the global oxidation resistance of the TiAlN coating.

Even more intensive local oxidation takes place at the sites of the growth defects due to the pores that are formed at the rim of the defect. During oxidation, they provide direct paths between the coating surface and the substrate, for transport of oxygen inward and substrate elements to the surface. The intensity of the oxidation process depends on the density of the pores and how open the pathways around the growth defect toward the substrate are. On the other hand, the number and size of pores at the rim of the growth defects depend on the shape and size of seeds or pits on the substrate surface that caused their formation. Thus, the shape of the seed largely determines whether the oxidation down to the substrate will occur or not. It should be noted that not all defects extend through the entire coating. The surface area covered by the growth defects is also relatively small (less than 1%). Although the early stage always starts at the sites of the growth defects, we believe that their influence on the global oxidation resistance of coatings is not decisive.

Author Contributions: Conceptualization, manuscript writing, P.P.; SEM and FIB analysis of growth defects and oxidized samples, manuscript review, A.D.; STEM analysis, G.D. All authors have read and agreed to the published version of the manuscript.

Funding: This work was funded by the Slovenian Research Agency (program P2-0082). We also acknowledge funding from the European Regional Development Funds (CENN Nanocenter, OP13.1.1.2.02.006).

Institutional Review Board Statement: Not applicable.

Informed Consent Statement: Not applicable.

Data Availability Statement: Data is contained within the article.

Conflicts of Interest: The authors declare no conflict of interest.

References

1. Jindal, P.C.; Santhanam, A.T.; Schleinkofer, U.; Shuster, A.F. Performance of PVD TiN, TiCN, and TiAlN coated cemented carbide tools in turning. *Int. J. Refract. Met. Hard Mater.* **1999**, *17*, 163–170. [CrossRef]
2. Chen, L.; Paulitsch, J.; Du, J.Y.; Mayrhofer, P.M. Thermal stability and oxidation resistance of Ti-Al-N coatings. *Surf. Coat. Technol.* **2012**, *206–318*, 2954–2960. [CrossRef]
3. Horling, A.; Hultman, L.; Oden, M.; Sjolen, J.; Karlsson, L. Thermal stability of arc evaporated high aluminium-content TiAlN thin films. *J. Vac. Sci. Technol.* **2002**, *20*, 1815–1823. [CrossRef]
4. Mayrhofer, P.H.; Horling, A.; Karlsson, L.; Sjolen, J.; Larsson, T.; Mitterer, C.; Hultman, L. Self-organized nanostructures in the Ti–Al–N system. *Appl. Phys. Lett.* **2003**, *83*, 2049. [CrossRef]
5. McIntyre, D.; Greene, J.E.; Håkansson, G.; Sundgren, J.E.; Münz, W.D. Oxidation of metastable single-phase polycrystalline $Ti_{0.5}Al_{0.5}N$ films: Kinetics and mechanisms. *J. Appl. Phys.* **1990**, *67*, 1542. [CrossRef]
6. Ichimura, H.; Kawana, A. High-temperature oxidation of ion-plated TiN and TiAlN films. *J. Mater. Res.* **1993**, *8*, 1093–1100. [CrossRef]
7. Feng, Z.; Zhang, L.; Ke, R.X.; Wn, Q.L.; Wang, Z.; Lu, Z.H. Thermal stability and oxidation behavior of AlTiN, AlCrN and AlCrSiWN coatings. *Int. J. Refract. Met. Hard Mater.* **2014**, *43*, 241–249. [CrossRef]
8. Joshi, A.; Hu, H.S. Oxidation behavior of titanium-aluminium nitrides. *Surf. Coat. Technol.* **1995**, *76–77*, 499–507. [CrossRef]
9. Lembke, M.I.; Lewis, D.B.; Münz, W.D. Localised oxidation defects in TiAlN/CrN superlattice structured hard coatings grown by cathodic arc/unbalanced magnetron deposition on various substrate materials. *Surf. Coat. Technol.* **2000**, *125*, 263–268. [CrossRef]
10. Pfeiler, M.; Zechner, J.; Penoy, M.; Michotte, C.; Mitterer, C.; Kathrein, M. Improved oxidation resistance of TiAlN coatings by doping with Si or B. *Surf. Coat. Technol.* **2009**, *203*, 3104–3110. [CrossRef]
11. Pfeiler, M.; Scheu, C.; Hutter, H.; Schnöller, J.; Michotte, C.; Mitterer, C.; Kathrein, M. On the effect of Ta on improved oxidation resistance of Ti-Al-Ta-N coatings. *J. Vac. Sci. Technol.* **2009**, *27*, 554. [CrossRef]
12. Riedl, H.; Holec, D.; Rachbauer, R.; Polcik, P.; Hollerweger, R.; Paulitsch, J.; Mayrhofer, P.H. Phase stability, mechanical properties and thermal stability of Y alloyed Ti-Al-N coatings. *Surf. Coat. Technol.* **2013**, *235*, 174. [CrossRef]
13. Asanuma, H.; Klimashin, F.F.; Polcik, P.; Kolozsvári, S.; Riedl, H.; Mayrhofer, P.H. Impact of lanthanum and boron on the growth, thermomechanical properties and oxidation resistance of Ti–Al–N thin films. *Thin Solid Film.* **2019**, *688*, 137239. [CrossRef]
14. Hollerweger, R.; Riedl, H.; Paulitsch, J.; Arndt, M.; Rachbauer, R.; Polcik, P. Origin of high temperature oxidation resistance of Ti–Al–Ta–N coatings. *Surf. Coat. Technol.* **2014**, *257*, 78–86. [CrossRef]

15. Wang, Q.M.; Garkas, W.; Renteria, A.F.; Leyens, C.; Kim, K.H. Oxidation Behaviour of Ti_2AlN Films Composed Mainly of Nanolaminated MAX Phase. *J. Nanosci. Nanotechnol.* **2011**, *11*, 10. [CrossRef] [PubMed]
16. Ikeda, T.; Satoh, H. Phase formation and characterization of hard coatings in the Ti-Al-N system prepared by the cathodic arc ion plating method. *Thin Solid Film.* **1991**, *195*, 99–110. [CrossRef]
17. Kim, B.J.; Kim, Z.C.; Nah, J.W.; Lee, J. High temperature oxidation of $(Ti_{1-X} Al_X)$ N coatings made by plasma enhanced chemical vapor deposition. *J. Vac. Sci. Technol.* **1999**, *17*, 133. [CrossRef]
18. Vaz, F.; Rebouta, L.; Andritschky, M.; da Silva, M.F.; Soares, J.C. Oxidation resistance of (Ti,Al,Si)N coatings in air. *Surf. Coat. Technol.* **1998**, *98*, 912–917. [CrossRef]
19. Peng, J.; Su, D.; Wang, C. Combined effect of aluminium content and layer structure on the oxidation performance of $Ti_{1-x}Al_xN$ based coatings. *J. Mater. Sci. Technol.* **2014**, *30*, 803–807. [CrossRef]
20. Hovsepian, P.E.; Ehiasarian, A.P. Six strategies to produce application tailored nanoscale multilayer structured PVD coatings by conventional and High Power Impulse Magnetron Sputtering (HIPIMS). *Thin Solid Film* **2019**, *688*, 137409. [CrossRef]
21. Chen, L.; Du, Y.; Xiong, X.; Chang, K.K.; Wu, M.J. Improved properties of Ti-Al-N coating by multilayer structure. *Int. J. Refract. Met. Hard Mater.* **2011**, *29*, 681–685. [CrossRef]
22. Hollerweger, R.; Riedl, H.; Arndt, M.; Rachbauer, R.; Kolozsvari, S.; Primig, S.; Mayrhofer, P.H. Guidelines for increasing the oxidation resistance of Ti-Al-N based coatings. *Thin Solid Film.* **2019**, *688*, 137290. [CrossRef]
23. Donohue, L.A.; Smith, I.J.; Münz, W.D.; Petrov, I.; Greene, J.E. Microstructure and oxidation-resistance of TiAlCrYN layers grown by combined steered-arc/unbalanced-magnetron-sputter deposition. *Surf. Coat. Technol.* **1997**, *95–96*, 226–231. [CrossRef]
24. Vannemann, V.; Stock, H.R.; Kohlscheen, J.; Rambadt, S.; Erkens, G. Oxidation resistance of titanium-aluminium-silicon nitride coatings. *Surf. Coat. Technol.* **2003**, *174–175*, 408–415. [CrossRef]
25. Moser, M.; Kiener, D.; Scheu, C.; Mayrhofer, P.H. Influence of Yttrium on the thermal stability of Ti-Al-N thin films. *Materials* **2010**, *3*, 1573–1592. [CrossRef]
26. Hovsepian, P.E.; Lewis, D.; Luo, Q.; Münz, W.-D.; Mayrhofer, P.; Mitterer, C.; Zhou, Z.; Rainforth, W. TiAlN based nanoscale multilayer coatings designed to adapt their tribological properties at elevated temperaures. *Thin Solid Film.* **2005**, *485*, 160–168. [CrossRef]
27. Wu, G.; Ma, S.; Xu, K.; Chu, P.K. Oxidation resistance of qintuple Ti-Al-Si-C-N coatings and associated mechanism. *J. Vac. Sci. Technol.* **2012**, *30*, 041508. [CrossRef]
28. Zhu, L.; Hu, M.; Ni, W.; Liu, Y. High temperature oxidation behavior of TiAlN coating and TiAlSiN coating. *Vacuum* **2012**, *86*, 1795–1799. [CrossRef]
29. Polcar, T.; Cavaleiro, A. High temperature behavior of nanolayered CrAlTiN coating: Thermal stability, oxidation and tribological properties. *Surf. Coat. Technol.* **2014**, *257*, 70–77. [CrossRef]
30. Xu, Y.X.; Riedl, H.; Holec, D.; Chen, L.; Du, Y.; Mayrhofer, P.H. Thermal stability and oxidation resistance of sputtered TiAlCrN hard coatings. *Surf. Coat. Technol.* **2017**, *324*, 48–56. [CrossRef]
31. Sui, X.; Li, G.; Zhou, H.; Zhang, S.; Yu, Y.; Wang, Q.; Hao, J. Evolution behavior of oxide scales of TiAlCrN coatings at high temperature. *Surf. Coat. Technol.* **2019**, *360*, 133–139. [CrossRef]
32. Glatz, S.A.; Hollerweger, R.; Polcik, P.; Rachbauer, R.; Paulitsch, J.; Mayrhofer, P.H. Thermal stability and mechanical properties of arc evaporated Ti–Al–Zr–N hard coatings. *Surf. Coat. Technol.* **2015**, *266*, 1–9. [CrossRef]
33. Chen, L.; He, L.; Xu, Y.; Zhou, L.; Pei, F.; Du, Y. Influence of ZrN on oxidation resistance of Ti–Al–N coating. *Surf. Coat. Technol.* **2014**, *244*, 87–91. [CrossRef]
34. Abadias, G.; Saladukhin, I.A.; Uglov, V.V.; Zlotski, S.V.; Eyidi, D. Thermal stability and oxidation behavior of quaternary TiZrAlN magnetron sputtered thin films: Influence of the pristine microstructure. *Surf. Coat. Technol.* **2013**, *237*, 187–195. [CrossRef]
35. Koller, C.M.; Hollerweger, R.; Sabitzer, C.; Rachbauer, R.; Kolozsvári, S.; Paulitsch, J.; Mayrhofer, P.H. Thermal stability and oxidation resistance of arc evaporated TiAlN, TaAlN, TiAlTaN, and TiAlN/TaAlN coatings. *Surf. Coat. Technol.* **2014**, *259*, 599–607. [CrossRef]
36. Rachbauer, R.; Holec, D.; Mayrhofer, P.M. Increased thermal stability of Ti-Al-N thin films by Ta alloying. *Surf. Coat. Technol.* **2012**, *211*, 98–103. [CrossRef]
37. Khetan, V.; Valle, N.; Duday, D.; Michotte, C.; Ogletree, M.P.D.; Choquet, P. Influence of temperature on oxidation mechanisms of fiber-textured AlTiTaN coatings. *Appl. Mat. Interfaces* **2014**, *6*, 4115–4125. [CrossRef]
38. Lembke, M.I. Oxidation Behaviour of TiAlN Based Nanolayered Hard Coatings. Available online: http://shura.shu.ac.uk/19951/ (accessed on 22 January 2021).
39. Fernandes, F.; Morgiel, J.; Polcar, T.; Cavaleiro, A. Oxidation and diffusion processes during annealing of TiSi(V)N films. *Surf. Coat. Technol.* **2015**, *275*, 120–126. [CrossRef]
40. Panjan, P.; Drnovšek, A.; Kovač, J.; Gselman, P.; Bončina, T.; Paskvale, S.; Čekada, M.; Kek-Merl, D.; Panjan, M. Oxidation resistance of CrN/(Cr,V)N hard coatings deposited by DC magnetron sputtering. *Thin Solid Film.* **2015**, *591*, 323–329. [CrossRef]
41. Panjan, P.; Drnovšek, A.; Kovač, J.; Čekada, M.; Panjan, M. Oxidation processes in vanadium-based single-layer and nanolayer hard coatings. *Vacuum* **2017**, *138*, 230–237. [CrossRef]
42. Panjan, P.; Drnovšek, A.; Gselman, P.; Čekada, M.; Panjan, P.; Boncina, T.; Kek Merl, D. Influence of Growth Defects on the Corrosion Resistance of Sputter-Deposited TiAlN Hard Coatings. *Coatings* **2019**, *9*, 511. [CrossRef]

43. Panjan, P.; Drnovšek, A.; Gselman, P.; Čekada, M.; Panjan, P. Review of Growth Defects in Thin Films Prepared by PVD Techniques. *Coatings* **2020**, *10*, 447. [CrossRef]
44. Jehn, H.A. Improvement of the corrosion resistance of PVD hard coating–substrate system. *Surf. Coat. Technol.* **2000**, *125*, 212–217. [CrossRef]
45. Park, H.S.; Kappl, H.; Lee, K.H.; Lee, J.J.; Jehn, H.A.; Fenker, M. Structure modification of magnetron-sputtered CrN coatings by intermediate plasma etching steps. *Surf. Coat. Technol.* **2000**, *133–134*, 176–180. [CrossRef]
46. Mattox, D.M. Atomistic Film Growth and Resulting Film Properties. Available online: http://www.htskorea.com/product/ambios/stress.pdf (accessed on 22 January 2021).
47. Münz, W.D. Titanium aluminum nitride films: A new alternative to TiN coatings. *J. Vac. Sci. Technol. A* **1986**, *4*, 2717–2725. [CrossRef]

Article

Laser-Assisted Surface Texturing of Ti/Zr Multilayers for Mesenchymal Stem Cell Response

Suzana Petrović [1,*], Davor Peruško [1], Evangelos Skoulas [2], Janez Kovač [3], Miodrag Mitrić [1], Jelena Potočnik [1], Zlatko Rakočević [1] and Emmanuel Stratakis [2]

1. Vinča Institute of Nuclear Sciences, University of Belgrade, P.O. Box 522, 11001 Belgrade, Serbia; dperusko@vin.bg.ac.rs (D.P.); mmitric@vin.bg.ac.rs (M.M.); jpotocnik@vin.bg.ac.rs (J.P.); zlatkora@vin.bg.ac.rs (Z.R.)
2. Institute of Electronic Structure and Laser (IESL), Foundation for Research and Technology (FORTH), N. Plastira 100, Vassilika Vouton, 70013 Heraklion, Crete, Greece; skoulasv@iesl.forth.gr (E.S.); stratak@iesl.forth.gr (E.S.)
3. Jožef Stefan Institute, Jamova 39, 1000 Ljubljana, Slovenia; Janez.kovac@ijs.si
* Correspondence: spetro@vin.bg.ac.rs; Tel.: +381-11-3408560

Received: 6 November 2019; Accepted: 10 December 2019; Published: 13 December 2019

Abstract: The formation of an ordered surface texture with micro and nanometer features on Ti/Zr multilayers is studied for better understanding and improvement of cell integration. Nanocomposite in form 30×(Ti/Zr)/Si thin films was deposited by ion sputtering on Si substrate for biocompatibility investigation. Surface texturing by femtosecond laser processing made it possible to form the laser-induced periodic surface structure (LIPSS) in each laser-written line. At fluence slightly above the ablation threshold, beside the formation of low spatial frequency-LIPSS (LSFL) oriented perpendicular to the direction of the laser polarization, the laser-induced surface oxidation was achieved on the irradiated area. Intermixing between the Ti and Zr layers with the formation of alloy in the sub-surface region was attained during the laser processing. The surface of the Ti/Zr multilayer system with changed composition and topography was used to observe the effect of topography on the survival, adhesion and proliferation of the murine mesenchymal stem cells (MSCs). Confocal and SEM microscopy images showed that cell adhesion and their growth improve on these modified surfaces, with tendency of the cell orientation along of LIPSS in laser-written lines.

Keywords: multilayer thin films; Ti-based alloy; ultrafast laser-modification; cell response

1. Introduction

Thin films and coatings are considered as very applicable for biomaterials, since surface properties are a key factor in the interaction of materials with the biological environment. Surface composition and morphology regulate surface bioactivity and other biofunctionalities, in terms of the adsorption of proteins on the material surface, which is determinant for the subsequent processes of cell growth, differentiation, and extracellular matrix formation [1–3]. Titanium-based materials are nowadays well integrated into the body, due to high specific strength, excellent corrosion resistance, and good biocompatibility [4,5]. Currently, one of the main tasks is development of the Ti-based alloys with a high concentration of β-stabilizer elements (β phase of titanium), and to provide compliance between the elasticity of the implant and the surrounding hard tissues (bones). The most suitable alloying elements to be added in these new alloys are niobium, tantalum, zirconium, and molybdenum, as they do not exhibit any cytotoxic reaction in contact with cells [6]. The admissible way for preparation of Ti-based alloy coatings could be deposition of multilayer structure with alternate distribution of Ti and other (Zr, Ta, Mo, Nb) components in thin films. The superior biocompatibility of Ti-based coatings originates in easy formation of outer Ti-oxide layer with a negative charge at physiological pH and protects against the metallic components dissolving in biological fluids [7]. One promising candidate for alloying is Zr, which act as a neutral element when dissolved in Ti and can improve mechanical properties of alloys [7]. From the biomedical point of view, Zr is fully soluble in both allotropic phases of Ti, improving the mechanical strength, corrosion resistance, and biocompatibility of Ti alloys [8]. Zirconium (Zr) has received special attention as an alloying element in Ti-based alloys, because it acts like non-toxic and non-allergenic elements, thereby avoiding stress shielding effects and implant failure [9]. Binary Ti–Zr alloy has exhibited a good combination of mechanical properties, with the advantage of easy manufacturing, in comparison to multicomponent, Ti-based alloys. Additionally, Zr with the lower elastic modulus (~80 GPa) can contribute to reducing stress shielding, which is the reason for the biomechanical incompatibility between the pure Ti-based implant and the bone [10].

Currently, more sophisticated and versatile methods/tools for surface engineering and synthesis of nanoscale facilities for biomedical applications are needed. In traditional chemical and physical methods, although the microstructure can be controlled to some extent, the specific shape and size in some particular applications cannot be precisely controlled. Most of the traditional methods have disadvantages, such as low efficiency, high cost and difficult-to-machine. By contrast, laser processing can easily form and control desired complicated topography with high resolution and high economic efficiency [11,12]. Techniques based on laser surface modification have advantages such as less debris contamination, reproducibility, precision and minimal heat-affected zone, which can create a controlled surface texture in specific, localized areas in a short time period [13]. Femtosecond laser texturing is a mask-free contactless technology, fully adaptable for 2D and 3D shapes on almost all materials [14]. The surface modification of the biomaterial has aimed to create specific chemical and physical properties that offer a favorable cellular response [15,16]. On the other hand, ultrafast laser surface modification is a unique method, which allows production of bioactive surface with formation of the desired oxide and alloy, creation of nano/micro textures and change wettability of the surface. Surface topographies with defined micro- and nanometer features generated by irradiation with linearly polarized laser radiation are well known as a laser-induced periodic surface structure (LIPSS). These surface structures have appeared in two forms as low spatial frequency LIPSS (LSFL) and high spatial frequency LIPSS (HSFL) [17]. The LSFLs usually have a spatial period close to the irradiation wavelength, oriented perpendicularly to the polarization vector. LSFL is generally accepted to originate from the interaction of the incident laser beam and electromagnetic wave scattered at the surface with the possibility to involve the excitation surface plasmon polariton. On the other hand, the HSFLs, oriented parallel to the polarization vector, are characterized with significantly lower periodicity than the irradiation wavelength. Several physical mechanisms have been proposed to explain the formation of HSFLs induced by laser pulses, such as self-organization, second harmonic generation, excitation of surface plasmon polaritons, and Coulomb explosion, but none have been proven fully yet [18]. The initial interactions between osteoblasts and nano-modified polymeric, ceramic and metallic substrates have indicated that nanoscale roughness can significantly affect cell adhesion, proliferation, and spreading. A few studies have indicated that increased osteoblast proliferation on the nanostructured surfaces coincided with an increase in alkaline-phosphatase synthesis, increased Ca-containing mineral deposition, and higher immunostaining of osteocalcin and osteopontin [19]. Recently, it has been demonstrated that osteointegration can be accelerated if the titanium surface is pre-treated to have a specific topography containing both micro- and nano-scale features [13]. In vivo experiments have shown that titanium and zirconium implants exhibit good osteointegration and that both elements have a high degree of bone-implant contact [20,21].

The aim of this paper is to study the relationship between laser processing and osteoblast-like cell response on titanium–zirconium multilayer thin films. The formation of an ordered surface line-texture with micro and nanometer features was achieved by optimizing the laser parameters, including applied laser fluences, number of pulses, and scanning rate. The micro/nano patterns distribution with changed composition and surface topography is evaluated in aim to determining their influence on cell adherence, cell morphology and the possibility of cell orientation along the laser-induced periodic structure.

2. Materials and Methods

The titanium–zirconium multilayer thin films were deposited in a Balzers Sputtron II system, using 1.3 keV argon ions and 99.9% pure Ti and Zr targets. Before deposition, the chamber was evacuated to the base pressure of 1×10^{-6} mbar, while the Ar partial pressure during deposition was 1×10^{-3} mbar. Silicon wafer Si (100) is used as substrate, which was cleaned by etching in HF and immersion in deionized water before mounting in the chamber. The deposition of multilayers was performed in a single vacuum run, at deposition rate of 0.17 nm s^{-1} for both Ti and Zr components, without heating of the substrates. The complete multilayer structure consisted of 30×(Ti/Zr) bilayers with total thickness of 1 µm, where thickness of individual Ti and Zr layers were about 17 nm.

Ultrafast laser modification of the multilayer 30×(Ti/Zr) thin films was performed with the Yb:KGW laser source Pharos SP from Light Conversion. The surface of thin films was irradiated by focused linearly p-polarized pulses with the following characteristics: repetition rate of 1 kHz, pulse duration of 160 fs, central wavelength of 1030 nm and 43 µm Gaussian spot diameter in focus. Samples were laser processed in an open air ambient environment and mounted on a motorized, computer-controlled, X-Y-Z translation stage, at normal incidence to the laser beam. The irradiations were conducted at identical conditions, creating lines on the surface of 5 mm × 5 mm at laser pulse energy of 60 µJ. In each line, energy per pulse was assumed to be constant, since the pulse energy deviation was less than 1%. The irradiation of the samples was performed in defocus mode by positioning the samples out of focus at 4 cm, thereby a pulse fluence of 0.4 J cm^{-2} with a spatial extension of the Gaussian beam profile was achieved. Indirectly, under the given laser irradiation conditions, the width of line with value of 75 µm was adjusted by defocusing of the laser beam, while distances between lines keep constant at value of 80 µm. The lines were direct laser writing at scan velocity of 2 mm s^{-1} (Figure 1).

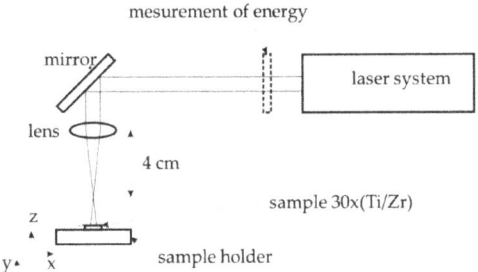

Figure 1. Schema of the experimental setup for laser processing.

Detailed surface morphology after irradiation was examined firstly by optical microscopy, and then by scanning electron microscopy (JEOL JSM-7500F, Tokyo, Japan) equipped with energy-dispersive X-ray spectroscopy (EDS) (High Wycombe, UK). A fracture cross section of the irradiated sample was performed by FEI SCIOS 2 microscopy, Hillsboro, OR, USA. The native cross-section of the sample was simply prepared by made fracture transversely to the laser-written lines. Phase composition and crystal structure analysis with X-ray diffraction (XRD), where the Cu Kα XRD pattern was collected by a Bruker D8 Advance Diffractometer, Karlsruhe, Germany. For these measurements, the Bragg–Brentano geometry with a step of 0.05° and the time interval of 30 s per step was used. The distribution of elements and interfaces of the multilayer 30×(Ti/Zr)/Si system were determined by time-of-flight SIMS instrument (Ion-TOF, Loughborough, UK). For composition depth profiling high energy Ga$^+$ pulsed primary source (25 keV) was combined with low energy sputter guns at 2 keV (Cs$^+$ and O^{2+}) at angle of 45° to sample surface.

Cell study under static in vitro conditions was performed with the mouse mesenchymal stem cells (MSCs) line (C57BL/6). MSCs cells were grown in cell culture flasks using Dulbecco's modified Eagle's medium [DMEM (Invitrogen, Grand Island, NY, USA) supplemented with 10% fetal bovine serum (FBS, Biosera, Sussex, UK)] in a 5% CO_2 incubator (Thermo Scientific, Waltham, MA, USA) at 37 °C. Laser-processed samples were autoclaved and transferred into sterile wells of 24 well plates (Sarstedt; Numbrecht, Germany). Culture medium with 2×10^4 cells were seeded onto the samples, where they were cultured in different time periods, ranging from one to three days in order to estimate the cell orientation, adhesion and proliferation. After each time points, the medium with MSCs cells were removed and the cultured samples were washed twice with 0.1 M sodium cacodylate buffer (SCB) and fixed with 2% glutaraldehyde (GDA) and 2% paraformaldehyde (PFA) in 0.1 M SCB for 30 min. Subsequently, samples were washed twice with 0.1 M SCB and dehydrated in increasing concentrations (from 30%–100%) of ethanol. Finally, before the determination of MSCs osteoblast cell morphology, samples were dried in a critical point drier (Baltec CPD 030, London, UK), sputter-coated with thin (10 nm) gold/palladium layer (Baltec SCD 050) and observed by scanning electron microscope (JEOL JSM-6390 LV). The cytoskeleton, focal adhesion points and nucleus of MSCs were stained for actin filaments, vinculin and DAPI. Specifically, after one and three days of cell cultured, the samples were fixed with 4% paraformaldehyde (PFA) for 15 min and permeabilized with 0.1% Triton X-100 in phosphate-buffered saline (PBS) for 5 min. The non-specific binding sites were blocked with 2% bovine serum albumen (BSA) in PBS for 30 min. Cell imaging was performed using a Leica SP8 inverted scanning confocal microscope (Jena, Germany) with ×40 objective.

3. Results and Discussion

The simultaneous creation of micro and nanometer morphological features on the surface of the multilayer Ti/Zr structure for biomedical application was achieved by laser writing of lines with a relatively wide laser beam (75 μm). Morphological shapes in micrometer dimensions consisted of 65 laser-written lines on a surface of 5 mm × 5 mm, whereby the width of each line was approximately 40 μm and the distance between them is also 40 μm (Figure 2a) as a consequence of the Gaussian shape of the laser beam profile. In each laser-written line, the creation of the laser-induced periodic surface structure (LIPSS) occurred, so that they were oriented normally to the direction of laser polarization (Figure 2b,c). These ripple structures are attributed to well-defined low spatial frequency LIPSS (LSFL) as a result of an interference of the incident laser beam with a surface electromagnetic wave excited during the laser irradiation with the possibility to involve the excitation surface plasmon polariton [22,23]. Applying a laser fluence of about 0.4 J cm^{-2}, which is slightly higher than the ablation threshold for the Ti/Zr system (0.22 J cm^{-2}), the LIPSS formation was accompanied by laser ablation of the multilayer 30×(Ti/Zr) thin films, as a consequence of the multi-pulse effect occurring between successive pulses [24]. LSFL ripples were not smooth, indicating the absence of any hydrodynamic effects during laser processing, but rather the dominant direct removal of thin film materials. The fragments were retained after laser ablation at the top of the LSFL ripples, where their appearance could indicate an unequal erosion of the layered structure (Figure 2c). In the zone of laser-written lines, the regular LSFL ripples appeared with a periodicity of approximately 800 nm. At given laser fluence, the ripple length ranges from quite small under 1 μm, to those whose length exceeds 7 μm (Figure 2c). In the cracks between LSFL ripples, the formation of high spatial frequency LIPSS (HSFL) was observed, which could be related to the decrease in energy distributed within the multilayer 30×(Ti/Zr) structure.

On the other hand, the HSFL ripples were formed in the zone between the two laser-written lines as a result of the action and overlapping of the edge Gaussian profile, where the energy is expected to be quite low, below the ablation threshold (Figure 2d,e). These HSFL ripples were oriented perpendicular to LSFL ripples with periodicity close to 200 nm, wherein the width of the individual HSFL ripple was below 100 nm. No ablation of the thin film materials was observed in this zone, so their formation can be attributed to the oxide phase upgrade or inducing the dewetting process [18,25,26].

On that occasion, the absorbed laser energy was induced the surface instabilities due to softening and perturbation of crystal and chemical bindings, as an endorsement of the model of self-organized structure formation. [18]. The solid-state dewetting can be induced during the laser processing as spontaneous regrouping of surface material into small-sized ripples (similar to nanorods) on the hot surface but at temperatures lower than the Ti melting point (Ti is top layer) [27]. Moving away from the laser-written line, the HSFLs gradually disappeared but not completely (Figure 1e), as they became pronounced again from the middle of unmodified area, approaching to the next line.

For the first rough check of the composition changes after laser treatment of 30×(Ti/Zr) system at the pulse fluence of 0.4 J cm^{-2}, the EDS method was used. Despite the used instrument analytical accuracy of $2\sigma = 2\%$, the obtained results should be taken as semi-quantitative because of the small thickness of 30×(Ti/Zr) multilayer system [23,24]. The elemental composition recorded by the EDS method at three different locations, in the center of the laser-written line, at the edge of this line and in the area between the lines, was compared in the spectra presented in Figure 3. Based on the obtained spectra, it can be concluded that there are no drastic changes in the concentrations for Ti, Zr, Si and O components, between the observed areas. It could be distinguished that laser ablation of the Ti/Zr multilayer was registered in the area of the laser-written line, whereby the zirconium was more removed than titanium and concentration of silicon increased in this area. On the other hand, surface oxidation could be observed as an effect of laser processing of the Ti/Zr system, which was reflected in the increasing of the oxygen concentration, especially in the area of the laser-written lines.

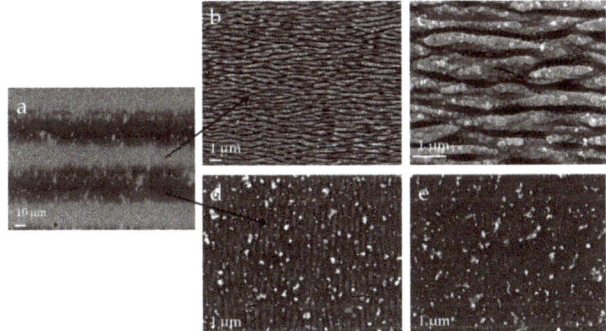

Figure 2. Scanning electron microscope (SEM) microphotographs of the 30×(Ti/Zr)/Si multilayer system after laser processing at laser fluence of 0.4 J cm^{-2}; (**a**) view of laser-written lines at magnification of 500×, (**b**,**c**) view of morphology inside of line with magnification 5000× and 20,000×, respectively and (**d**,**e**) view of morphology between lines at 20,000× magnification.

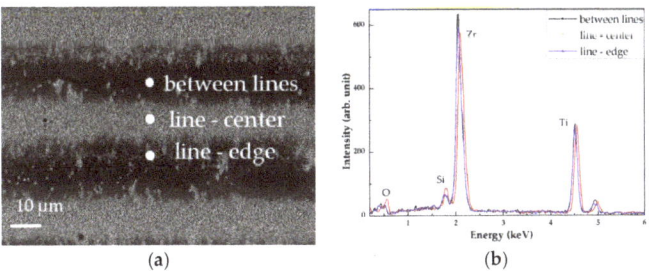

Figure 3. SEM microphotographs and energy-dispersive X-ray spectroscopy (EDS) analysis of the 30×(Ti/Zr)/Si multilayer system after laser processing at laser fluence of 0.4 J cm^{-2}; (**a**) view of lase-written lines at magnification of 500× with the positions of the recorded EDS spectra, and (**b**) EDS spectra.

A cross-section view of 30×(Ti/Zr)/Si multilayer system was obtained with SEM microscopy on the broken sample, including the unmodified part and the irradiated area in a place normal to the laser-written lines (Figure 4). Inside the as-deposited 30×(Ti/Zr) multilayer thin film were very well separated and alternately arranged Ti and Zr layers with almost identical thicknesses of individual layers (Figure 4a). The SEM cross-section view of 30×(Ti/Zr)/Si multilayer after laser processing in the zone of the laser-written lines (Figure 4b–d), confirms that the creation of periodical structure in form of LSFL ripples was accompanied by ablation of the material. However, laser ablation was quite unequal by making the number of the removed layers variable and ranged in interval of 8–15 layers (130–250 nm) when viewed at the top of the ripples. The depth of cracks between LSFL ripples were about 20 layers (350 nm), while their repetition period was about 800 nm (Figure 4b), which is consistent with the periodicity of the formed LSFLs. The spatial distribution of Ti and Zr components inside the laser-modified area retained a layered structure, with barely noticeable expansion of the interfaces between the Ti and Zr layers. Also, slight deformation or corrugating of the inner Ti and Zr layers could be observed, most likely by inducing internal stresses during the laser irradiation (Figure 4d).

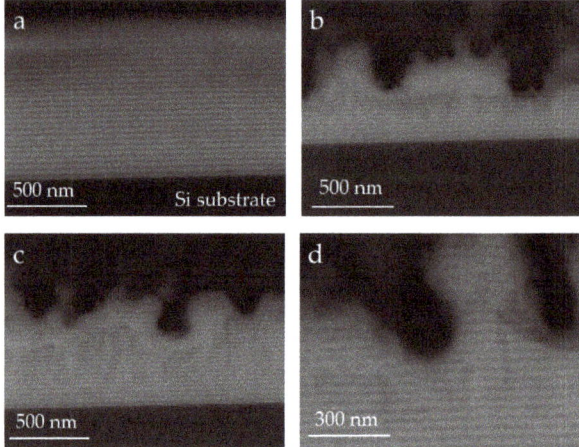

Figure 4. SEM cross-section view of the native broken 30×(Ti/Zr)/Si multilayer sample: (**a**) unmodified part at magnification of 120,000×, (**b**) and (**c**), the irradiated area in a place normal to the laser-written lines at magnification of 120,000×, and (**d**) part of laser-modified sample at 200,000× magnification.

The concentration depth profiles recorded by SIMS technique for the as-deposited and laser processed multilayer 30×(Ti/Zr)/Si system are presented in Figure 5. Results obtained by SIMS analysis of almost half of the as-deposited Ti/Zr multilayer structure showed that the Ti and Zr layers were very well separated with clearly defined interfaces between them (Figure 5a). After femtosecond laser processing of Ti/Zr multilayer structure, the distribution of the components could not be accurately determined, since the roughness of the laser-modified surface was quite high. The differences in height greater than 300 nm were reached, especially at the crack locations between LSFL ripples. In these positions the Si substrate was relatively close to the surface, which can be reason for the appearance of Si component in the spectrum (Figure 5b). In addition, it can be assumed that there was a possible diffusion of Si atoms from the substrate into a thin layer during laser irradiation [25]. The main components Ti and Zr were well intermixed, which is inconsistent with the result obtained by the SEM cross-section analysis, where the layered structure is retained after modification (Figure 5b). These differences can be attributed to the fact that the signal in SIMS analysis was taken from a large sample surface, regardless of the morphological characteristics on it. However, the presence of oxide phases after laser modification could be determined by SIMS analysis with high accuracy. Laser-induced surface oxidation was reflected in the formation of ultra-thin oxide layers composed from both Ti and Zr oxide phases. The sub-surface layers of least the first few layers (4–5 layers based on the SEM cross-section image) were intermixed and both oxide phases (Ti-oxide and Zr-oxide) existed on the contact surface, which was important for studying the cellular response. On the other hand, the pure oxygen signal was not registered in the spectrum, indicating that all the penetrated oxygen in the multilayer structure was in the form of compounds (oxides).

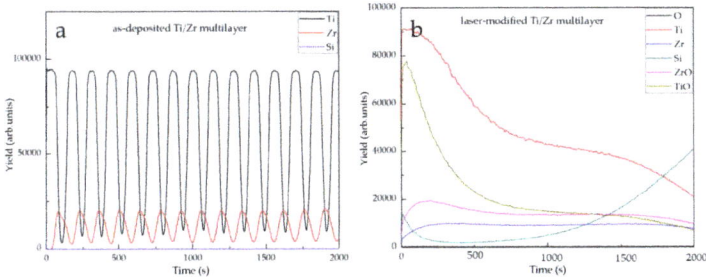

Figure 5. SIMS spectra before (**a**) and after (**b**) laser modification of 15×(Ti/Zr)/Si multilayer thin film.

Phase composition and crystal structure for as-deposited and laser-treated multilayer 30×(Ti/Zr)/Si system were determined by the XRD technique, as presented in Figure 6. All diffraction lines for Ti and Zr components were overlapped, because these elements were situated in the same group of periodic system one under another. In XRD pattern for as-deposited 30×(Ti/Zr)/Si multilayer system was identified both α-Ti and β-Ti phases (Figure 6a) at the following crystalline orientations α-Ti(100), β-Ti(110), α-Ti(102) and β-Ti(200) [28]. In this case, zirconium played a role as a β-stabiliser element, which induced the formation of β-Ti phases during the deposition of thin Ti and Zr layers [4]. Comparing the XRD patterns obtained for as-deposited and after laser processing of Ti/Zr multilayer thin film, it was observed that diffraction lines did not change positions. However, the intensities of these diffraction lines are changed, indicating that laser processing favored texture with dominant α-Ti(102) and β-Ti(200) crystalline orientation (Figure 6b).

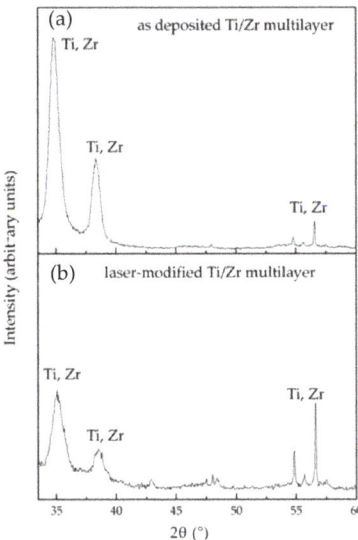

Figure 6. X-ray diffraction (XRD) patterns before (**a**) and after (**b**) laser modification of 15×(Ti/Zr)/Si multilayer thin film.

Before cell seeding on laser-processed 30×(Ti/Zr)/Si systems, the wettability was estimated by measuring the contact angle for as-deposited and laser-modified samples (Figure 7). In the experiment, estimation of the contact angle was determined the angle between the tangent of the solid surface and the tangent to the liquid (water) at the contact line among the three phases. The contact angle measurement included repeating the measurement five times, with an accuracy of 5%, for each surface just before seeding the cell culture. The contact angle for laser-modified Ti/Zr multilayer was greatly increased up to 136° value, which contributed to the achievement of a moderately hydrophobic surface. This increase in contact angle from the value of 82° for as-deposited sample was attributed to the formed surface topography in form of LIPSS with micro- and nanometer features, but also to the formation of Ti and Zr oxides on the contact surface [29]. Surface wettability can influence protein adsorption by controlling the total amount of proteins bound to the material, as well as their conformation and orientation after adsorption. The interaction between cells and Ti-based alloy can be considered through the dynamics of proteins when adsorbed onto the material surfaces. Proteins are structurally and chemically asymmetric, and during their adsorption the specific orientation and conformation were determined, defining the domain of the molecule that will interact and attach with the material [30].

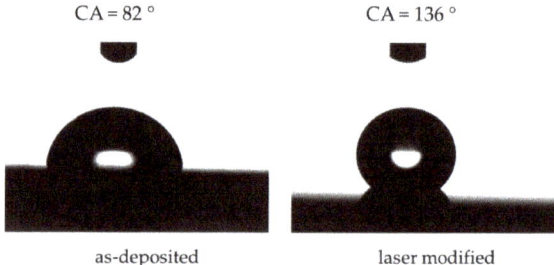

Figure 7. Contact angle measurements for as-deposited and laser modified surfaces of 30×(Ti/Zr) multilayer thin film.

The biocompatibility of the laser-modified Ti/Zr multilayers is reflected in cell adhesion and proliferation through contact, adhesion, and spreading in the initial phase of the cell-material interaction. Therefore, the laser-created surface topography with nano- and micro features plays a key role in further cellular behavior [31]. Morphological characteristics of MSCs cell proliferation on unmodified and laser-modified 30×(Ti/Zr)/Si multilayer surfaces was estimated by SEM and confocal analysis after one and three days' cultivation. After one-day cultivation on the surface of both unmodified and laser patterned Ti/Zr multilayers, the MSCs cells showed very good adhesion (Figure 8). On the as-deposited 30×(Ti/Zr)/Si sample with flat surfaces, there were easily visible cell groups an arbitrary cell growth occurring in all directions (Figure 8a,b), after one-day cultivation. On the other hand, under the same experimental conditions, on the 30×(Ti/Zr)/Si multilayer with laser-induced surface ripple morphology, the MSCs cells adhered, with a tendency for growth along the ripples' orientation (Figure 8c,d). In addition to the evident directed growth of osteoblast-like cells on a laser-created surface topography, it can be observed that the ripple morphology induced better communication between cells, due to significantly elongated cell groups with the aim to connect (Figure 8c,d). Statistically, a significantly larger number of cells per surface (250 μm × 250 μm) were attached on the laser-processed Ti/Zr multilayer (~80 cells) than to the flat as-deposited sample (~25 cells).

The proliferation of MSCs cells was achieved significantly, especially after three days when the number of cells was increased. Where almost whole surfaces were covered by cells for both as-deposited (flat surface) and laser-modified (ripple morphology) 30×(Ti/Zr)/Si samples after three-days cultivation (Figure 9). The number of cells for laser-modified 30×(Ti/Zr)/Si samples were about 220 cells per surface of 250 μm × 250 μm, while a slightly smaller number of cells about 170 cells were present on a flat surface. Cell metabolic activities, determined by the MTT method including the measurement of the optical density at 545 nm, were found to result in approximately similar but statistically significant values (~42%) for both laser-modified and unmodified samples. It could be observed that the surface topography including micrometer sized laser-written lines with nanometer sized ripples did not influence on the cell proliferation rate. These facts can contribute to the conclusion that Ti/Zr multilayer systems have satisfactory biocompatibility regardless of surface conditions (Figure 9a,c). However, after three-day cultivation, the laser-modified surface was covered with a slightly larger number of smaller cells compared to as-deposited 30×(Ti/Zr)/Si multilayer, which was visible based on the cell cytoskeleton (Figure 9b,d). One of the most interesting findings is that osteoblast-like cells were oriented in all directions; they even had some kind of radial orientation in certain areas, in case of the as-deposited sample (Figure 9b). However, on the surface topography with ripples had a cell tendency to grow towards the ripples direction, with clear proliferation between one to three days (Figure 9d). Moreover, this study of the 30×(Ti/Zr)/Si multilayer demonstrated that in micro-line features, the ridge width is commonly larger than or equal to the size of a single cell, permissive for cell attachment and migration, as well as cell alignment following the geometrical guidance. In contrast, nano-ripple features are similar to the ECM (extracellular matrix) architectures and typically much smaller than a single cell, consequently inducing cell alignment in a more fundamental way such as mimicking or signaling the cell membrane receptors [32].

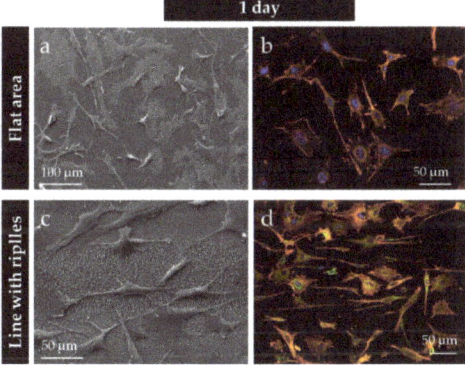

Figure 8. SEM and fluorescent images of mesenchymal stem cells (MSCs) osteoblast cultivated on the as-deposited (**a**,**b**), and laser-processed 30×(Ti/Zr)/Si multilayer thin film (**c**,**d**), for one-day cultivation, respectively.

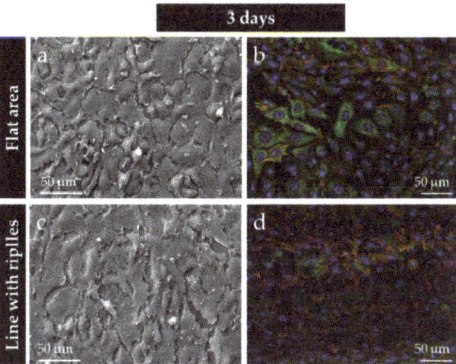

Figure 9. SEM and fluorescent images of MSCs osteoblast cultivated on the as-deposited (**a**,**b**), and laser-processed 30×(Ti/Zr)/Si multilayer thin film (**c**,**d**), for three-day cultivation, respectively.

4. Conclusions

The surface functionalization of the Ti-based alloy in terms of improving the osteoblast cell response was achieved by the deposition of a Ti/Zr multilayer structure and ultrafast laser processing. Adjusting the morphology and composition of the physical vapor-deposited 30×(Ti/Zr)/Si multilayer structure was achieved at optimal combination of laser parameters. The micro- and nanometer morphological features were obtained by laser writing micrometer sized lines on a relatively large 5 mm × 5 mm surface. In each laser-written line were created the laser-induced periodic surface structure defined as a low spatial frequency LIPSS (LSFL) with the periodicity close to 800 nm oriented normal to the laser polarization. While the space between the lines was filled with high spatial frequency LIPSS (HSFL) oriented perpendicular to LSFL ripples with periodicity of about 200 nm. The desired chemical composition required for good biocompatibility of Ti-based alloys was provided by the formation of a very thin oxide layer composed of Zr and Ti oxides, as well as intermixing of these components in the sub-surface region.

The biocompatibility of the laser-processed 30×(Ti/Zr)/Si multilayers was confirmed by the cultivation of a MSC-established adherent mouse osteoblast cell line. The osteoblast cells adhered and proliferated on the 30×(Ti/Zr)/Si multilayers, regardless of the fact that the samples were pre-laser treated to form specific surface topographies. However, after one- and three-day cultivation, the osteoblast cells showed a growth along ripples with a tendency to connection via their elongated parts, in the case of laser-patterned samples. Bioactivation of this specific 30×(Ti/Zr)/Si multilayer system with laser surface texturing and adjusting of surface composition could be potentially useful for tissue engineering and the application of this material as an implant.

Author Contributions: The contributions by author are as follows: conceptualization, S.P.; methodology, D.P., E.S. (Evangelos Skoulas), and Z.R.; formal analysis, J.K., M.M., J.P. and Z.R.; investigation, S.P. and D.P.; writing—original draft preparation, S.P.; writing—review and editing, S.P.; supervision, E.S. (Emmanuel Stratakis).

Funding: This research was funded by the EU-H2020 research and innovation program under grant agreement N 654360 having benefitted from the access provided by Foundation for Research and Technology Hellas (FORTH) access provider (Institute of Electronic Structure and Lasers i.e., Institution) in Heraklion, Crete, Greece within the framework of the NFFA-Europe Transnational Access Activity. This research was sponsored by the Ministry of Education, Science and Technological Development of the Republic Serbia through projects No. OI 171023, III 45016 and III45005.

Conflicts of Interest: The funders had no role in the design of the study; in the collection, analyses, or interpretation of data; in the writing of the manuscript, or in the decision to publish the results.

References

1. Othman, Z.; Cillero Pastor, B.; Van Rijt, S.; Habibovic, P. Understanding interactions between biomaterials and biological systems using proteomic. *Biomaterials* **2018**, *167*, 191–204. [CrossRef] [PubMed]
2. Su, Y.; Luo, C.; Zhang, Z.; Hermawan, H.; Zhu, D.; Huang, J.; Liang, Y.; Li, G.; Ren, L. Bioinspired surface functionalization of metallic biomaterials. *J. Mech. Behav. Biomed. Mater.* **2018**, *77*, 90–105. [CrossRef] [PubMed]
3. Bose, S.; Robertson, S.F.; Bandyopadhyay, A. Surface modification of biomaterials and biomedical devices using additive manufacturing. *Acta Biomater.* **2018**, *66*, 6–22. [CrossRef] [PubMed]
4. Correa, D.R.N.; Kuroda, P.A.B.; Lourenço, M.L.; Fernandes, C.J.C.; Buzalaf, M.A.R.; Zambuzzi, W.F.; Grandini, C.R. Development of Ti-15Zr-Mo alloys for applying as implantable biomedical devices. *J. Alloy. Compd.* **2018**, *749*, 163–171. [CrossRef]
5. Rack, H.J.; Qazi, J.I. Titanium alloys for biomedical applications. *Mater. Sci. Eng. C* **2006**, *26*, 1269–1277. [CrossRef]
6. Donato, T.A.G.; De Almeida, L.H.; Nogueira, R.A.; Niemeyer, T.C.; Grandini, C.R.; Caram, R.; Schneider, S.G.; Santos, A.R., Jr. Cytotoxicity study of some Ti alloys used as biomaterial. *Mater. Sci. Eng. C* **2009**, *29*, 1365–1369. [CrossRef]
7. Zhao, D.; Chen, C.; Yao, K.; Shi, X.; Wang, Z.; Hahn, H.; Gleiter, H.; Chen, N. Designing biocompatible Ti-based amorphous thin films with no toxic element. *J. Alloy. Compd.* **2017**, *707*, 142–147. [CrossRef]

8. Hsu, H.C.; Wu, S.C.; Sung, Y.C.; Ho, W.F. The structure and mechanical properties of as-cast Zr-Ti alloys. *J. Alloy. Compd.* **2009**, *488*, 279–283. [CrossRef]
9. Li, Y.; Yang, C.; Zhao, H.; Qu, S.; Li, X.; Li, Y. New developments of Ti-based alloys for biomedical applications. *Materials* **2014**, *7*, 1709–1800. [CrossRef]
10. Aristizabala, M.; Jamshidib, P.; Sabooric, A.; Coxd, S.C.; Attallah, M.M. Laser powder bed fusion of a Zr-alloy: Tensile properties and biocompatibility. *Mater. Lett.* **2020**, *259*, 126897. [CrossRef]
11. Yu, Z.; Yang, G.; Zhang, W.; Hu, J. Investigating the effect of picosecond laser texturing on microstructure and biofunctionalization of titanium alloy. *J. Mater. Process. Tech.* **2018**, *255*, 129–136. [CrossRef]
12. Sartinska, L.L.; Barchikovski, B.; Wagenda, N.; Rut, B.M.; Timofeeva, I.I. Laser induced modification of surface structures. *Appl. Surf. Sci.* **2007**, *253*, 4296–4299. [CrossRef]
13. Frostevarg, J.; Olsson, R.; Powell, J.; Palmquist, A.; Branemark, R. Formation mechanisms of surfaces for osseointegration on titanium using pulsed laser spattering. *Appl. Surf. Sci.* **2019**, *485*, 158–169. [CrossRef]
14. Berg, Y.; Kotler, Z.; Shacham-Diamand, Y. Holes generation in glass using large spot femtosecond laser pulses. *J. Micromech. Microeng.* **2018**, *28*, 035009. [CrossRef]
15. Jenko, M.; Gorensek, M.; Godec, M.; Hodnik, M.; Setina-Batic, B.; Donik, C.; Grant, J.T.; Dolinar, D. Surface chemistry and microstructure of metallic biomaterials for hip and knee endoprostheses. *Appl. Surf. Sci.* **2018**, *427*, 584–593. [CrossRef]
16. Simitzi, C.; Ranella, A.; Stratakis, E. Controlling the morphology and outgrowth of nerve and neuroglial cells: The effect of surface topography. *Acta Biomater.* **2017**, *51*, 21–52. [CrossRef]
17. Gregorcic, P.; Sedlacek, M.; Podgornik, B.; Reif, J. Formation of laser-induced periodic surface structures (LIPSS) on tool steel by multiple picosecond laser pulses of different polarizations. *Appl. Surf. Sci.* **2016**, *387*, 698–706. [CrossRef]
18. Varlamova, O.; Costache, F.; Reif, J.; Bestehorn, M. Self-organized pattern formation upon femtosecond laser ablation by circularly polarized light. *Appl. Surf. Sci.* **2006**, *252*, 4702–4706. [CrossRef]
19. Gittens, R.A.; Lachlan, T.M.; Olivares-Navarrete, R.; Cai, Y.; Berner, S.; Tannenbaum, R.; Schwartz, Z.; Sandhage, K.H.; Boyan, B.D. The effects of combined micron-/submicron-scale surface roughness and nanoscale features on cell proliferation and differentiation. *Biomaterials* **2011**, *32*, 3395–3403. [CrossRef]
20. Stanciuc, A.M.; Flamant, Q.; Sprecher, C.M.; Alini, M.; Anglada, M.; Peroglio, M. Femtosecond laser multi-patterning of zirconia for screening of cell-surface interactions. *J. Eur. Ceram. Soc.* **2018**, *38*, 939–948. [CrossRef]
21. Ohtsu, N.; Kozuka, T.; Yamane, M.; Arai, H. Surface chemistry and osteoblast-like cell response on a titanium surface modified by a focused Nd: YAG laser. *Surf. Coat. Technol.* **2017**, *309*, 220–226. [CrossRef]
22. Bonse, J.; Koter, R.; Hartelt, M.; Spaltmann, D.; Pentzien, S.; Hohm, S.; Rosenfeld, A.; Krüger, J. Tribological performance of femtosecond laser-induced periodic surface structures on titanium and a high toughness bearing steel. *Appl. Surf. Sci.* **2015**, *336*, 21–27. [CrossRef]
23. Kirnera, S.V.; Wirth, T.; Sturm, H.; Krüger, J.; Bonse, J. Nanometer-resolved chemical analyses of femtosecond laser-induced periodic surface structures on titanium. *J. Appl. Phys.* **2017**, *122*, 104901. [CrossRef]
24. Može, M.; Zupančič, M.; Hočevar, M.; Golobič, I.; Gregorčič, P. Surface chemistry and morphology transition induced by critical heat flux incipience on laser-textured copper surfaces. *Appl. Surf. Sci.* **2019**, *490*, 220–230. [CrossRef]
25. Petrović, S.; Peruško, D.; Kovač, J.; Panjan, P.; Mitrić, M.; Pjević, D.; Kovačević, A.; Jelenković, B. Design of co-existence parallel periodic surface structure induced by picosecond laser pulses on the Al/Ti multilayers. *J. Appl. Phys.* **2017**, *122*, 115302. [CrossRef]
26. Oh, H.; Pyatenko, A.; Lee, M. Laser dewetting behaviors of Ag and Au thin films on glass and Si substrates: Experiments and theoretical considerations. *Appl. Surf. Sci.* **2019**, *475*, 740–747. [CrossRef]
27. Abbott, W.M.; Corbett, S.; Cunningham, G.; Petford-Long, A.; Zhang, S.; Donegan, J.F.; McCloskey, D. Solid state dewetting of thin plasmonic films under focused cw-laser irradiation. *Acta Mater.* **2018**, *145*, 210–219. [CrossRef]
28. JCPDS. *International Centre for Diffraction Data (ICDD), PCPDFWIN v. 2.00*; ICDD: Newtown Square, PA, USA, 1998.
29. Bizi-Bandoki, P.; Benayoun, S.; Valette, S.; Beaugiraud, B.; Audouard, E. Modifications of roughness and wettability properties of metals induced by femtosecond laser treatment. *Appl. Surf. Sci.* **2011**, *257*, 5213–5218. [CrossRef]

30. Tang, L.; Thevenot, P.; Hu, W. Surface chemistry influence implant biocompatibility. *Curr. Top. Med. Chem.* **2008**, *8*, 270–280. [CrossRef]
31. Anselme, K. Osteoblast adhesion on biomaterials. *Biomaterials* **2000**, *21*, 667–681. [CrossRef]
32. Babaliari, E.; Kavatzikidou, P.; Angelaki, D.; Chaniotaki, L.; Manousaki, A.; Siakouli-Galanopoulou, A.; Ranella, A.; Stratakis, E. Engineering Cell Adhesion and Orientation via Ultrafast Laser Fabricated Microstructured Substrates. *Int. J. Mol. Sci.* **2018**, *19*, 2053. [CrossRef] [PubMed]

© 2019 by the authors. Licensee MDPI, Basel, Switzerland. This article is an open access article distributed under the terms and conditions of the Creative Commons Attribution (CC BY) license (http://creativecommons.org/licenses/by/4.0/).

Article

Metallurgical Soldering of Duplex CrN Coating in Contact with Aluminum Alloy

Pal Terek [1], Lazar Kovačević [1], Aleksandar Miletić [1,2], Branko Škorić [1,*], Janez Kovač [3] and Aljaž Drnovšek [3]

1. Faculty of Technical Sciences, University of Novi Sad, Trg Dositeja Obradovića 6, 21000 Novi Sad, Serbia; palterek@uns.ac.rs (P.T.); lazarkov@uns.ac.rs (L.K.); aleksandar.miletic@polymtl.ca (A.M.)
2. Department of Engineering Physics, Polytechnique Montreal, Montreal, QC H3T 1J4, Canada
3. Jožef Stefan Institute, Jamova 39, 1000 Ljubljana, Slovenia; janez.kovac@ijs.si (J.K.); aljaz.drnovsek@ijs.si (A.D.)
* Correspondence: skoricb@uns.ac.rs; Tel.: +381-21-485-2342

Received: 17 February 2020; Accepted: 20 March 2020; Published: 24 March 2020

Abstract: Coatings deposited by physical vapor deposition (PVD) significantly reduce the wear of high pressure die casting tools; however, cast alloy soldering still has a strong negative effect on production efficiency. Although a lot of research has been already done in this field, the fundamental understanding of aluminum alloy soldering toward PVD coatings is still scarce. Therefore, in this work the performance of CrN duplex coatings with different roughness is evaluated by a modified ejection test performed with delayed (DS) and conventional casting solidification (CS). After the ejection tests, sample surfaces and layers were subjected to comprehensive characterizations of their morphological and chemical characteristics. Considerably lower values of the ejection force were recorded in DS experiments than in CS experiments. Surface roughness played an important role in the CS experiments, while samples with different surface topographies in the DS experiments performed in a similar fashion. The decrease in the ejection force, observed in DS tests, is attributed to the formation of a thick Cr–O layer on CrN coating which reduced soldering and sliding friction against thick Al–O casting scale. The Cr–O layer formed in DS experiments suffered from diffusion wear by cast alloy. The observed oxidation phenomena of nitride coatings may be utilized in a design of non-sticking coatings.

Keywords: die casting; aluminum; tool life; soldering; ejection test; CrN; surface roughness; oxidation; diffusion wear

1. Introduction

High pressure die casting (HPDC) is a technology used for the mass production of near-net shape parts of non-ferrous alloys, with thin walls and smooth surfaces. Due to the ever-increasing application of lightweight components in automotive products and other products, the use of HPDC technology for the production of aluminum alloy castings is constantly expanding. This kind of large volume production is economically justified only by highly efficient production of high-quality parts.

During operation die (tool) surfaces are exposed to wear by: erosion, corrosion and soldering, thermal cycling fatigue [1,2], and adhesion [3]. These processes affect the tool life and casting quality, but more importantly they increase the production costs due to increased: machine down times, number of rejected castings [3,4], energy and materials consumption. Aluminum alloys have the highest affinity toward iron contained in die steel materials and consequently induce the most pronounced soldering. To avoid cast alloy soldering before every casting cycle die-lubricant is sprayed on tool surfaces. The demands in terms of casting surface quality have been constantly increasing. In order to achieve this high surface quality, specific casting processing conditions, which significantly increase tool wear due to the increased erosion and soldering effects, are required.

A prominent approach in the reduction of tool wear and improvement of tool performance is the application of thin ceramic coatings which are deposited by physical vapor deposition (PVD) techniques [1–3,5]. When ceramic coating is applied, the tool erosion is suppressed due to coating high hot hardness; corrosion and soldering are reduced due to high inertness and thermal stability of ceramic coating materials [1,6,7]; and finally, resistance to thermal fatigue cracking is improved by applying coatings of increased toughness, such as coatings of nanocomposite and nanolayer design [1,5,8]. In order to attain all these properties in one coating, the system of coating layers have to be adequately designed, as proposed by Lin et al. in [1]. Owing to its relatively high hardness, high oxidation and corrosion resistance, thermal stability [9] and low stress [10], CrN is still one of the commercially most used coatings for protection of tools HPDC of aluminum alloys. However, in recent years chromium-based coatings such as CrAlN [1,11], AlCrN [5,12], CrN/AlN [13], Cr_2O_3 [1,13], AlCrSiN [5,14] of different design received a considerable attention in protection of HPDC tools.

Nowadays, the improvement of casting release (ejection) from a die and reduction of lubricant consumption have been catching a significant attention in HPDC industry [12,15]. Additionally, the reduction of cast alloy soldering on long die-cores is still a great technological challenge. Therefore, in recent years, numerous studies have been focused on a topic of application of PVD coatings on HPDC tools [5,6,12,16]. Paiva et al. [5] evaluated the performance of AlCrN and two nanocomposite coatings (AlTiN/Si_3N_4; AlCrN/Si_3N_4) in high temperature tribological tests and in real HPDC industrial production. They showed that the tool life was considerably improved by application of nanocomposite coatings. Nunes et al. [16] investigated $Ti_xAl_{x-1}N$ coatings with different aluminum content, they compared the coating behavior obtained in tribological tests with the behavior obtained in HPDC industrial trials. They revealed that $Ti_xAl_{x-1}N$ coating with higher aluminum content provides better protection for HPDC tools. Bobzin et al. [6] employed a rotating immersion test (laboratory tests) and HPDC trials for evaluation of a CrN/AlN/Al_2O_3 coating system and two commercial coatings. CrN/AlN/Al_2O_3 coating system exhibited very good behavior in both kinds of tests, however, its top Al_2O_3 layer suffered from thermal cracking caused by phase change. Wang et al. [12] investigated the soldering performance of several nitride coatings using the aluminum adhesion test (laboratory test) and HPDC plant trails without application of die lubricants. Based on their findings, they proposed AlCrN coating as the most optimal candidate for the lubricant-free die casting. They measured the lowest force for separation from a casting and revealed the lowest soldering tendency for this coating.

Generally, the cast alloy soldering phenomena are divided into mechanical soldering (sticking) and metallurgical soldering [17]. Since many PVD coatings used in HPDC industry are inert to molten aluminum alloys [6,7], metallurgical soldering effects are easy to overlook and are generally not well recognized by scientific community. Formation of a built-up layer in the contact of a cast alloy and coated tool surfaces is usually attributed to mechanical soldering [3,18]. Nevertheless, evidences of metallurgical phenomena are present in literature. For example, corrosion of underlying substrate in a contact with molten aluminum alloy occurs through coating growth defects [19,20]. In such a process, a casting hooks the coating which hampers the casting ejection and consequently causes coating detachment. To prevent this, Abusuilik B. Saleh proposed intermediate coating treatment [19], while sealing of the coating defects by atomic layer deposition of thin Al_2O_3 layer [20] is also a promising approach. Besides soldering through coating defects, cast alloy can firmly bond in a form of a built-up layer to smooth coated surfaces which is often called sticking. Such remnants of the cast alloy on coated surfaces indicate hampered casting ejection. We believe that this process occurs due to the chemical compatibility or inter-diffusion between paired materials, and it should be classified as coatings metallurgical soldering. In few recent works [3,12,16] this phenomenon was observed, however, answers about fundamental mechanisms behind it were neither given nor discussed. Metallurgical soldering has not even been addressed for CrN coating which is the most investigated coating for protection of HPDC tools. Further development of soldering (sticking) resistant PVD coatings requires thorough understanding of processes involved in interaction of the cast alloy with the coating. This was our primary motivation for performing the investigation presented herein.

Evaluation of performance of coating materials through practical experiments is time consuming, expensive, less controllable, and is highly limited in terms of studying of isolated, specific wear mechanisms (soldering, erosion, thermal fatigue) [21]. On the other hand, laboratory experiments are simple, quick, they provide isolation of a single wear mechanism, have high repeatability, and allow quantification of wear. One group of the laboratory tests used for evaluation of the soldering tendency, are separation tests [21]. These tests involve metal casting process for production of sample-casting assembly and a mechanical process of separation of a sample from a casting. In this way, most important processes that lead to cast alloy soldering and formation of a built-up layer (galling) are simulated. Therefore, these tests are considered the most appropriate for soldering evaluation.

A separation test used for evaluation of cylindrical pin samples is known as an ejection test. The improved ejection test, developed in our previous work [21], is simple to perform, however, it simulates only one casting cycle. In this test, the time cast alloy spends in contact with a pin sample is very short and therefore mainly mechanical soldering effects are simulated [21].

One approach to induce metallurgical soldering mechanisms is repetition of ejection tests in hundreds, or thousands of casting cycles which is highly impractical. The other, less time consuming and less expensive approach is extension of the time a molten metal stays in contact with coated surfaces. With the goal to enhance metallurgical soldering effects, and to have more comprehensive understanding of processes occurring in the contact of coated tool surfaces with molten aluminum alloys, in this investigation the ejection test was modified by adopting the latter approach, i.e., by introducing a delayed casting solidification. The performance of CrN duplex coatings in the contact with Al–Si–Cu alloy was studied. CrN was chosen as a model coating for this investigation because it is commonly applied in the field, it is simple for evaluation, and can help in understanding the soldering wear of other Cr-N-based coatings. CrN coating was prepared to different degrees of surface roughness. The obtained results are compared to results of our previous study in which conventional solidification methodology was used.

2. Materials and Methods

Soldering performance of duplex CrN coated core pins in contact with Al–Si–Cu alloy was evaluated by ejection test performed in two configurations. A standard configuration involved sting solidification, while the newly proposed configuration involved delayed casting solidification, performed with two periods of delay. Surface characterization of samples from both experiments is performed to explain the trends observed in quantitative data for both tests. In addition, coated samples were annealed in a separate experiment in order to have better understanding of heat-induced changes inside a CrN coating layer.

2.1. Samples Preparation

Substrates used in this investigation were produced from a quenched and double tempered EN X27CrMoV51 hot-working tool steel. Substrates were heated to 1000 °C and kept for 30 min for austenitizing, which was followed by oil quenching and double tempering (1 h at 620 °C and for 1 h at 500 °C). Substrates were machined by applying a sequence of procedures regularly used in production of HPDC tool parts, namely by turning, grinding and polishing. Samples were produced in two shapes, disc samples and cylindrical samples. Disc-shaped samples, with the dimensions ϕ25 × 5 mm, were produced for characterization of materials properties and annealing experiment. Cylindrical pin-shaped samples, with the dimensions of the working part ϕ15 × 100 mm, were used in ejection tests for evaluation of soldering tendency. For detailed drawing of pin samples please refer to our previous work [3]. Hardness obtained after quenching and double tempering of steel pin samples was 455 ± 41 HV_{30}.

CrN duplex composite layers were produced by plasma nitriding of steel substrates with subsequent deposition of CrN coating. Plasma nitriding was performed in a unit equipped with a pulsed plasma generator (ION-25I, IonTech, Sofia, Bulgaria). The nitriding process was performed in atmosphere with gas ratio of $H_2:N_2$ = 3:1 during a 12 h long processing cycle with 0.6 duty cycle.

In order to maximize adhesion of CrN layer a compound layer formed during nitriding was removed by polishing. The polishing procedure was performed using diamond paste with 3 µm and 6 µm granulations. As a result, two groups of samples (surfaces) of different roughness, were formed. Rough samples (R) were obtained by polishing with 6 µm paste, while smooth samples (S) were obtained by two-step polishing with 6 and 3 µm diamond paste.

Prior to coating deposition samples were degreased and cleaned in ultrasonic baths with deionized water and detergents and dried in hot air. CrN coating was deposited in an industrial thermionic arc ion plating system (BAI 730M, Balzers, Balzers, Liechtenstein) with the 2-fold samples rotation employed. More information on the employed deposition system might be found in [22]. The coating preparation process begun by heating samples in plasma to 450° which was followed by ion etching employing 200 V DC bias for 15 min. A thin Cr layer was deposited on substrates to improve CrN coating adhesion. During deposition, the bias voltage of −125 V was applied, while nitrogen partial pressure was kept constant. The typical deposition rate was around 50 nm/min. After the deposition, a group of smooth samples was submitted to additional polishing with 3 µm diamond paste. As a result, post deposition polished (PP) samples were obtained. Designation of all samples used in this study, procedures performed in their production and their roughness are presented in Table 1.

Table 1. Samples designations, production procedures and roughness parameters.

Group of Samples	Rough	Smooth	Post Polished
Sample name	CrN-R	CrN-S	CrN-PP
Nitriding	×	×	×
Substrate polishing after nitriding 3 µm	×	×	×
Substrate polishing after nitriding 6 µm		×	×
CrN coating	×	×	×
Polishing after coating deposition 6 µm			×
R_a [µm]	0.145	0.032	0.027
R_{sk}	−0.179	0.491	−1.162

2.2. Soldering Evaluation–Ejection Test

The employed ejection test is explained in detail in our previous works [3,21]. In this test a cylindrical pin-shaped sample is cast-in in an Al–Si–Cu alloy casting. In this way a pin-casting assembly is obtained. In the next step, the pin sample is ejected from the casting. During the ejection process a force-displacement diagram is recorded, with the maximum force representing a quantitative measure of the bonding strength between paired materials (pin and casting materials), i.e., their soldering tendency.

In this study, pin-casting assemblies were produced in two configurations (methods). In the first configuration the casting was performed by gravity pouring molten EN AC-46200 aluminum alloy, at temperature of 730 °C, into a specially designed steel die [3,21], preheated to temperature of 320 °C. After the die was filled, the casting is allowed to solidify. In the rest of the article this configuration will be referred as conventional solidification method. The second configuration is a modification of the previous, it was performed using the same experimental die. Before casting, the die with mounted sample was preheated in a furnace for 40 min to achieve a target temperature of 600 °C. After cast metal was poured, the die was placed into a furnace heated to 700 °C to delay the casting solidification for a predetermined time (5 min and 20 min). After the predetermined time, the die was taken out of the furnace and the casting was allowed to solidify. This test configuration is called delayed solidification method and it is illustrated in Figure 1. This procedure was developed to intensify and extend corrosion processes that occur between a casting and a pin sample.

The conventional solidification experiments were conducted with three times repetitions while the delayed solidification experiments were performed without repetitions. Abbreviations used for designation of different test configurations are: CS—conventional solidification, DS 5—delayed solidification for 5 min and DS 20—delayed solidification for 20 min.

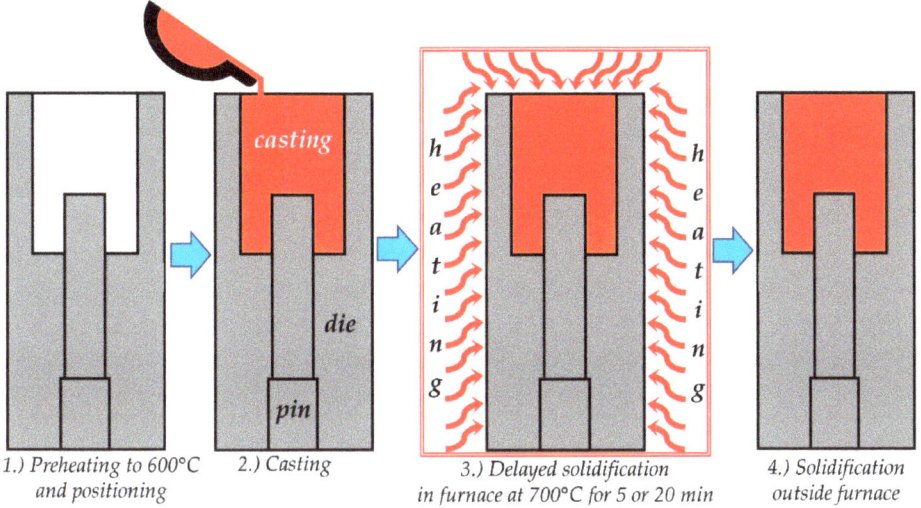

Figure 1. Schematic illustration of the casting method with the delayed solidification.

The temperature of die and sample surfaces were, for chosen conditions, determined in trial experiments by using exposed junction K-type thermocouples made from 0.25 mm wires (Omega Engineering Inc., Norwalk, Connecticut, USA) and by infrared thermometer. After taking the filled die out of the furnace, a casting cooling curve was recorded by immersing a K-type thermocouple casting's thermal axis.

A tensile testing machine (ZDM 5/91, VEB, Leipzig, Germany) was used for ejection of pins out of castings. More details can be found in our previous works [3,21]. During the ejection, a force-displacement curve was recorded (ejection curve). The force recorded during the test represents the soldering tendency of a cast alloy toward pin material. In the present study, the highest force recorded during the ejection test was chosen as a quantitative parameter for comparing the behavior of samples subjected to different experimental conditions.

2.3. Annealing Experiment

In order to evaluate the effect of high temperature on the structure and chemical composition of CrN duplex coating, annealing experiment was conducted. Disc shaped sample was placed in a laboratory tubular furnace preheated to 650 °C and kept there for 75 min in ambient air. The annealing time was chosen to equal the total time samples spend at high temperature in the 20 min delayed solidification test. In that test, samples are preheated for 40 min, solidification is delayed for 20 min, casting solidification lasted 15 min, which in total gives period of 75 min. The temperature of 650 °C was chosen as an average temperature to which the samples were exposed in DS 20 test.

2.4. Samples Characterization

Surface roughness was acquired by a stylus profilometer (Talysurf, Taylor Hobson, Leicester, United Kingdom). Instrumented hardness tester (H100C, Fischerscope, Windsor, CT, USA) was used for the determination of mechanical properties of nitrided layer and CrN coating. During the indentation tests, the load of 50 mN and 100 mN was applied. A hardness profile was acquired by making indentations with loads of 100 mN along the thickness of the plasma nitrided layer. This hardness depth profile was used to determine the nitrided case thickness, in compliance with EN ISO 2639:2002, and the maximal layer hardness. After the ejection tests, examination of sample surfaces and cross sections provided additional information about the material behavior and about the soldering processes. Focused ion beam (FIB) (Helios Nanolab 650i, Fei, Hillsboro, Oregon, OR, USA) equipped with energy dispersive spectroscopy (EDS) was used for sample surface and cross-sectional analyses. In order to precisely identify the composition of very thin layers, samples were subjected to time of flight secondary ion mass spectroscopy (ToF-SIMS). ToF–SIMS instrument (ToF–SIMS 5, IONTOF, GmbH, Münster, Germany) equipped with a Bi liquid metal ion gun with a kinetic energy of 30 keV was used. The SIMS spectra were measured by scanning Bi+ ion beam over an area of 100 × 100 μm in size. SIMS depth profiles were measured in a dual beam depth profiling mode using a 2 keV Cs$^+$ ion beam rastering over an area of 0.4 × 0.4 mm for sputtering. Etching rate was estimated to be 0.20 nm/s.

3. Results

3.1. Materials and Layers Properties

Plasma nitriding process resulted in 90 ± 10 μm thick nitriding layer, which consisted of 87 μm thick diffusion layer and 3 μm thick compound layer. Hardness of the nitrided layer after the compound layer was removed was 1300 ± 75 $HV_{0.01}$.

Figure 2 presents results of the cross-sectional FIB and EDS analysis of CrN coating in its initial state. CrN coating was 2.7 ± 0.25 μm thick and exhibited hardness of 2735 ± 235 $HV_{0.05}$. The FIB cross-sectional analysis revealed a fine-grained compact microstructure of CrN single-layer coating, Figure 2a. Grains of relatively even size are uniformly distributed in the whole coating layer. According to EDS line analysis all chemical elements are uniformly distributed across the coating layer, Figure 2b. EDS analysis shows CrN coating is under-stoichiometric which is in agreement with results published by our team members in [23,24], where CrN coating was produced in the same way and where it was shown CrN coating was of sub-stoichiometric Cr_2N composition. Considering that values in EDS analysis are presented in number of counts these results should be used only for qualitative evaluations and comparisons. Roughness of samples from different groups are presented in Table 1. The highest average roughness (R_a) was measured on samples from the rough group. Low and similar average roughness was measured on samples from the smooth and post polished groups. Although later two groups have been characterized by similar average roughness, skewness roughness parameter (R_{sk}) clearly indicates that surface topography of samples from these groups differ significantly.

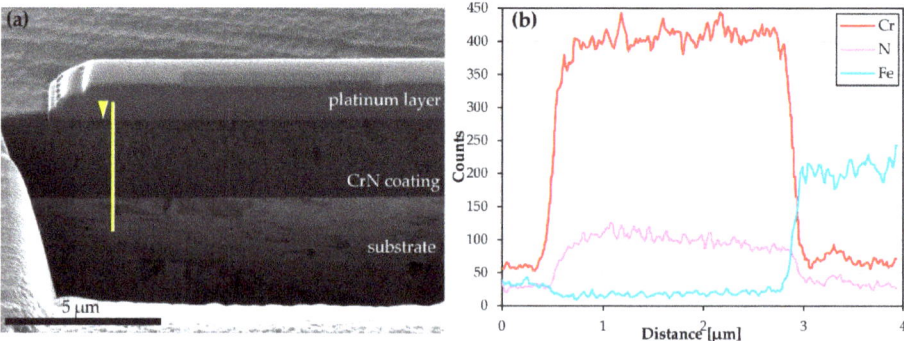

Figure 2. CrN coating in initial state, (**a**) ion induced secondary electron image of CrN coating cross-section, (**b**) chemical composition obtained by EDS analysis along the line depicted in the image (**a**).

3.2. Results of the Ejection Tests

Values of the maximal ejection force, recorded in all ejection tests, are jointly presented with the surface roughness (R_a) in Figure 3. Several trends are observable in the presented graph. First, the average ejection force in CS experiment increases with the decreased roughness. This was reported in our previous study [3]. Second, the ejection force in DS experiments was considerably lower than in CS experiments. Third, the longer the delay in solidification the lower the ejection force. Fourth, similar ejection force was measured in DS experiments on samples of different roughness. Altogether, the ejection force in DS experiments is reduced when compared to CS experiment, and there is a clear trend between the level of reduction and surface roughness, with the highest reduction observed for the post-polished sample. For better comprehension, the level of reduction of the ejection force is presented in percentages in Table 2.

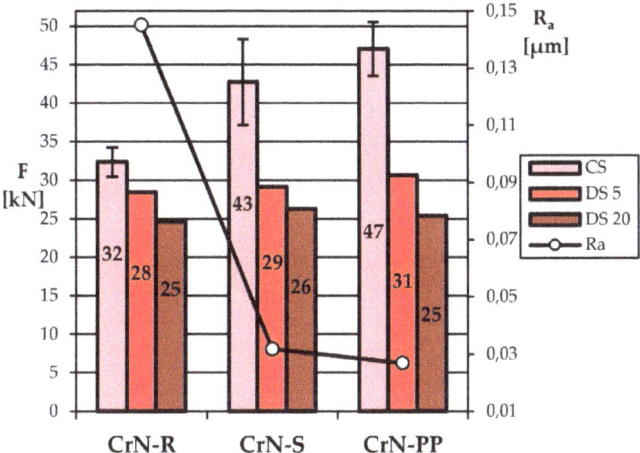

Figure 3. Maximal ejection force obtained for CrN pins with the different roughness in different experimental setups, error bars represent the ±1, 95% confidence interval (CI).

Table 2. Percentage (%) of the reduction in ejection force in delayed casting solidification (DS) experiments in comparison to values obtained in CS experiments.

Sample Name	CrN-R	CrN-S	CrN-PP
DS 5	12.5	32.5	34
DS 20	22	39.5	47

3.3. Cross Sectional Analysis of Coated Samples After Ejection Tests

Results of FIB analysis performed on CrN-R sample subjected to CS experiment are presented in Figure 4. Cross sectional analysis was performed on a location with a cast alloy built-up layer present in one typical micro groove (Figure 4a,b). This location was in a contact with the cast alloy with part of it left on the coating surface after the ejection process. The cross-sectional image indicates that CrN layer stayed intact. The layer was not damaged by tribological (mechanical) processes and its initial fine-grained microstructure was not changed (Figure 4b). Contrast observed in the built-up layer (Figure 4b) shows that different Al–Si–Cu cast alloy phases are present in the layer. Such a contrast in images formed by ion induced secondary electrons arises due to differences in ion channeling in phases with diverse crystalline structures and orientations.

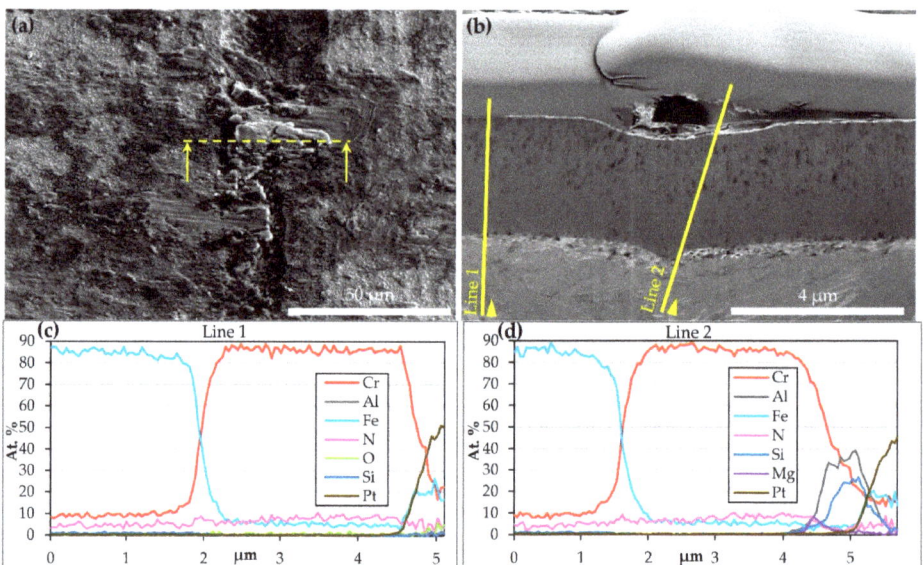

Figure 4. Focused ion beam (FIB) analysis of the CrN-R pin sample with aluminum built-up after casting solidification (CS) experiment, (**a**) secondary electron image, line and arrows indicate FIB milling line and the direction of cross section observation, (**b**) cross-sectional ion induced secondary electron image, (**c**) and (**d**) chemical composition obtained by EDS analysis conducted on cross section along lines drawn in the image (b).

Results of a cross-sectional EDS line analysis are presented in Figure 4c,d. The analysis was performed at two locations, at location without a built-up layer (line 1) and at location with the built-up layer (line 2). The chemical composition of CrN coating layer measured on two locations is similar. The Cr content is very high and almost constant, while the N content is low but also constant across the coating layer. Although a thin bright layer is observed on the top of the coating (Figure 4b), the expected increase of O content was not detected in neither of two tested locations. EDS lines presented in Figure 4d show that intermetallic phases inside the built-up layer consist of Al, Si, and Mg.

Results of FIB analysis performed on CrN-R sample subjected to DS 20 experiment are presented in Figure 5. The cross-sectional analysis was performed on a location where thin cast alloy built-up layer was present, Figure 5a. The CrN coating was not damaged mechanically, however there were changes in coating microstructure and phase constitution, Figure 5b. The bottom layer of CrN coating appears darker than the rest of the coating, which suggests his layer is of different chemical and/or phase composition. Crystallographic (ion channeling) contrast of the same region shown in Figure 5c, indicates that the bottom layer has large, elongated, crystals aligned perpendicular to the interface. On the top of CrN coating, a very thin surface layer can be seen, Figure 5b,c. The cast alloy built-up layer in the analyzed region is very thin and consists of phases (possibly intermetallic) with different crystallographic orientation.

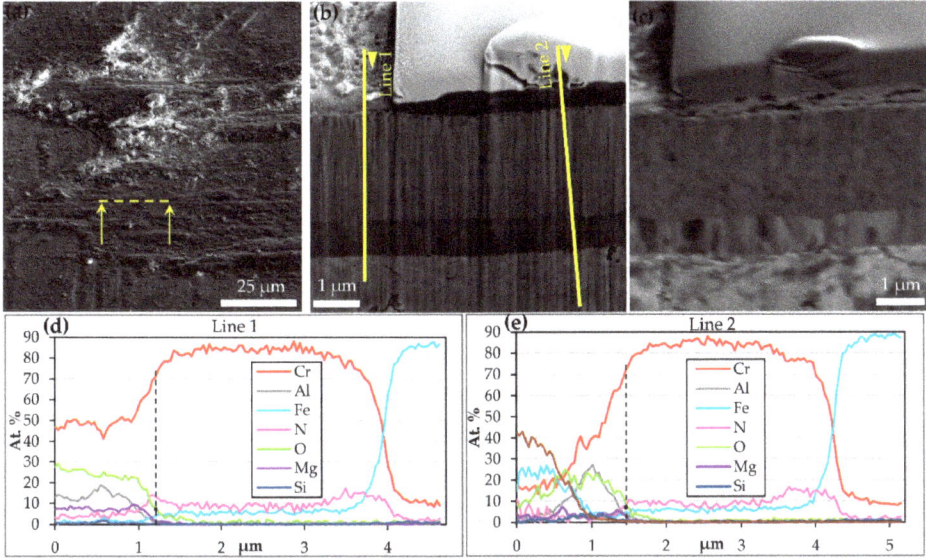

Figure 5. FIB analysis of the CrN-R pin sample with aluminum built-up after DS 20 experiment, (**a**) secondary electron image, line and arrows indicate FIB milling line and the direction of cross section observation, (**b**) cross-sectional secondary electron image, (**c**) cross-sectional ion induced secondary electron image, (**d**) and (**e**) chemical composition obtained by EDS analysis conducted on cross section along lines drawn in the image (**b**).

Results of EDS analysis performed on a cross section of CrN-R sample subjected to DS 20 experiment are presented in Figure 5d,e. The N content is a bit higher (lower Cr/N atomic ratio) in a top thin layer and substantially higher in a bottom thicker layer. The bottom layer of increased N content corresponds to the darker layer with elongated crystals, seen in Figure 5b,c. Increased values of O, Al, and Mg were detected in top regions of both CrN coating and built-up layer. It is worth to note that the line of O content does not follow the line of Al and Mg content. At certain depths (indicated by dashed lines in Figure 5d,e) the O content is fairly high, while the content of Al and Mg is low and insignificant. This means that the top of CrN layer is oxidized (Cr–O formed) and that above it a layer of Al and Mg oxides formed, out of constituents of the casting material.

3.4. ToF-SIMS Analysis of Samples after Ejection Tests

In order to determine more accurately the phases that constitute different layers observed in FIB images, ToF-SIMS analysis was engaged. Depth profiling was performed on pin locations which were not exposed to a cast alloy and on locations which were exposed to a cast alloy. ToF-SIMS depth profiles obtained for the most representative samples are presented in Figures 6 and 7. Signals of CrO^-, AlO^- and CrN^- secondary ions are presented, since they are of the most importance for the present study.

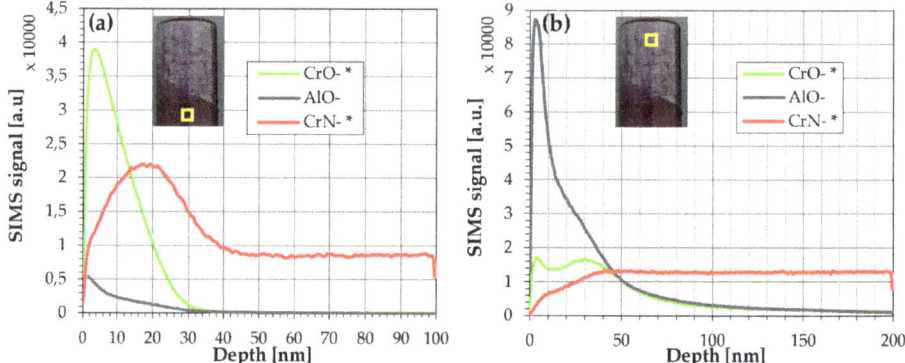

Figure 6. ToF-SIMS depth spectra of the CrN-R sample after CS experiment: (**a**) profile at the location not exposed to cast alloy; (**b**) profile at the location exposed to cast alloy, with built-up layer. The intensity of signals marked with * is multiplied five times. Figure inserts show the investigated pin surfaces and the locations of the depth profiling.

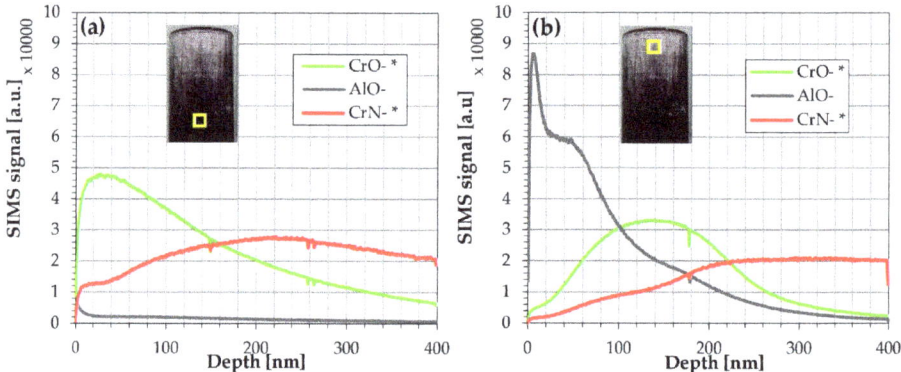

Figure 7. ToF-SIMS depth spectra of the CrN-R sample after DS 20 experiment: (**a**) profile at the location not exposed to cast alloy; (**b**) profile at the location exposed to cast alloy, with built-up layer. The intensity of signals marked with * is multiplied 5 times. Figure inserts show the investigated pin surfaces and the locations of the depth profiling.

A very thin surface layer of Cr–O (15–20 nm) is found at both analyzed location of CrN-R sample from CS experiment. This layer formed on the top of CrN coating, Figure 6. A peak of CrN⁻ near the surface region is not realistic, it is an artefact of the applied measuring technique. It formed due to the matrix effect caused by presence of an oxide [25]. At the location which was not exposed to a cast alloy a smaller quantity of Al–O was found in a thin surface layer, Figure 6a. On the other side, at the location which was exposed to a cast alloy a 15 nm thick layer with higher quantity of Al–O was detected, Figure 6b. It has to be noted that the thin layer with high Al–O content also has lower content of Cr–O.

Figure 7 shows ToF-SIMS depth profiles of a CrN-R sample subjected to the DS 20 experiment. A quite thick Cr–O layer (~150 nm), in which very small quantity of Al–O is identified, is present at the location which was not exposed to a cast alloy. In the top surface layer, the content of CrN is low, it increases with the depth as the content of Cr–O decreases. A top layer of Al–O with a low content of Cr–O and CrN is present at the location which was exposed to a cast alloy. Deeper in the layer, the content of Al–O decreases, while contents of Cr–O and CrN increases. Cr–O content reaches its maximum at the lower depth than CrN which means that Cr–O layer lies over the CrN. The thickness of the Cr–O layer in this region is approximately 150 nm. As the content of Cr–O declines the content of CrN rises and reaches its maximum. Similar findings were obtained from ToF-SIMS analysis of samples subjected to DS 5 experiments.

3.5. Cross-sectional Analysis of Samples Subjected to Annealing Test

FIB cross-sectional analysis of the CrN sample annealed in air at 650 °C is presented in Figure 8. The analysis is performed at location where a nodular defect was present in the coating. In the middle of the coating microstructure is unchanged. Similar grain size and distribution are observed as in as-deposited CrN coating presented in Figure 2a. On the other side, microstructure of CrN top and bottom layer transformed during the annealing experiment. A thin bright layer formed in the top part of the coating with a layer consisting of larger elongated crystals beneath it, Figure 8a. A similar layer, consisting of large elongated crystals is present in the bottom part of CrN coating. Such crystals are comparable to those observed in the bottom part of CrN-R sample subjected to DS 20 experiment, presented in Figure 5. Results of a cross-sectional EDS line analysis are shown in Figure 8b. Note that these results are presented in the number of counts which means that they can be used for qualitative analysis. A top thin layer (~200 nm) has increased content of O. Considering that in this top layer the content of Cr matches with the content of O, this is probably Cr–O layer. The EDS analysis showed that both (top and bottom) layers with large elongated crystals have increased content of N and somewhat lower content of Cr.

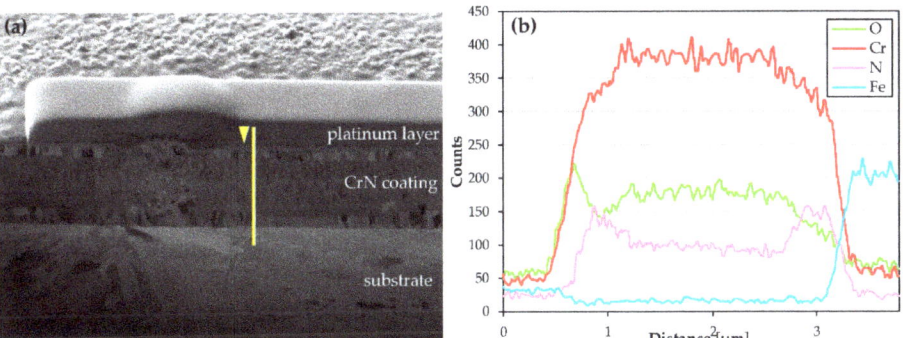

Figure 8. CrN coating in annealed state, (**a**) ion induced secondary electron image of CrN coating cross-section in the annealed state, (**b**) chemical composition obtained by EDS analysis along the line depicted in the image (**a**).

4. Discussion

4.1. Initial Coating Properties

Qualitative results of EDS analysis indicate that the investigated CrN coating has under-stoichiometric composition (Cr_2N). Its chemical composition is comparable to the chemical composition of CrN coating from previous studies, which were produced in similar conditions using the same deposition chamber [23,24]. High hardness for CrN coating is attributed to its dense, fine-grained microstructure and under-stoichiometric coating composition [26]. Besides high hardness, such a microstructure ensures a high cracking resistance. PVD coatings usually consist of columnar grains where cracks propagate easily between the grains [27,28]. However, when fine grains constitute the microstructure, as in our CrN, cracks have to bend around the grains which results in increased cracking resistance. High cracking resistance is of high importance because coatings applied on HPDC tools are subjected to thermal fatigue caused by alternating heating and cooling cycles inherent in the casting process [2]. Additionally, the underlying plasma nitrided layer has very important role in the composite system. It has high hardness (1300 $HV_{0.1}$) and therefore it provides high load bearing capacity for CrN coating.

4.2. Chemical Composition and Microstructure After Casting Experiments

Compared to CS experiments, DS casting experiments significantly altered coating chemical composition and microstructure. The observed changes have specific impact on the behavior of coated samples during the ejection tests. Therefore, to explain the differences observed in the ejection tests, changes in coating chemical and microstructural properties have to be addressed first in detail.

CrN coating was not significantly changed after it was exposed to testing conditions in CS experiment. Microstructure was unchanged, while chemical composition was altered only in a very thin layer at the top of the coating. A thin (15–20 nm) Cr–O layer was formed. This layer should be Cr_2O_3 [26,29], which formed during the casting process when the pin surfaces were shortly exposed to temperatures up to ~600 °C.

It is known that Cr_2O_3 layer forms on CrN coating in these conditions [30]. ToF-SIMS analysis (Figure 6) shows that oxide layer forms more intensively at pin locations which were not exposed to a cast alloy, i.e., it forms more readily at locations which were exposed to air in die. A thin Al–O layer (detected only by ToF-SIMS), is the casting oxide scale which forms when molten aluminum alloy comes in contact with air. However, considering the standard Gibbs free energy of formation, from Ellingham diagram [31], the Al–O layer observed in this study could also form by reduction of the Cr–O top layer by Al from the casting alloy. The time that a liquid cast alloy stays in a contact with a pin sample is very short in CS experiment [21]. This explains why the Al–O layer on the cast alloy is not so thick. The changes observed in CrN layer after conducting the CS experiments are not detrimental to coating integrity. In the industrial use of the coating on HPDC tools, the observed changes would appear after the first few hundreds, or thousand, of casting cycles.

Two types of changes were observed in samples after the DS experiments. The first type involved formation of oxide top layer and its interaction with the cast alloy. The second involved changes in microstructure and chemical composition just beneath the top oxide layer and at the bottom of CrN coating layer. Oxidation of CrN coating occurred during preheating stage in which samples could reach temperatures above 650 °C. Formation of oxide at the top of the coating was also confirmed by EDS study conducted on the annealed samples, Figure 8. Oxide layer in CrN coatings forms due to outward diffusion of Cr and inward diffusion of O through the formed oxide layer [9,26,29]. For under-stoichiometric CrN (Cr_2N) coating, oxidation process forms a layer of Cr_2O_3 [26,29]. This layer suppresses the outward diffusion of N [26,29], as a consequence the Cr/N ratio right beneath the oxide layer is decreased, Figure 5. As a result, stoichiometric CrN forms beneath the oxide layer [26]. Such a layer is favorable for applications on HPDC tools because it oxidizes slower and forms denser microstructure than the initial under-stoichiometric CrN [9,26]. However, large elongated crystals present in the CrN layer beneath the oxide does not promise high mechanical properties.

Both EDS and ToF-SIMS analyses (Figures 5 and 7) showed that the cast alloy built-up layer on CrN coatings subjected to DS experiments consists of a relatively thick Al–O layer. This Al–O layer formed in a reaction of Al–Si–Cu cast alloy with the top Cr–O layer. This analysis showed that during DS experiments coating is subjected to diffusion wear by cast alloy, which should be classified as the coating metallurgical soldering mechanism. Thickness of Al–O layer depends on two factors, the first is the thickness of an initial Cr–O layer, and the second is the time a coated sample spends in contact with a molten alloy. Both are more pronounced in DS tests than in CS tests, i.e., thickness of Cr–O layer is larger and the exposure time to molten alloy is longer (~35 min, [32]), which explains why Al–O layer is thicker after DS tests. Upon casting solidification, the pin-casting contact is basically established between Cr–O and Al–O layers (Cr–O/Al–O pair).

On the bottom of CrN coating, an N-rich layer formed due to outward diffusion of N from the underlying nitrided steel, which is expected to occur at temperatures higher than 550 °C [33]. Due to the increased N content, a stoichiometric CrN was formed. Large, elongated grains aligned perpendicularly to the interface constitute this layer, which is similar to microstructure observed in the layer present in the top area just beneath the oxide layer. If the coating would be exposed to similar casting conditions for a longer time, the two N-rich layers would eventually meet and the whole CrN coating would consist of large elongated grains. Although in a real industrial production die surfaces can be exposed to temperatures as high as 650 °C, since production cycles are short such N-rich layers cannot form in a single cycle or after smaller number of cycles. However, after exposing the coating material to high temperatures in tens of thousands of casting cycles N-rich layers are likely to occur.

Changes observed in annealed sample agree very well with the above discussed changes observed after DS tests. This implies that changes induced during DS tests appear as a consequence of heating and exposure to an oxidizing environment. The only difference in results obtained in these two experiments is the thickness of oxide and "nitrogen-rich" layers. Both layers were thinner in CrN coatings subjected to DS experiments. Oxide layer is thinner due to the shorter period of exposure to air and quite possibly due to the reduction of Cr–O by Al contained in a casting alloy.

4.3. Ejection Force

Although DS experiments were performed without repetition, results of the ejection test should be regarded as reliable. Sequence of steps performed in a DS experiment is almost equal to sequence performed during a CS experiment. Two experiments differ only in selection of parameters, such as preheating temperature and the delay of the solidification process, which were precisely controlled. Therefore, we postulated that the expected variation in the ejection force (coefficient of variation) in DS experiments should be similar to variation obtained for CS experiments, approximately 10% of the measured value [21]. Small variations in the ejection force obtained for different samples in DS 5 and DS 20 experiments (Figure 3) corroborate this statement.

In the next paragraphs we will discuss on values of the ejection force measured for different samples, in different testing conditions.

High values of the ejection force were measured in CS experiments. In CS experiment, the pin-casting contact is established between a Cr–O coating layer and a very thin layer of Al–O contained in a casting. Formation of these layers was discussed in detail in the previous subsection. In the initial stage of the ejection process, the Al–O layer is easily removed because the cast alloy is very soft, and as such it does not provide enough support for very thin Al–O layer. As a consequence, pin samples were mostly sliding against surface of Al–Si–Cu cast alloy. In rare publications on this topic, it is documented that such sliding occurs with a considerable friction, because aluminum alloy increases the adhesive wear component [34]. In the cited study, high friction coefficient of 0.5 was measured between Cr–O and Al in a pin-on-plate test [34]. Even higher friction coefficient (>1.4) was recorded in cross-cylinder tests [35,36]. Therefore, high values of the ejection force obtained in CS experiments are in agreement with high coefficient of friction characteristic for Cr–O/Al pair [34–36].

Values of the ejection force recorded in both DS 5 and DS 20 experiments were substantially lower than in CS experiment. In DS experiments the pin-casting contact is established between thick a Cr–O coating layer and a thick layer of Al–O formed on casting surface. Considering that these layers have substantial hardness, they did not wear off during the pin ejection process. Due to their high hardness [37] and high chemical inertness, these layers in tribological contact impede adhesive wear and high friction [34,38]. For example, for Cr–O/Al_2O_3 pair in a pin-on-plate test friction coefficient of 0.4 was measured, while in work [39] friction coefficient was between 0.35 and 0.4. Therefore, lower values of the ejection force recorded in DS experiments are attributed to lower coefficient of friction of materials in contact.

Surface roughness strongly affected the ejection force in CS experiment, where the ejection force increased with the decreased roughness. A detailed study on the mechanical soldering mechanisms responsible for such a trend was provided in one of our previous studies [3]. On the other hand, an almost equal ejection force was measured in DS experiments for samples of different roughness. These observations along with the above provided discussion suggest that in DS experiments effects of surface chemistry are more dominant than the effects of surface topography, i.e., that metallurgical mechanisms are more dominant than mechanical mechanisms.

4.4. Coating Changes in Delayed Solidification Experiments and Implications on its Performance in HPDC Process

The observed transformation of CrN duplex layer has several drawbacks. Formation of large elongated (columnar) grains in stoichiometric CrN layers might be detrimental for application on HPDC tools. Such layers have lower mechanical properties [28], lower resistance to crack propagation, and are prone to intergranular sliding during severe mechanical loading [27,28]. Additionally, a columnar grain structure creates diffusion paths between grains which enhance diffusion of O [9].

Results of this study unambiguously show that formation of oxide layer on CrN coating is highly beneficial for reduction of ejection force and cast alloy soldering. This is in agreement with findings published in [1,40], in which the application of Cr_2O_3 as a working layer of HPDC tools is highly recommended. However, diffusion wear of Cr–O layer observed in this study can induce negative effects in a long run of HPDC production. The issue is the cyclic exchange of oxidation and diffusion wear processes, which significantly contribute to the overall wear of a coating. If the coating diffusion wear would be slowed down, or brought under the control, the benefits of Cr–O application would prevail. This might be achieved either by application of thick Cr–O working layers or by nanolayer coating design. We propose a coating of a nanolayer design where nanolayers of Cr–O and O-diffusion barrier material would be alternatively deposited. Besides low ejection forces and reduced soldering effect, such a design would suppress the oxygen diffusion out of the coating layer. Application of coatings with Cr–O ("sacrificial") layer allows design of die-cores without drafts. In this way the technological limitation of the casting design is overcome, which is extremely beneficial in HPDC technology.

5. Conclusions

Soldering performance of duplex CrN coating was evaluated by modified ejection test with the delayed casting solidification (DS). Solidification was delayed for 5 min and 20 min. Obtained results were compared to the results from our previous study [3], in which the same coating was evaluated in the conventional solidification (CS) experiment.

In both experiments with the delayed casting solidification the ejection force, required for separation of CrN coated pin samples and Al-alloy castings, was considerably reduced compared to the conventional solidification tests. Depending on sample roughness, the ejection force was reduced from 20% to 50% of the value obtained in conventional solidification experiments.

In order to understand these differences and to study the wear mechanisms acting in different experimental setups, thorough characterization of sample surfaces covered with built-up layers was conducted. It was found that in both conventional and delayed solidification experiments Cr–O formed at the coating side while Al–O formed at the casting side. Cr–O oxides formed when CrN coating was exposed to air at high temperatures, while Al–O formed in Al–Si–Cu alloy by reduction of Cr–O by Al from the alloy. Since the time CrN coating was exposed to both air and liquid cast alloy was longer in the DS experiments, both Cr–O and Al–O layers were substantially thicker in the DS tests. Considering that reduction of Cr–O occurred by diffusion of O into the cast alloy, this kind of wear should be regarded as coating metallurgical soldering. Such a wear mechanism has not been reported in the literature from the field, so far.

Besides oxidation, microstructural and compositional changes were found in top and bottom layers of CrN coating after DS experiments. Two N-rich layers formed, one just beneath the Cr–O oxide, and the other just above the nitrided steel. The first layer formed as a consequence of CrN oxidation, and the second as a consequence of N outward diffusion from the underlying nitrided layer. In both layers elongated grain structure was found.

During ejection of coated pins from castings formed in CS experiments, at the beginning Cr–O layer was in the contact with a very thin Al–O layer. This thin Al–O layer was easily removed, and the sliding mainly occurred between Cr–O and Al–Si–Cu alloy. High adhesion between these materials resulted in high ejection force. On the other hand, during ejection in DS experiments a thick Al–O casting scale inhibited the contact of Al–Si–Cu alloy with a thick Cr–O layer. As a consequence, soldering and friction between a coated pin and a casting was reduced and lower ejection force was recorded.

It was found that pin samples with different surface roughness in DS experiments exhibited approximately the same values of the ejection force. This is contrary to CS experiments where roughness played important role. Differences arise because the pin-casting material pair formed in DS experiments (Cr–O/Al–O) greatly reduces adhesion and galling which both increases with the decrease in surface roughness. These observations show that in DS experiments the effect, of surface chemistry is more dominant than the effect of surface roughness.

This study showed that modification of ejection test by introduction of the delayed casting solidification is appropriate approach for introduction of severe soldering and corrosion conditions which are required for appropriate evaluation of metallurgical soldering performance of coating materials. Such a test configuration allowed us to recognize several mechanisms of CrN coating soldering wear, which were not addressed in the literature so far.

In order to confirm the findings presented herein, while excluding the negative effects of DS experiments, for future investigations we suggest performing the CS experiment with previously oxidized CrN coating. The other important point is determination of the intensity of coatings diffusion wear in contact with aluminum alloy castings. In such a way it can be determined whether the CrN oxidation prior to application on HPDC tools is beneficial. Additionally, the soldering performance and wear of Al_2O_3 coating and/or Al_2O_3-forming (i.e., TiAlN, TiAlSiN) coatings should be compared with Cr_2O_3 coating and/or Cr_2O_3-forming coatings (i.e., CrN, CrAlN).

Author Contributions: Conceptualization, P.T. and L.K.; methodology, P.T., L.K., A.M. and B.Š.; validation, L.K. and A.M.; investigation, P.T., L.K., J.K. and A.D.; resources, B.Š. and J.K.; writing—original draft preparation, P.T.; writing—review and editing, P.T., L.K., A.M. and B.Š.; visualization, A.M.; supervision, B.Š.; project administration, B.Š.; funding acquisition, B.Š. All authors have read and agreed to the published version of the manuscript.

Funding: This research was funded by Serbian-Slovenian bilateral project (2018–2019) grant 48. This work was also funded by the Slovenian Research Agency (program P2-0082) and European Regional Development Funds (CENN Nanocenter, OP13.1.1.2.02.006).

Acknowledgments: Special thanks to Peter Panjan, Institute "Jožef Stefan" (Ljubljana, Slovenia), for help in samples characterization and fruitful discussions about the results obtained in this study. Additionally, we are very grateful to Miha Čekada, Institute "Jožef Stefan" (Ljubljana, Slovenia) for all the help and for providing the resources needed for the investigation. Authors also gratefully acknowledge Termometal d.o.o. (Ada, Serbia) for samples and tools production.

Conflicts of Interest: The authors declare no conflict of interest. The funders had no role in the design of the study; in the collection, analyses, or interpretation of data, in the writing of the manuscript, or in the decision to publish the results.

References

1. Lin, J.; Carrera, S.; Kunrath, A.O.; Myers, S.; Mishra, B.; Ried, P.; Moore, J.J.; Zhong, D. Design methodology for optimized die coatings: The case for aluminum pressure die-casting. *Surf. Coat. Technol.* **2006**, *201*, 2930–2941. [CrossRef]
2. Panjan, P.; Čekada, M.; Kirn, R.; Soković, M. Improvement of die-casting tools with duplex treatment. *Surf. Coat. Technol.* **2004**, *180*, 561–565. [CrossRef]
3. Terek, P.; Kovačević, L.; Miletić, A.; Panjan, P.; Baloš, S.; Škorić, B.; Kakaš, D. Effects of die core treatments and surface finishes on the sticking and galling tendency of Al–Si alloy casting during ejection. *Wear* **2016**, *356–357*, 122–134. [CrossRef]

4. Duarte, A.; Oliveira, F.J.; Costa, F.M. Characterisation of interface formed at 650°C between AISI H13 steel and Al–12Si–1Cu aluminium melt. *Int. J. Cast Metals Res.* **2010**, *23*, 231–239. [CrossRef]
5. Paiva, J.M.; Fox-Rabinovich, G.; Junior, E.L.; Stolf, P.; Ahmed, Y.S.; Martins, M.M.; Bork, C.; Veldhuis, S. Tribological and wear performance of nanocomposite PVD hard coatings deposited on aluminum die casting tool. *Materials* **2018**, *11*, 358. [CrossRef] [PubMed]
6. Bobzin, K.; Brögelmann, T.; Hartmann, U.; Kruppe, N.C. Analysis of CrN/AlN/Al2O3 and two industrially used coatings deposited on die casting cores after application in an aluminum die casting machine. *Surf. Coat. Technol.* **2016**, *308*, 374–382. [CrossRef]
7. Tentardini, E.K.; Aguzzoli, C.; Castro, M.; Kunrath, A.O.; Moore, J.J.; Kwietniewski, C.; Baumvol, I.J.R. Reactivity between aluminum and (Ti,Al)N coatings for casting dies. *Thin Solid Films* **2008**, *516*, 3062–3069. [CrossRef]
8. Torres, E.; Ugues, D.; Brytan, Z.; Perucca, M. Development of multilayer coatings for forming dies and tools of aluminium alloy from liquid state. *J. Phys. D Appl. Phys.* **2009**, *42*, 105306. [CrossRef]
9. Lin, J.; Zhang, N.; Sproul, W.D.; Moore, J.J. A comparison of the oxidation behavior of CrN films deposited using continuous dc, pulsed dc and modulated pulsed power magnetron sputtering. *Surf. Coat. Technol.* **2012**, *206*, 3283–3290. [CrossRef]
10. Navinšek, B.; Panjan, P.; Milošev, I. Industrial applications of CrN (PVD) coatings, deposited at high and low temperatures. *Surf. Coat. Technol.* **1997**, *97*, 182–191. [CrossRef]
11. Lugscheider, E.; Bobzin, K.; Barimani, C.; Bärwulf, S.; Hornig, T. PVD hard coatings protecting the surface of thixoforming tools. *Adv. Eng. Mater.* **2000**, *2*, 33–37. [CrossRef]
12. Wang, B.; Bourne, G.R.; Korenyi-Both, A.L.; Monroe, A.K.; Midson, S.P.; Kaufman, M.J. Method to evaluate the adhesion behavior of aluminum-based alloys on various materials and coatings for lube-free die casting. *J. Mater. Process. Technol.* **2016**, *237*, 386–393. [CrossRef]
13. Bobzin, K.; Brögelmann, T.; Brugnara, R.H.; Kruppe, N.C. CrN/AlN and CrN/AlN/Al2O3 coatings deposited by pulsed cathodic arc for aluminum die casting applications. *Surf. Coat. Technol.* **2015**, *284*, 222–229. [CrossRef]
14. Veprek, S. Recent search for new superhard materials: Go nano! *J. Vac. Sci. Technol. A* **2013**, *31*, 050822. [CrossRef]
15. Kyler, M.; Blowers, M.; Rakita, A.; Koehler, D.; Lee, M.; Landa, Q.; Han, C.; Daugherty, D.; Gettinger, A.H. Optimizing Die Cooling Using Pulsed Spray/Lube Residuals on Die Surfaces Using Spray Methods. In Proceedings of the Die Casting Congress: West Coast, Long Beach, CA, USA, 11–12 January 2018.
16. Nunes, V.; Silva, F.J.G.; Andrade, M.F.; Alexandre, R.; Baptista, A.P.M. Increasing the lifespan of high-pressure die cast molds subjected to severe wear. *Surf. Coat. Technol.* **2017**, *332*, 319–331. [CrossRef]
17. Han, Q.; Viswanathan, S. Analysis of the mechanism of die soldering in aluminum die casting. *Metall. Mater. Trans. A* **2003**, *34*, 139–146. [CrossRef]
18. Shankar, S.; Apelian, D. Die soldering: Mechanism of the interface reaction between molten aluminum alloy and tool steel. *Metall. Mater. Trans. B* **2002**, *33*, 465–476. [CrossRef]
19. Abusuilik, S.B. Pre-, intermediate, and post-treatment of hard coatings to improve their performance for forming and cutting tools. *Surf. Coat. Technol.* **2015**, *284*, 384–395. [CrossRef]
20. Panjan, P.; Drnovšek, A.; Gselman, P.; Čekada, M.; Panjan, M.; Bončina, T.; Merl, D.K. Influence of growth defects on the corrosion resistance of sputter-deposited TiAlN hard coatings. *Coatings* **2019**, *9*, 511. [CrossRef]
21. Terek, P.; Kovačević, L.; Miletić, A.; Kukuruzović, D.; Baloš, S.; Škorić, B. Improved ejection test for evaluation of soldering tendency of cast alloy to die core materials. *J. Mater. Process. Technol.* **2019**, *266*, 114–124. [CrossRef]
22. Kadlec, S.; Maček, M.; Wouters, S.; Meert, B.; Navinšek, B.; Panjan, P.; Quaeyhaegens, C.; Stals, L.M. Plasma diagnostics of triode ion-plating systems by energy-resolved mass spectroscopy and comparison of TiN film properties. *Surf. Coat. Technol.* **1999**, *116–119*, 1211–1218. [CrossRef]
23. Čekada, M.; Maček, M.; Kek Merl, D.; Panjan, P. Properties of Cr(C,N) hard coatings deposited in Ar-C2H2-N2 plasma. *Thin Solid Films* **2003**, *433*, 174–179. [CrossRef]
24. Navinšek, B.; Panjan, P.; Urankar, I.; Cvahte, P.; Gorenjak, F. Improvement of hot-working processes with PVD coatings and duplex treatment. *Surf. Coat. Technol.* **2001**, *142–144*, 1148–1154. [CrossRef]
25. Stevie, F.A. *Secondary Ion Mass Spectrometry: Applications for Depth Profiling and Surface Characterization*; Momentum Press: New York, NY, USA, 2016; ISBN 978-1-60650-589-2.

26. Qi, Z.B.; Liu, B.; Wu, Z.T.; Zhu, F.P.; Wang, Z.C.; Wu, C.H. A comparative study of the oxidation behavior of Cr2N and CrN coatings. *Thin Solid Films* **2013**, *544*, 515–520. [CrossRef]
27. Rzepiejewska-Malyska, K.; Parlinska-Wojtan, M.; Wasmer, K.; Hejduk, K.; Michler, J. In-situ SEM indentation studies of the deformation mechanisms in TiN, CrN and TiN/CrN. *Micron* **2009**, *40*, 22–27. [CrossRef]
28. Lin, J.; Moore, J.J.; Sproul, W.D.; Mishra, B.; Wu, Z.; Wang, J. The structure and properties of chromium nitride coatings deposited using dc, pulsed dc and modulated pulse power magnetron sputtering. *Surf. Coat. Technol.* **2010**, *204*, 2230–2239. [CrossRef]
29. Huber, E.; Hofmann, S. Oxidation behaviour of chromium-based nitride coatings. *Surf. Coat. Technol* **1994**, *68–69*, 64–69. [CrossRef]
30. Panjan, P.; Navinšek, B.; Cvelbar, A.; Zalar, A.; Milošev, I. Oxidation of TiN, ZrN, TiZrN, CrN, TiCrN and TiN/CrN multilayer hard coatings reactively sputtered at low temperature. *Thin Solid Films* **1996**, *281–282*, 298–301. [CrossRef]
31. Reed, T.B. *Free Energy of Formation of Binary Compounds An Atlas of Charts for High-Temperature Chemical Calculations*; MIT Press: Cambridge, MA, USA, 1972; ISBN 9780262180511.
32. Terek, P. Application of surface engineering technologies for improvement of die casting tools quality. Ph.D. Thesis, University of Novi Sad, Novi Sad, Serbia, 2016.
33. Djellal, R.; Saker, A.; Bouzabata, B.; Mekki, D.E. Thermal stability and phase decomposition of nitrided layers on 316L and 310 austenitic stainless steels. *Surf. Coat. Technol.* **2017**, *325*, 533–538. [CrossRef]
34. Wang, L.; Nie, X. Effect of annealing temperature on tribological properties and material transfer phenomena of CrN and CrAlN coatings. *J. Mater. Eng. Perform.* **2014**, *23*, 560–571. [CrossRef]
35. Jerina, J.; Kalin, M. Aluminium-alloy transfer to a CrN coating and a hot-work tool steel at room and elevated temperatures. *Wear* **2015**, *340–341*, 82–89. [CrossRef]
36. Kalin, M.; Jerina, J. The effect of temperature and sliding distance on coated (CrN, TiAlN) and uncoated nitrided hot-work tool steels against an aluminium alloy. *Wear* **2015**, *330–331*, 371–379. [CrossRef]
37. Wang, D.-Y.; Lin, J.-H.; Ho, W.-Y. Study on chromium oxide synthesized by unbalanced magnetron sputtering. *Thin Solid Films* **1998**, *332*, 295–299. [CrossRef]
38. Polcar, T.; Martinez, R.; Vítů, T.; Kopecký, L.; Rodriguez, R.; Cavaleiro, A. High temperature tribology of CrN and multilayered Cr/CrN coatings. *Surf. Coat. Technol.* **2009**, *203*, 3254–3259. [CrossRef]
39. Urgen, M.; Ezirmik, V.; Senel, E.; Kahraman, Z.; Kazmanli, K. The effect of oxygen content on the temperature dependent tribological behavior of Cr-O-N coatings. *Surf. Coat. Technol.* **2009**, *203*, 2272–2277. [CrossRef]
40. Ho, W.Y.; Huang, D.-H.; Huang, L.-T.; Hsu, C.-H.; Wang, D.-Y. Study of characteristics of Cr2O3/CrN duplex coatings for aluminum die casting applications. *Surf. Coat. Technol.* **2004**, *177–178*, 172–177. [CrossRef]

© 2020 by the authors. Licensee MDPI, Basel, Switzerland. This article is an open access article distributed under the terms and conditions of the Creative Commons Attribution (CC BY) license (http://creativecommons.org/licenses/by/4.0/).

Article

Properties of Tool Steels and Their Importance When Used in a Coated System

Bojan Podgornik *, Marko Sedlaček, Borut Žužek and Agnieszka Guštin

Institute of Metals and Technology, SI-1000 Ljubljana, Slovenia; marko.sedlacek@imt.si (M.S.); borut.zuzek@imt.si (B.Ž.); agnieszka.gustin@imt.si (A.G.)
* Correspondence: bojan.podgornik@imt.si; Tel.: +386-1-4701-930

Received: 29 January 2020; Accepted: 10 March 2020; Published: 12 March 2020

Abstract: The introduction of new light-weight high-strength materials, which are difficult to form, increases demands on tool properties, including load-carrying capacity and wear resistance. Tool properties can be improved by the deposition of hard coatings but proper combination and optimization of the substrate properties are required to prepare the tool for coating application. The aim of this paper is to elaborate on tool steel substrate properties correlations, including hardness, fracture toughness, strength and surface quality and how these substrate properties influence on the coating performance. Results show that hardness of the steel substrate is the most influential parameter for abrasive wear resistance and load-carrying capacity, which is true for different types of hard coatings. However, high hardness should also be accompanied by sufficient fracture toughness, especially when it comes to very hard and brittle coatings, thus providing a combination of high load-carrying capacity, good fatigue properties and superior resistance against impact wear. Duplex treatment and formation of a compound layer during nitriding can be used as an additional support interlayer, but its brittleness may result in accelerated coating cracking and spallation if not supported by sufficient core hardness. In terms of galling resistance, even for coated surfaces substrate roughness and topography have major influence when it comes to hard ceramic coatings, with reduced substrate roughness and coating post-polishing providing up to two times better galling resistance.

Keywords: tool steel substrate; coatings; hardness; fracture toughness; load-carrying capacity; wear

1. Introduction

In forming applications of modern metallic materials, including die casting, stamping, forging and rolling, tool lifetime is limited due to very demanding working conditions. These include mechanical, thermal and impact loading [1,2]. Under such complex working conditions, comprising high contact stresses as well as abrasive and adhesive wear [3–5], tool surfaces are attacked by different wear and fatigue mechanisms [6–8]. By the current demands on reducing mass and size of components, lowering fuel consumption and CO_2 emission, increasing recycling and improving overall strength and safety [9,10]—especially when talking about transportation and energy sector— tools are exposed to new and more severe requirements and demands. This is related to design of the tool, selection of material and heat treatment [11], and surface engineering, where substrate preparation is essential [12].

Increased demands mean strengthened requirements in terms of the tool material properties, including temperature resistance, strength, shock and fatigue resistance, impact and sliding wear resistance, etc. Ductility and fracture toughness are among the main tool properties essential for the majority of forming applications and the influencing of tool resistance [13,14]. Properties of the tool core material, i.e., tool steel, depend on its chemical composition and production process, but primarily on the heat treatment parameters. Heat treatment defines final microstructure and corresponding properties. In general, steel properties, mainly focused on strength and hardness are provided by a well-defined heat treatment process, consisting of an austenitization treatment with a subsequent quenching and a multiple tempering [13,15]. Typically, a trade-off between toughness and hardness is required [16]. On the other hand, through optimized heat treatment procedures, involving vacuum hardening, selection of tempering and austenitizing temperature, and inclusion of deep cryogenic treatment [17], fine-grained microstructure with homogeneous carbides distribution can be obtained [15]. Thus, improved toughness and fatigue resistance is obtained while maintaining high strength and hardness [18].

The typical property used for the planning of heat treatment for tool steels is hardness. High hardness is related to abrasive wear resistance and resistance to plastic deformation. However, beside hardness there are also other material properties, which are based on the application and surface engineering techniques, which become more important as we use tools that are more complex. These include fracture toughness, compressive and bending strength, creep and wear resistance, machinability, etc. [2]. Different material properties are defined and determined by different standards and test methods. However, each standard and test method uses specific test specimens with different geometries. Different geometries relate to different heat treatment conditions, and thus in microstructure deviation [19] and problematic properties correlation. On the other hand, circumferentially notched and fatigue-pre-cracked tensile bar (CNPTB) specimens used for measuring fracture toughness of more brittle materials, i.e., tool steels have been found as the best alternative, allowing determination and correlation of many different properties [20]. The advantage of the CNPTB specimen is in its radial symmetry and uniform microstructure through the whole volume.

Introduction of light-weight high-strength materials, being very difficult to form also sets more demanding properties requirements on tool properties, especially its surface and wear resistance [21]. One way of improving wear properties of the tool is application of different heat and diffusion treatments, i.e., plasma nitriding, thus modifying surface microstructure and properties [22,23]. Another way, proven in cutting tool applications is deposition of wear resistant coatings [24]. However, limited load-carrying capacity, adhesion and topographical characteristics of the substrate restrict successful application of hard wear-resistant coatings on forming tools. Thus, proper combination and optimization of the substrate properties are needed in order to prepare tools for coating deposition and provide improve performance of the tool [23–27].

The aim of this research work, carried out by using CNPTB test specimen configuration was to determine correlations between different tool steel properties, including fracture toughness, hardness, compressive and bending strength, wear resistance and surface quality and how these substrate properties influence on the coating performance.

2. Materials and Methods

2.1. Materials

Two tool steels being the most common in forming and tooling industry have been included in this investigation. First one, aimed at studying the effect of heat treatment parameters as well as additional plasma nitriding on the tool steel substrate properties, their correlation and influence on the load-carrying capacity and sliding wear resistance was conventional AISI H11 type hot-work tool steel (H11), produced through casting, electro slag re-melting, forging and annealing. The second one was high fatigue strength P/M cold work tool steel (PM-CW), aimed at determining the influence of hardness and fracture toughness as well as surface preparation on the load-carrying capacity as well as impact and galling wear resistance when coated with different type of hard coatings. Chemical composition of the investigated tool steels is given in Table 1.

Table 1. Chemical composition of the investigated tool steels (wt %).

No.	Tool Steel	% C	% Si	% Mn	% Cr	% Mo	% V	% W	% Co	% Fe
1	H11	0.37	<1.0	0.27	5.16	1.28	0.41	–	–	balance
2	PM-CW	0.85	0.54	0.39	4.36	2.79	2.11	2.54	4.52	balance

2.2. Heat Treatment and Coatings

Heat treatment of hot work tool steel (H11) included vacuum heat treatment performed in Ipsen VTC 324-R horizontal vacuum furnace (Ipsen, Kleve, Germany). The specimens machined from soft annealed material were preheated to 850 °C and then progressively heated at 10 °C/min to the austenitizing temperature of 990–1000 and 1030 °C, respectively, soaked for 20 min and quenched in N_2 gas flow at a cooling speed of 3 °C/s. After quenching specimens were double tempered for 2 h. First tempering was always performed at 540 °C, immediately followed by second tempering at six different temperatures, varied from 550 to 630 °C.

Specimens planned to investigate the effect of heat treatment parameters on the load-carrying capacity and sliding wear were further plasma nitrided in a Metaplas Ionon HZIW 600/1000 reactor (Metaplas Ionon, Bergisch Gladbach, Germany), surface polished and coated. Plasma nitriding was performed at 540 °C for 20 h using different gas mixtures. One group of test specimens was treated in 25% N_2:75% H_2 gas mixture, resulting in diffusion zone of about 260 μm and approx. 5 μm thick top compound layer. Another group was treated in 5% N_2:95% H_2 gas mixture, providing ~230 μm thick diffusion zone without any compound layer. After nitriding all specimens were polished to a mirror-like finish (R_a = 0.1 μm) and coated with the commercial TiN/TiB_2 nanocomposite multilayer coating. Coating consisted of a primary TiN monolayer, a multilayer zone (TiN/Ti-B-N) with a lamella thickness of 85 nm and a top TiB_2 overcoat. TiN/TiB_2 coating with total thickness of ~2 μm and hardness of 3000 HV was deposited by a bipolar-pulsed glow discharge PACVD. Processing temperature was 530 °C and pressure 200 Pa.

In the case of PM cold work tool steel (PM-CW) three groups of vacuum heat treatment parameters were used and combined with the process of deep cryogenic treatment (Table 2). First group (A1) providing maximum hardness was quenched from high austenitizing temperature (1130 °C) and triple tempered at low tempering temperatures (520/520/490 °C). In order to increase fracture toughness but still maintain hardness above 64 HRC second group (A2) was austenitized at 1100 °C and tempered at 500 °C (500/500/470 °C). The last group (A3) was hardened from 1070 °C and triple tempered at increased tempering temperature (585/585/565 °C), thus providing high fracture toughness. In the cases when vacuum heat treatment was combined with deep cryogenic treatment—DCT (groups B1–B3), DCT was performed immediately after quenching and followed by a single 2 h tempering. DCT consisted of a controlled immersion of the test specimens in liquid nitrogen for 25 h (Table 2). After heat treatment specimens were mirror polished (R_a = 0.10 μm), sputter cleaned and coated. Investigation included three representative PVD coatings These were monolayer TiAlN coating (~3300 HV), AlTiN/TiN multi-layer coating (~3500 HV) with lamellas thickness of ~50 nm and ~80 nm, respectively, and (Ti,Si)N nano-composite coating (~3800 HV). All three coatings were about 2 μm thick and deposited by magnetron sputtering process at the substrate temperature of ~450 °C. Details of the deposition process are provided in Reference [28,29].

The effect of substrate roughness on galling performance was evaluated by preparing A1 group of specimens with four different procedures; coarse grinding and polishing with a 20 μm industrial polishing paste (A1-1; R_a = 0.5 μm), coarse grinding and double polishing with 20 μm and 10 μm polishing paste (A1-2; R_a = 0.3 μm), fine grinding (A1-3; R_a = 0.15 μm) and polishing (A1-4; R_a = 0.1 μm). Afterwards specimens were coated with commercial TiN monolayer and W-doped DLC multilayer coating [30].

Table 2. Vacuum heat treatment and deep-cryogenic treatment parameters for P/M cold work tool steel.

Group	Austenitizing		Deep-Cryogenic Treatment		Tempering	
	Temp. [°C]	Time [min]	Temp. [°C]	Immersion Time [h]	Temp. [°C]	Time [h]
A1	1130	6	–	–	520/520/500	2/2/2
A2	1100	20	–	–	500/500/480	2/2/2
A3	1070	20	–	–	585/585/565	2/2/2
B1	1130	6	−196	25	520	2
B2	1100	20	−196	25	500	2
B3	1070	20	−196	25	585	2

2.3. Mechanical Properties

For each material CNPTB specimens [20] (Figure 1) were machined and used for further investigation. Fracture toughness was measured by pre-cracking CNPTB specimen under rotating-bending loading (400 N, 4500 cycles). After pre-cracking specimens were subjected to tensile load at the cross-head speed of 1.0 mm/min until fracture. Measuring the load at fracture (P) and diameter of the fractured area (d) fracture toughness is calculated according to the Equation (1) [31,32]. Details of the CNPTB specimen and fracture toughness measurement technique are given in Ref. [20]. For each group of treat treated specimens at least 12 samples were characterized in order to provide reliable results.

$$K_{Ic} = \frac{P}{d_0^{3/2}} \cdot \left(-1.27 + 1.72 \frac{d_0}{d}\right) \quad (1)$$

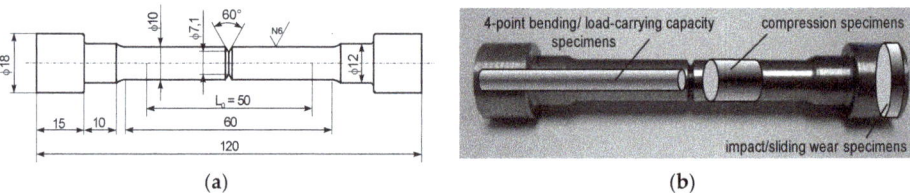

Figure 1. (a) Circumferentially notched and fatigue-pre-cracked tensile bar (CNPTB) specimen and (b) extraction of 4-point bending, load-carrying capacity, compression and sliding/impact wear test specimens.

Rockwell-C hardness measurements, performed circumferentially (×3) on each CNPTB specimen, were carried out on Willson-Rockwell B2000 machine (Buehler, Esslingen, Germany) and then average value calculated.

One half of the fractured CNPTB specimen was used to machine ϕ 10 mm × 12.5 mm cylinder for compression test (Figure 1) and ϕ 18 mm × 8 mm disc for sliding and impact wear testing. Compression tests were performed according to ASTM E9-09 standard [33] and used to determine yield strength, maximum compression strength and strain hardening exponent. Strain hardening exponent was defined between yield and maximum compression stress on a log-log plot of true stress-true strain. The other part of the fractured CNPTB specimen was used to prepare 4-point bending test specimen (ϕ 5 mm × 60 mm) or load-carrying capacity test specimens (ϕ 10 mm × 60 mm) (Figure 1). After high-speed machining, bending and load-carrying capacity test specimens were ground and polished (R_a = 0.1 µm). 4-point bending tests at room temperature were performed according to ASTM E290-09 standard [34], using support span of 40 mm and load span of 16 mm.

Effect of heat treatment on the machinability was analyzed by measuring surface roughness of as machined 4-point bending specimens. High-speed machining involved pre-turning with standard cutting inserts (Sandvik-Coromant DNMG11 R0.4, Sandviken, Sweden) at a cutting speed of 100 m/min, depth of cut of 0.3 mm and feed rate of 0.12 mm/rev, followed by final turning (VBMT 16 04 cutting inserts) at the same cutting speed, depth of cut of 0.2 mm and feed rate of 0.08 mm/rev. Each specimen was machined with a new cutting insert and surface roughness analyzed (ISO 4287:1997 standard [35]) by Alicona InfiniteFocus G4 microscope (Alicona, Raaba, Austria).

2.4. Load-Carrying Capacity and Wear Testing

Load-carrying capacity was evaluated by load-scanning test rig, Figure 2a. The test configuration consists of two crossed cylinders (10 mm) which are sliding against each other at fixed speed, but progressive loading. Thus, each position of the wear scar corresponds to a unique load without any loading history [36]. In this investigation coated tool steel cylinder was loaded against polished WC cylinder (R_a = 0.05 µm, 2200 HV), using dry sliding conditions, room temperature, fixed sliding speed of 0.01 m/s and normal load in the range of 400–4000 N (p_H = 2.8–6.1 GPa). Load-carrying capacity is determined by defining critical loads at which first cracks in the coating are observed and when coating starts to flake [4,36].

Sliding wear tests were done under dry sliding conditions using ball on disc contact configuration and reciprocating sliding (Figure 2b). Polished 20 mm diameter Al_2O_3 ball (R_a = 0.05 µm) was used as a counter material in order to simulate abrasive wear mode and focus all the wear on the investigated disc material. Test were done under normal room conditions (RT and 45% RH) and elevated temperature of 150 °C applying different contact conditions; loads corresponding to contact pressure between 800 and 1300 MPa and sliding speeds between 0.01 and 0.1 m/s, obtained by changing oscillating frequency from 1 to 15 Hz. All tests were performed up to the total sliding distance of 100 m (up to two hours), with the average coefficient of friction being analyzed and wear volume measured using 3D confocal microscope.

Impact wear tests were performed on servo hydraulic fatigue testing machine, with the coated disc being repetitively impacted against a WC ball (φ 32 mm; Figure 2c). Testing machine is position controlled during testing, which includes continuous monitoring of the impact force. Impact wear tests were performed at the frequency of 30 Hz, initial impacting distance of 0.5 mm and maximum impacting load of 5.5 kN (p_H = 3.5 GPa). New WC ball was used for each test and testing specimens cleaned with ethanol. Adhesive wear of the WC ball was prevented by applying a thin layer of lithium grease on the disc surface.

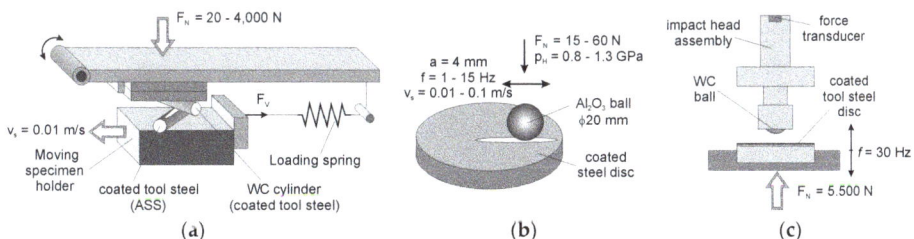

Figure 2. Load-carrying capacity and wear testing setups; (a) load scanner, (b) reciprocating sliding wear test, (c) impact wear test.

Effect of substrate roughness and surface quality on resistance against galling was also evaluated by a load-scanning test rig (Figure 2a). In this case tempered austenitic stainless steel (ASS) cylinder (AISI 304, 335 HV, R_a = 0.2 µm, φ 10 mm) was used as a moving counter cylinder. Galling tests were performed dry, using sliding speed of 0.01 m/s and normal load from 20 to 1300 N. Results were then evaluated by analyzing wear tracks after sliding and determining critical loads for galling initiation and gross galling formation [36,37].

3. Results

3.1. Tool Steel Substrate Properties Correlation

Diagram displaying fracture toughness and hardness of vacuum heat treated AISI H11 type hot work tool steel as depending on the tempering and austenitizing temperature is shown in Figure 3. Fracture toughness obtained by hardening from 1000 °C followed by double tempering at 630 °C was 87 MPa√m. It was reduced to less than 30 MPa√m by reducing tempering temperature to 550 °C. Hardness, on the other hand, increased from 40 HRC to almost 50 HRC. Further hardness increase is provided by increased Si content and austenitizing temperature. In the case of low Si content, increase in austenitizing temperature to 1030 °C results in about 5% higher hardness (up to 52 HRC) and for high Si content even more, especially at low tempering temperatures, up to 54 HRC. However, for low Si hot work tool steel increase in austenitizing temperature also provided higher fracture toughness, being between 45 and 115 MPa√m, while for high Si content fracture toughness has been reduced at elevated austenitizing temperature, ranging between 25 and 80 MPa√m, as shown in Figure 3.

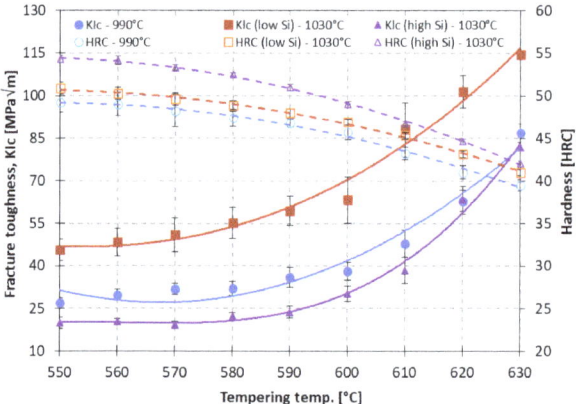

Figure 3. Effect of Si content and heat treatment temperatures on fracture toughness and hardness of AISI H11 hot work tool steel.

Yield and ultimate compression strength of the investigated hot work tool steel (AISI H11) are between 1200 and 1850 MPa, and 1450 and 2130 MPa, respectively. Similar to hardness, ultimate compression strength and yield strength are increasing when increasing austenitizing temperature (for 5%) and Si content, but decreasing with tempering temperature, as shown in Figure 4a. On the other hand, material ductility being analyzed by measuring strain hardening exponent was found mainly independent on the austenitizing and tempering temperature. For the tempering temperatures up to 610 °C it shows more or less constant value, 0.45 for low Si content and 0.4 for high Si content.

In the case of bending test, maximum and yield strength at the austenitizing temperature of 990 °C changed from 1860 and 3260 MPa to 2600 and 4550 MPa, respectively, when reducing tempering temperature from 630 to 550 °C. Further increase was obtained by increasing austenitizing temperature to 1030 °C. In this case and low Si content, yield and maximum bending strength reached peak values (tempering at 550 °C) of 2700 and 4780 MPa, respectively, and even up to 2800 and 4950 MPa, respectively, for high Si content (Figure 4b).

When analyzing correlations, strong correlation between hardness and strength of tool steel substrate has been observed. In agreement with well-established correlations [38,39] compression and bending strength increase linearly with hardness, but dropping digressively with fracture toughness, as shown in Figure 5. On the other hand, strain hardening exponent has no direct correlation with hardness but it shows rising trend with increased fracture toughness (Figure 6).

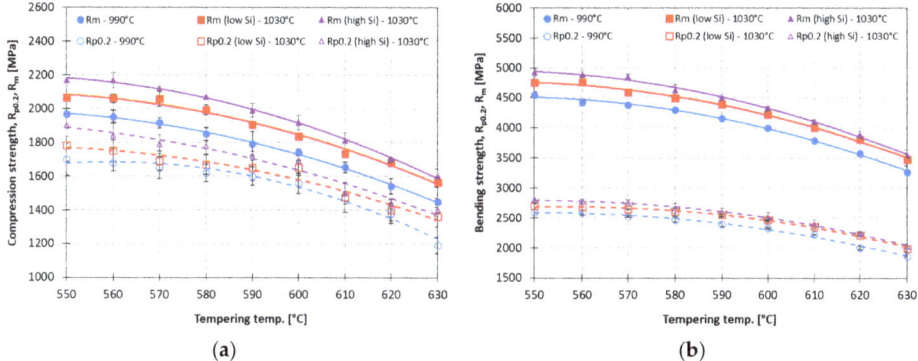

Figure 4. (**a**) Compression and (**b**) bending strength tempering diagram.

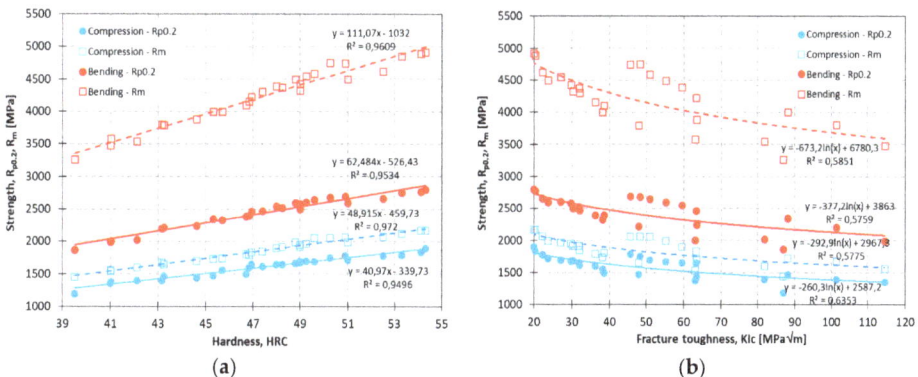

Figure 5. Strength vs. (**a**) hardness and (**b**) fracture toughness correlation.

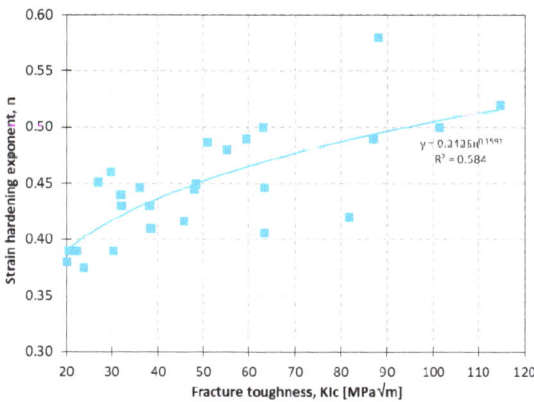

Figure 6. Strain hardening exponent vs. fracture toughness correlation.

Surface roughness analysis of machined 4-point bending and load-carrying capacity specimens revealed deteriorated surface quality with higher average roughness and intensified tearing component when increasing tempering as well as austenitizing temperature [39]. As shown in Figure 7, lower average roughness values (R_a), lower kurtosis (R_{ku}) representing less sharp surface profile and zero skewness (R_{sk}) indicating symmetric profile, with all indicating improved machinability and better surface quality of tool steel substrate, are obtained when increasing hardness (above 45 HRC) and having fracture toughness below 60 MPa√m. However, when material becomes too hard, above 50 HRC or too tough (>80 MPa√m) surface quality quickly deteriorates (Figure 7) and becomes too rough for coating deposition [40].

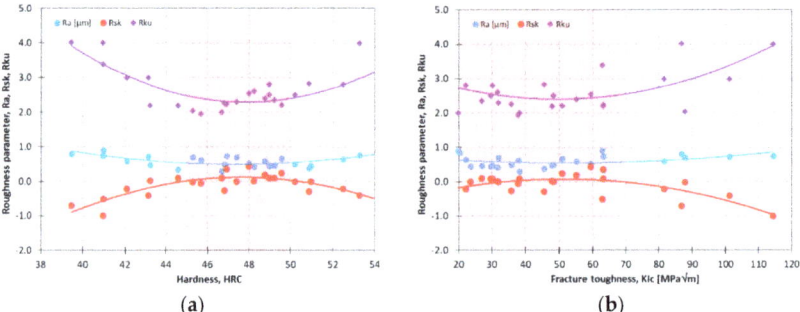

Figure 7. Effect of (**a**) hardness and (**b**) fracture toughness on surface quality.

In terms of tribological properties, coefficient of friction for AISI H11 tool steel was found largely independent on the austenitizing and tempering temperature, displaying average value of about 0.75, which is well in agreement with many tribological investigations on tool steels. On the other hand, wear volume was found to increase with tempering temperature and being dependent also on austenitizing temperature. Furthermore, the form and rate of increase was dependent on the contact conditions used, as shown in Figure 8. In the case of low load-low sliding speed and high load-low sliding speed (Figure 8a) conditions wear volume shows linear increase with tempering temperature, with the higher austenitizing temperature giving faster increase rate and higher wear, especially for high loads. By increasing the sliding speed (low load-high sliding speed and high load-high sliding speed—Figure 8b) effect of austenitizing temperature on wear has been reversed. In these cases, wear shows exponential increase with tempering temperature but drop in values for higher austenitizing temperature, which is more pronounced in the low tempering range (Figure 8b). Furthermore, the best wear resistance and the lowest wear was observed for mid-tempering range between 570 and 590 °C. This indicates that for low sliding speeds higher fracture toughness obtained by lower austenitizing temperature is dominating over hardness, while for high sliding speeds hardness prevails.

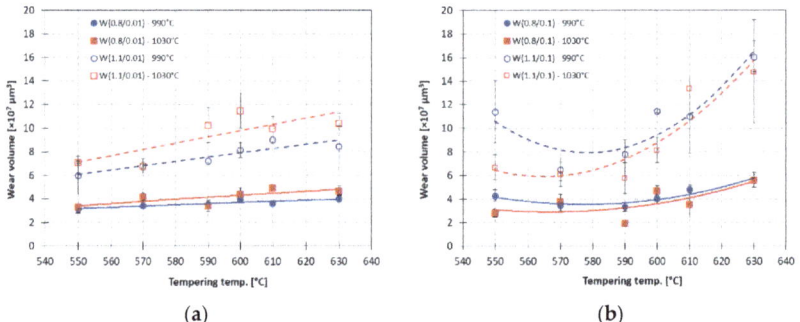

Figure 8. Wear of AISI H11 tool steel as a function of heat treatment temperatures; (**a**) low sliding speed and (**b**) high sliding speed case.

When it comes to friction under abrasive wear conditions, steady-state coefficient of friction for tool steel was found largely unaffected by fracture toughness and hardness when operating within working hardness range (42–52 HRC). On the other hand, as shown in Figure 9, wear volume and wear rate, defined as wear volume divided by load and sliding distance show direct dependency on hardness and fracture toughness. Under the abrasive wear conditions wear rate increases with fracture toughness and is reduced with hardness, with the hardness being found as the most influencing parameter. For the best abrasive wear resistance tool steel hardness should be above 48 HRC (strength above 1900 MPa) and fracture toughness below 55 MPa√m, although not too low (Figure 9).

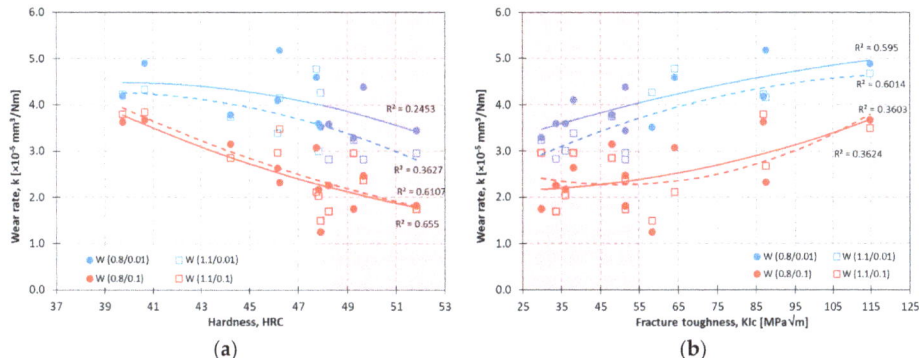

Figure 9. Effect of (**a**) hardness and (**b**) fracture toughness on hot work tool steel wear resistance.

3.2. Thermo-Chemical Treatment vs. Coating

Based on Archard law [41] abrasive wear resistance of materials is in general dependent on their hardness. Hardness increase and thus better wear resistance of steels can be achieved by increasing austenitizing and decreasing tempering temperature [42]. However, even higher surface hardness is provided by thermo-chemical processes, i.e., nitriding and deposition of wear resistant coatings [12]. As shown in Figure 10 plasma nitriding in 5% N_2:95% H_2 gas mixture, providing nitrided surface without compound layer and a hardness of 1100 $HV_{0.05}$ reduces wear of hot work tool steel by about 25%. However, although operating under abrasive wear mode and wear being concentrated within the nitride zone of just 250 μm [4] steel core microstructure and properties, especially hardness and fracture toughness play an important role in terms of surface wear resistance. For lower austenitizing temperature with the hardness in the range of 47–50 HRC and fracture toughness above 35 MPa√m wear rate of nitrided surface increases as the hardness is reduced (higher tempering temperature). On the other hand, increase in austenitizing temperature, providing higher core hardness (51–53 HRC) but much lower toughness (<30 MPa√m; Figure 3) resulted in about 30% higher wear of the investigated AISI H11 tool steel, both at room and high temperature sliding. In this case wear resistance of plasma nitrided tool steel has been found more or less unaffected by core hardness, but mainly defined by reduced fracture toughness, which is further escalated by nitriding and formation of hard brittle surface zone [43].

Although nitriding improves wear resistance of metallic surfaces [44] it cannot match wear resistance provided by PVD and CVD coatings. As shown in Figure 10, coating of tool steel by hard protective TiN/TiB_2 coating can improve surface wear resistance by two orders of magnitude. Furthermore, when load is carried entirely by the top coating and substrate deformation is within elastic range wear rate of the coated surface becomes independent on the substrate preparation and properties, including heat treatment temperatures and/or plasma nitriding used [4].

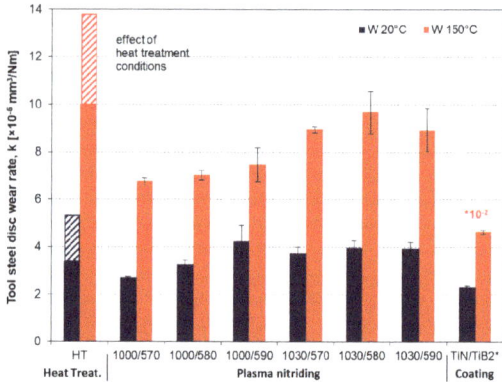

Figure 10. Effect of heat treatment, plasma nitriding and coating deposition on wear resistance of hot work tool steel.

In the case of coated surface load-carrying capacity is the primary requirement. As shown in Figure 11, it doesn't depend just on core hardness but also on the subsurface properties determined by the eventual thermo-chemical process [4]. In the case of plasma nitriding, producing few microns thick compound layer (25% N_2:75% H_2) on the hot work tool steel surface, increase in tempering temperature and corresponding drop in core hardness of less than 5 HRC led to about 30% lower critical loads for TiN/TiB$_2$ coating cracking and flaking or spallation. Even though compound layer may provide additional load support [45], it is quite brittle and thus its cracking resistance dependent on the core hardness. Any crack starting in the compound layer will propagate directly into the substrate but mainly into and through the top coating [46], Figure 12. By avoiding formation (5% N_2:95% H_2) or removing the intermediate compound layer load-carrying capacity drops further. However, it is influenced by the combined effect of fracture toughness and core hardness. For low hardness values (50 HRC) and fracture toughness above 30 MPa$\sqrt{m}$ (austenitizing @1000 °C) increase in fracture toughness obtained by higher tempering temperatures prevails over the reduction in core hardness providing better coating resistance to cracking and load-carrying capacity. Positive effect of core fracture toughness is present also for lower toughness values (<30 MPa$\sqrt{m}$) but it fades away as soon as core hardness drops to about 50 HRC, as shown in Figure 11.

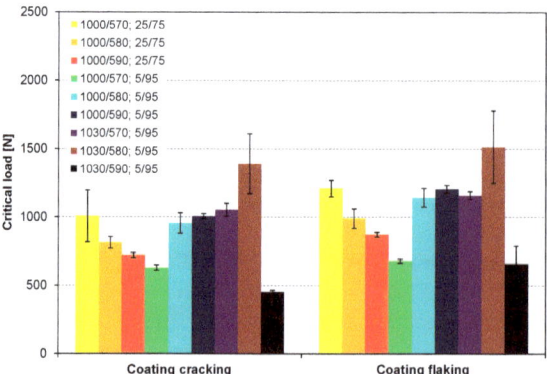

Figure 11. Dependency of load-carrying capacity on substrate heat treatment and plasma nitriding conditions [4]. Reprinted with permission from [4]. Copyright 2015 Elsevier.

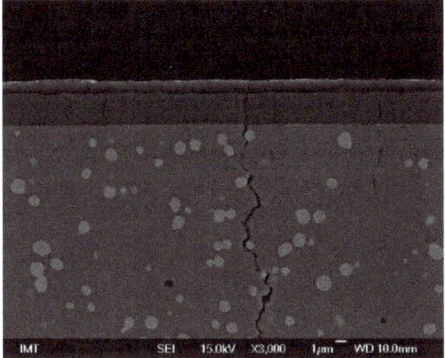

Figure 12. Crack pattern in TiN/TiB$_2$ coated and plasma nitrided hot work tool steel with intermediate compound layer.

3.3. Effect of Substrate Hardness and Fracture Toughness on Coating Performance

Coatings represent only one part of the coated system and require proper substrate giving sufficient support. Thus, beside coating type its properties and performance greatly depend on the substrate, mainly its hardness and fracture toughness. Hardness provides resistance to plastic deformation while fracture toughness is responsible for hindering crack initiation and propagation. As shown in Table 3, by combining different heat treatment parameters (tempering and austenitizing temperature) with eventual deep cryogenic treatment different combinations of fracture toughness and hardness for cold work tool steel can be obtained.

Table 3. Fracture toughness and hardness of P/M cold work tool steel using different heat treatment procedures.

Substrate Treatment	Fracture Toughness K_{Ic} [MPa$\sqrt{m}$]	Hardness HRC	K_{Ic}/HRC
A1	6.1 ± 1.2	65.8 ± 0.2	0.09
A2	10.2 ± 2.0	64.0 ± 0.2	0.159
A3	12.7 ± 0.7	59.3 ± 0.1	0.214
B1	10.4 ± 0.8	65.0 ± 0.3	0.160
B2	12.4 ± 0.7	64.2 ± 0.4	0.193
B3	14.2 ± 0.4	59.5 ± 0.1	0.239

Load-carrying capacity results for three different coatings (monolayer, multilayer, nano-composite) obtained for different core hardness vs. fracture toughness values are shown in Figure 13. In the case of a TiAlN monolayer coating deposited on the hardest tool steel substrate (66 HRC; Group A1) first signs of coating cracking are observed at the critical load of about 3.1 kN. Increase in fracture toughness from 6 to 10 MPa$\sqrt{m}$ at the same time resulting in reduced substrate hardness (66→64 HRC; Group A2) leads to loss in load-carrying capacity, reducing critical load for coating cracking for about 10%, down to 2.8 kN. Further drop in hardness (60 HRC; Group A3), in spite of providing high fracture toughness values results in additional 10% drop in load-carrying capacity (L_C <2.5 kN). However, when maintaining high hardness level (above 65 HRC) any increase in fracture toughness, obtained by combining conventional heat treatment with deep cryogenic treatment [17,27] will provide better crack initiation and propagation resistance and thus higher load-carrying capacity (Group B1).

In the case of AlTiN/TiN multilayer coating with improved cracking resistance, substrate hardness above or equal 64 HRC provides comparable load-carrying capacity, regardless of the K_{Ic}/HRC ratio and level of the fracture toughness obtained (Figure 13). However, with the drop in substrate hardness below 60 HRC (Group A3) critical loads for the beginning of coating cracking are reduced, indicating about 20% lower load-carrying capacity. This time increase in fracture toughness (Group B3) provides some load-carrying capacity improvement, although not reaching the same level as with the harder substrates.

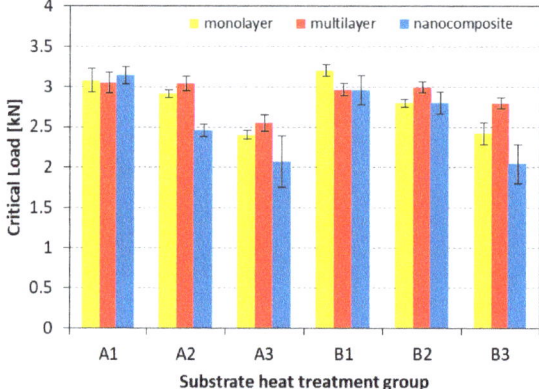

Figure 13. Load-carrying capacity of different coating types as a function of substrate heat treatment and properties.

The most marked effect of substrate properties on load-carrying capacity is found for the most brittle and hardest (Ti,Si)N nano-composite coating. Again, the best load-carrying capacity is provided by the hardest substrate, but very low for too soft one, irrespectively of the fracture toughness level obtained. On the other hand, fracture toughness becomes important for the intermediate working hardness values, as shown in Figure 13. Combination of high fracture toughness (>12 MPa$\sqrt{m}$; Group B2) and working hardness of about 64 HRC guarantees load-carrying capacity similar to hardest substrates but at greatly improved fatigue resistance.

Another coating property, altered by substrate properties is its impact wear resistance. In agreement with high load-carrying capacity the best impact wear resistance for different types of hard coatings is achieved when applied on steel substrate with the highest hardness (Group A1; Figure 14). In this case coatings are removed through abrasive wear mechanism, without any evident coating delamination or cracking. Even the smallest drop in hardness, even though accompanied by increased fracture toughness (Groups B1 and A2) leads to increased coating impact wear and beginning of coating delamination. However, if hardness of about 64 HRC is paired with the fracture toughness above 12 MPa$\sqrt{m}$ (Group B2) coating impact wear can be reduced for up to 30%, closely matching harder but more brittle substrate case. Improved fracture toughness retards crack initiation and propagation while high hardness guarantees high load support. Further rise in fracture toughness, meaning drop in hardness below 60 HRC (Group A3 and B3) has clear negative effect. Low load-carrying capacity of the steel substrate results in excessive deformations and thus in high impact wear of the coating. However, for the most brittle and the hardest coatings (i.e., (Ti,Si)N) substrate hardness is found as the dominant factor in terms of impact wear resistance. In this case, the steel substrate with the highest hardness is the most suitable (66 HRC; Group A1). Use of any softer substrate results in coating cracking and flaking with wear exceeding coating thickness (Figure 14).

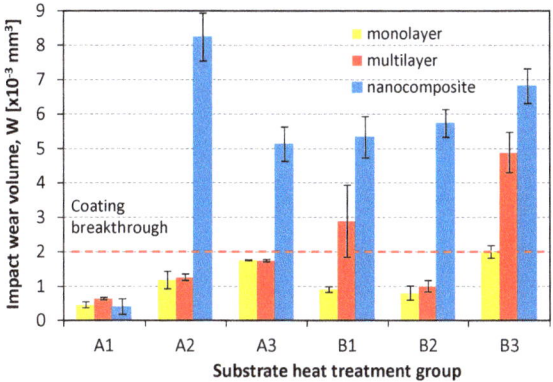

Figure 14. Impact wear volume for different coating types and substrate pre-treatments.

3.4. Effect of Substrate Roughness

If tool is coated its galling resistance depends on the material to be formed and coating type, as well as on the substrate roughness and any additional surface treatment, i.e., post-polishing [25,47]. In the case of hard nitride-based coatings, i.e., TiN, which generally show lower galling resistance than tool steels [48], substrate roughness is even more important. For rough, coarse ground substrates (A1-1 & A1-2) coating gives higher friction (0.4) and about 25% lower galling resistance as indicated by reduced critical loads for stainless steel transfer, as compared to uncoated tool steel (Figure 15). However, by reducing roughness of the substrate (A1-3 and A1-4) coated surface provides comparable resistance to galling and material transfer. Even more, when post-polished (A1-3 + post-polishing) to Ra values of 0.1, use of coated surface can provide up to two times higher resistance to galling. On the other hand, low friction and excellent galling resistance against ASS are obtained by low-friction DLC coating, regardless of the substrate roughness and post-polishing procedure (Figure 15), with the critical loads for galling exceeding 1 kN even under dry sliding [48].

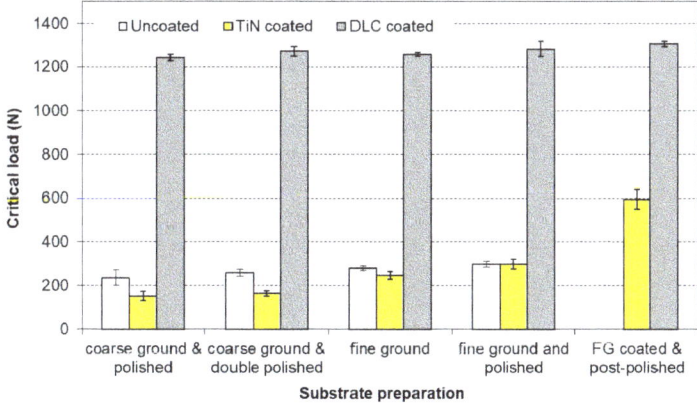

Figure 15. Effect of substrate roughness and post-polishing of coated surface on galling resistance.

4. Conclusions

Effect of steel substrate properties on coating performance can be summarized in the following conclusions:

- Yield and maximum compression and bending strength of the tool steel substrate show rising linear dependency on hardness and reduced trend with fracture toughness. On the other hand, strain hardening exponent has no direct correlation with hardness but shows rising trend with increased fracture toughness.
- Thermo-chemical treatment, i.e., plasma nitriding, provides up to 25% better tool steel wear resistance. However, even when plasma nitrided wear resistance depends on combination of fracture toughness and core hardness. Higher hardness of the core material improves abrasive wear resistance of the surface, but sufficient fracture toughness level needs to be provided. On the other hand, hard wear-resistant coatings outperform all other surface engineering techniques, providing up to two orders of magnitude better tool abrasive wear resistance.
- In the case of coated applications steel substrate must provide sufficient load-carrying capacity and support for the coating. A compound layer can be used as an additional interlayer, but its brittleness results in accelerated coating cracking as the core hardness is reduced. Even for cases without compound layer, high level of steel core hardness (above 50 HRC) is mandatory in order to provide good load support for the top coating. However, high hardness must also be supported by proper level of the fracture toughness (above 30 MPa$\sqrt{m}$).
- Surface roughness and topography have major influence on galling resistance in forming, with smoother surfaces and plateau-like topography providing better results. This is further escalated for coated surfaces, where galling resistance depends on substrate roughness level, coating type and material to be formed. In the case of typical hard ceramic coatings post-polishing of the coated surface and use of smoothened substrate gives about 2 times better galling resistance. On the other hand, for carbon-based low friction coatings post-polishing and roughness of the substrate have very limited effect on the tool resistance against galling and work material transfer.

Workshop practice recommendations:

- Abrasive wear resistance as well as surface quality of hot work tool steel mainly depend on hardness but also on fracture toughness. Good machinability and the best surface quality are obtained when hardness is between 45 and 50 HRC, and high abrasive wear resistance for hardness above 48 HRC. Fracture toughness, however, should be below 55–60 MPa$\sqrt{m}$ to get best performance. On the other hand, coefficient of friction is independent on heat treatment parameters and mainly depends on contact conditions.
- In the case of cold work tool steel, hardness of over 64 HRC is required to obtain sufficient load-carrying capacity of the coated surface, regardless of the coating type used. However, required level of the fracture toughness is dependent also on the coating type. In the case of monolayer coatings, the main parameter is hardness of the substrate with higher the better. For typical multilayer coatings, having improved resistance to crack initiation and propagation, substrate hardness of about 64 HRC is sufficient and high fracture toughness only required at low hardness. However, for very brittle coatings combination of working hardness and high fracture toughness (above 10 MPa$\sqrt{m}$) gives superior results.
- Substrate hardness is the most influential parameter also when it comes to impact wear resistance, which is true for different types of hard coatings. However, except for very brittle coatings fracture toughness of the steel substrate should be above 12 MPa$\sqrt{m}$ and hardness in the range of 64–65 HRC, thus providing combination of high load-carrying capacity, good fatigue properties and superior resistance against impact wear.

Author Contributions: Conceptualization, B.P.; methodology, B.P.; validation, B.P., M.S. and B.Ž.; formal analysis, M.S. and B.Ž.; investigation, M.S. and A.G.; writing—original draft preparation, B.P.; supervision, B.P. All authors have read and agreed to the published version of the manuscript.

Funding: This research was funded by Slovenian Research Agency (research core funding No. P2-0050 and research project L2-9211).

Acknowledgments: Authors would like to acknowledge help from Miha Čekada from Institute Jozef Stefan for the deposition and supply of coatings.

Conflicts of Interest: The authors declare no conflict of interest. The funders had no role in the design of the study; in the collection, analyses, or interpretation of data; in the writing of the manuscript, or in the decision to publish the results.

References

1. Behrens, B.-A.; Doege, E.; Reinsch, S.; Telkamp, K.; Daehndel, H.; Specker, A. Precision forging processes for high-duty automotive components. *J. Mater. Process. Technol.* **2007**, *185*, 139–146. [CrossRef]
2. Podgornik, B.; Leskovšek, V. Experimental evaluation of tool and high-speed steel properties using multi-functional K_{Ic}-test specimen. *Steel Res. Int.* **2013**, *84*, 1294–1301. [CrossRef]
3. Pérez, M.; Belzunce, F.J. The effect of deep cryogenic treatments on the mechanical properties of an AISI H13 steel. *Mater. Sci. Eng. A* **2015**, *624*, 32–40. [CrossRef]
4. Podgornik, B.; Leskovšek, V.; Tehovnik, F.; Burja, J. Vacuum heat treatment optimization for improved load carrying capacity and wear properties of surface engineered hot work tool steel. *Surf. Coat. Technol.* **2015**, *261*, 253–261. [CrossRef]
5. Shanbhag, V.V.; Rolfe, B.F.; Griffin, J.M.; Arunachalam, N.; Pereira, M.P. Understanding galling wear initiation and progression using force and acoustic emissions sensors. *Wear* **2019**, *436–437*, 202991. [CrossRef]
6. Mellouli, D.; Haddar, N.; Köster, A.; Ayedi, H.F. Hardness effect on thermal fatigue damage of hot-working tool steel. *Eng. Fail. Anal.* **2014**, *45*, 85–95. [CrossRef]
7. Norström, L.-Å.; Svensson, M.; Öhrberg, N. Thermal-fatigue behaviour of hot-work tool steels. *Met. Technol.* **1981**, *8*, 376–381. [CrossRef]
8. Leskovšek, V.; Šuštaršič, B.; Jutriša, G. The influence of austenitizing and tempering temperature on the hardness and fracture toughness of hot-worked H11 tool steel. *J. Mater. Process. Technol.* **2006**, *178*, 328–334. [CrossRef]
9. Shirgaokar, M. Technology to Improve Competitiveness in Warm and Hot Forging: Increasing Die Life and Material Utilization. Ph.D. Thesis, Ohio State University, Columbus, OH, USA, 2008.
10. Eller, T.; Greve, L.; Andres, M.; Medricky, M.; Meinders, V.T.; Boogaard, A.V.D.; Boogaard, A.H.V.D. Determination of strain hardening parameters of tailor hardened boron steel up to high strains using inverse FEM optimization and strain field matching. *J. Mater. Process. Technol.* **2016**, *228*, 43–58. [CrossRef]
11. Roberts, G.A.; Kennedy, R.; Krauss, G. *Tool Steels*, 5th ed.; ASM International: Materials Park, OH, USA, 1998.
12. Podgornik, B.; Leskovšek, V. Wear mechanisms and surface engineering of forming tools. *Mater. Tehnol.* **2015**, *49*, 313–324. [CrossRef]
13. Eser, A.; Broeckmann, C.; Simsir, C. Multiscale modeling of tempering of AISI H13 hot-work tool steel–Part 1: Prediction of microstructure evolution and coupling with mechanical properties. *Comput. Mater. Sci.* **2016**, *113*, 280–291. [CrossRef]
14. Li, J.-Y.; Chen, Y.-L.; Huo, J.-H. Mechanism of improvement on strength and toughness of H13 die steel by nitrogen. *Mater. Sci. Eng. A* **2015**, *640*, 16–23. [CrossRef]
15. Ramezani, M.; Pasang, T.; Chen, Z.; Neitzert, T.; Au, D. Evaluation of carbon diffusion in heat treatment of H13 tool steel under different atmospheric conditions. *J. Mater. Res. Technol.* **2015**, *4*, 114–125. [CrossRef]
16. Lerchbacher, C.; Zinner, S.; Leitner, H. Direct or indirect: Influence of type of retained austenite decomposition during tempering on the toughness of a hot-work tool steel. *Mater. Sci. Eng. A* **2013**, *564*, 163–168. [CrossRef]
17. Podgornik, B.; Paulin, I.; Zajec, B.; Jacobson, S.; Leskovšek, V. Deep cryogenic treatment of tool steels. *J. Mater. Process. Technol.* **2016**, *229*, 398–406. [CrossRef]
18. Telasang, G.; Majumdar, J.D.; Padmanabham, G.; Manna, I. Structure–property correlation in laser surface treated AISI H13 tool steel for improved mechanical properties. *Mater. Sci. Eng. A* **2014**, *599*, 255–267. [CrossRef]
19. Dossett, J.L.; Totten, G.E. *ASM Handbook Volume 4A: Steel Heat Treating Fundamental and Processes*; ASM International: Materials Park, OH, USA, 2013.
20. Podgornik, B.; Žužek, B.; Leskovšek, V. Experimental evaluation of tool steel fracture toughness using circumferentially notched and precracked tension bar specimen. *Mater. Perform. Charact.* **2014**, *3*, 20130045. [CrossRef]

21. Groche, P.; Christiany, M. Evaluation of the potential of tool materials for the cold forming of advanced high strength steels. *Wear* **2013**, *302*, 1279–1285. [CrossRef]
22. Leskovšek, V.; Podgornik, B. Vacuum heat treatment, deep cryogenic treatment and simultaneous pulse plasma nitriding and tempering of P/M S390MC steel. *Mater. Sci. Eng. A* **2012**, *531*, 119–129. [CrossRef]
23. Podgornik, B.; Leskovsek, V.; Arh, B. The effect of heat treatment on the mechanical, tribological and load-carrying properties of PACVD-coated tool steel. *Surf. Coat. Technol.* **2013**, *232*, 528–534. [CrossRef]
24. Caliskan, H.; Panjan, P.; Kurbanoglu, C. 3.16 hard coatings on cutting tools and surface finish. In *Comprehensive Materials Finishing*; Hashmi, M.S.J., Ed.; Elsevier: Oxford, UK, 2017; pp. 230–242.
25. Podgornik, B.; Jerina, J. Surface topography effect on galling resistance of coated and uncoated tool steel. *Surf. Coat. Technol.* **2012**, *206*, 2792–2800. [CrossRef]
26. Podgornik, B.; Zajec, B.; Bay, N.O.; Vižintin, J. Application of hard coatings for blanking and piercing tools. *Wear* **2011**, *270*, 850–856. [CrossRef]
27. Podgornik, B.; Sedlaček, M.; Čekada, M.; Jacobson, S.; Zajec, B. Impact of fracture toughness on surface properties of PVD coated cold work tool steel. *Surf. Coat. Technol.* **2015**, *277*, 144–150. [CrossRef]
28. Miletić, A.; Panjan, P.; Skoric, B.; Čekada, M.; Dražić, G.; Kovac, J. Microstructure and mechanical properties of nanostructured Ti–Al–Si–N coatings deposited by magnetron sputtering. *Surf. Coat. Technol.* **2014**, *241*, 105–111. [CrossRef]
29. Panjan, M.; Gunde, M.K.; Panjan, P.; Čekada, M. Designing the color of AlTiN hard coating through interference effect. *Surf. Coat. Technol.* **2014**, *254*, 65–72. [CrossRef]
30. Podgornik, B.; Hogmark, S.; Sandberg, O.; Leskovsek, V. Wear resistance and anti-sticking properties of duplex treated forming tool steel. *Wear* **2003**, *254*, 1113–1121. [CrossRef]
31. Wei, S.; Zhao, T.; Gao, D.; Liu, D.; Li, P.; Qui, X. Fracture toughness measurement by cylindrical specimen with ring-shaped crack. *Eng. Fract. Mech.* **1982**, *16*, 69–82.
32. Gdoutos, E. *Fracture Mechanics Criteria and Aslications*; Kluwer Academic Publishers: London, UK, 1990.
33. *ASTM E9-09 Standard Test Methods of Compression Testing of Metallic Materials at Room Temperature*; ASTM International: West Conshohocken, PA, USA, 2009.
34. *ASTM E290-09 Standard Test Methods for Bend Testing of Material for Ductility*; ASTM International: West Conshohocken, PA, USA, 2009.
35. *ISO 4287:1997 Geometrical Product Specifications (GPS)—Surface Texture: Profile Method—Terms, Definitions and Surface Texture Parameters*; International Organization for Standardization: Geneva, Switzerland, 1997.
36. Podgornik, B.; Hogmark, S.; Pezdirnik, J. Comparison between different test methods for evaluation of galling properties of surface engineered tool surfaces. *Wear* **2004**, *257*, 843–851. [CrossRef]
37. Podgornik, B.; Kafexhiu, F.; Nevosad, A.; Badisch, E. Influence of surface roughness and phosphate coating on galling resistance of medium-grade carbon steel. *Wear* **2020**, *446–447*, 203180. [CrossRef]
38. Pavlina, E.J.; Van Tyne, C.; Van Tyne, C. Correlation of yield strength and tensile strength with hardness for steels. *J. Mater. Eng. Perform.* **2008**, *17*, 888–893. [CrossRef]
39. Podgornik, B.; Puš, G.; Žužek, B.; Leskovšek, V.; Godec, M. Heat treatment optimization and properties correlation for H11-type hot-work tool steel. *Met. Mater. Trans. A* **2017**, *49*, 455–462. [CrossRef]
40. Wiklund, U.; Gunnars, J.; Hogmark, S. Influence of residual stresses on fracture and delamination of thin hard coatings. *Wear* **1999**, *232*, 262–269. [CrossRef]
41. Zmitrowicz, A. Wear patterns and laws of wear—A review. *J. Theor. Appl. Mech.* **2006**, *44*, 219–253.
42. Podgornik, B. Fracture toughness and wear resistance of hot work tool steel related to heat treatment conditions. In Proceedings of the National Conference on Vacuum Heat Treatment and Heat Treatment of Tools, Puchov, Slovakia, 20–21 November 2018; p. 10.
43. Fernandes, F.; Heck, S.; Picone, C.; Casteletti, L. On the wear and corrosion of plasma nitrided AISI H13. *Surf. Coat. Technol.* **2020**, *381*, 125216. [CrossRef]
44. Béjar, M.; Schnake, W.; Saavedra, W.; Vildósola, J. Surface hardening of metallic alloys by electrospark deposition followed by plasma nitriding. *J. Mater. Process. Technol.* **2006**, *176*, 210–213. [CrossRef]
45. Podgornik, B.; Vižintin, J.; Wänstrand, O.; Larsson, M.; Hogmark, S. Wear and friction behaviour of duplex-treated AISI 4140 steel. *Surf. Coat. Technol.* **1999**, *120*, 502–508. [CrossRef]

46. Podgornik, B.; Leskovšek, V.; Steiner-Petrovic, D.; Jerina, J. Load-carrying capacity of coated high-speed steel at room and elevated temperatures. In Proceedings of the 2nd Workshop on High Temperature Tribology-Mechanisms and Control of Friction & Wear, Luleå, Sweden, 18 March 2013; pp. 23–24.
47. Podgornik, B.; Hogmark, S.; Sandberg, O. Proper coating selection for improved galling performance of forming tool steel. *Wear* **2006**, *261*, 15–21. [CrossRef]
48. Podgornik, B.; Hogmark, S. Surface modification to improve friction and galling properties of forming tools. *J. Mater. Process. Technol.* **2006**, *174*, 334–341. [CrossRef]

 © 2020 by the authors. Licensee MDPI, Basel, Switzerland. This article is an open access article distributed under the terms and conditions of the Creative Commons Attribution (CC BY) license (http://creativecommons.org/licenses/by/4.0/).

Article

Distribution of the Deposition Rates in an Industrial-Size PECVD Reactor Using HMDSO Precursor

Žiga Gosar [1,2], Denis Đonlagić [3], Simon Pevec [3], Bojan Gergič [3], Miran Mozetič [4,5], Gregor Primc [4,5], Alenka Vesel [4,5,*] and Rok Zaplotnik [4,5]

[1] Development and Technology Department, Elvez Ltd, Ulica Antona Tomšiča 35, 1294 Višnja Gora, Slovenia; ziga.gosar@elvez.si
[2] Ecotechnology Programme, Jozef Stefan International Postgraduate School, Jamova Cesta 39, 1000 Ljubljana, Slovenia
[3] Faculty of Electrical Engineering and Computer Science, University of Maribor, Koroška Cesta 46, 2000 Maribor, Slovenia; denis.donlagic@um.si (D.Đ.); simon.pevec@um.si (S.P.); bojan.gergic@um.si (B.G.)
[4] Department of Surface Engineering, Jozef Stefan Institute, Jamova Cesta 39, 1000 Ljubljana, Slovenia; miran.mozetic@ijs.si (M.M.); gregor.primc@ijs.si (G.P.); rok.zaplotnik@ijs.si (R.Z.)
[5] Research and Development Department, Plasmadis Ltd., Teslova Ulica 30, 1000 Ljubljana, Slovenia
* Correspondence: alenka.vesel@guest.arnes.si

Citation: Gosar, Ž.; Đonlagić, D.; Pevec, S.; Gergič, B.; Mozetič, M.; Primc, G.; Vesel, A.; Zaplotnik, R. Distribution of the Deposition Rates in an Industrial-Size PECVD Reactor Using HMDSO Precursor. *Coatings* **2021**, *11*, 1218. https://doi.org/10.3390/coatings11101218

Academic Editors: Alessio Lamperti and Alessandro Patelli

Received: 26 August 2021
Accepted: 30 September 2021
Published: 5 October 2021

Publisher's Note: MDPI stays neutral with regard to jurisdictional claims in published maps and institutional affiliations.

Copyright: © 2021 by the authors. Licensee MDPI, Basel, Switzerland. This article is an open access article distributed under the terms and conditions of the Creative Commons Attribution (CC BY) license (https://creativecommons.org/licenses/by/4.0/).

Abstract: The deposition rates of protective coatings resembling polydimethylsiloxane (PDMS) were measured with numerous sensors placed at different positions on the walls of a plasma-enhanced chemical vapor deposition (PECVD) reactor with a volume of approximately 5 m^3. The plasma was maintained by an asymmetric capacitively coupled radiofrequency (RF) discharge using a generator with a frequency 40 kHz and an adjustable power of up to 8 kW. Hexamethyldisiloxane (HMDSO) was leaked into the reactor at 130 sccm with continuous pumping using roots pumps with a nominal pumping speed of 8800 m^3 h^{-1} backed by rotary pumps with a nominal pumping speed of 1260 m^3 h^{-1}. Deposition rates were measured versus the discharge power in an empty reactor and a reactor loaded with samples. The highest deposition rate of approximately 15 nm min^{-1} was observed in an empty reactor close to the powered electrodes and the lowest of approximately 1 nm min^{-1} was observed close to the precursor inlet. The deposition rate was about an order of magnitude lower if the reactor was fully loaded with the samples, and the ratio between deposition rates in an empty reactor and loaded reactor was the largest far from the powered electrodes. The results were explained by the loss of plasma radicals on the surfaces of the materials facing the plasma and by the peculiarities of the gas-phase reactions typical for asymmetric RF discharges.

Keywords: HMDSO; PECVD; deposition rate; uniformity of deposition; polymerization; organosilicon thin films

1. Introduction

Many materials should be coated with a thin protective layer to provide an adequate surface finish and stability in harsh environments [1–5]. A variety of techniques have been proposed, and a few have also been commercialized [6–10]. One technique for depositing compact and hydrophobic films similar to polydimethylsiloxane (PDMS) is plasma polymerization. A suitable monomer is provided and partially dissociated and ionized under plasma conditions [11,12]. The radicals adhere to the surface of any object exposed to the plasma and form a thin film. The structure and composition of the coating depend on the type of precursor, plasma parameters and specifics of the discharge used for sustaining gaseous plasma [13–17]. The growth kinetics is complex and difficult to control because of the large number of radicals formed in the gaseous plasma. An early report of the kinetics was presented by Bourreau et al. [18]. The authors used different sources to deposit protective coatings rich in silicon oxides: silane (SiH$_4$), hexamethyl disiloxane (HMDSO) and tetraethoxysilane (TEOS). They correlated the evolution of the coverage with the deposition kinetics and compared the growth rates. The profiles were independent of the substrate temperature or the deposition rate when silane was used as a precursor. In the case of organic precursors, however, the deposition rate decreased with an increase in the deposition temperature. They found the adsorption–desorption phenomena to be important factors for the coverage evolution. At low deposition temperatures, the film growth rate was sensitive to ion surface bombardment and resulted in a non-conformal deposit even in compounds with high surface mobility.

Theirich et al. [19] studied the gas-phase reactions in HMDSO/O_2 mixtures and pressures between 20 and 70 Pa. Plasma was characterized by mass spectrometry and infrared spectroscopy. They found the film homogeneity dominated by the precursor content and its spatial distribution in the gas or plasma phase. Three reactive intermediate species were proposed to act as a precursor for silica-like film growth, all having a mass of 148 Da, so the authors concluded that further work should be performed to distinguish between the radicals.

In their classic paper, Hegemann et al. [20] studied the deposition rate and three-dimensional uniformity of capacitively coupled radio-frequency (RF) plasma useful for depositing protective layers using HMDSO as a precursor. The deposition rate increased with monomer gas flow, whereas it was independent of pressure. Large differences in the deposition rates at different positions of the samples were reported, as well as the influence of the dimensions of the samples on the growth kinetics. In another paper [21], the same group investigated the deposition rate in symmetrical and asymmetrical electrode configurations and found that the deposition rate depended on the so-called reaction parameter (power input per gas flow of the monomer).

More recently, Ropcke's [22] group performed a detailed characterization of the HMDSO plasma by optical emission spectroscopy (OES) in the visible spectral range and infrared laser absorption spectroscopy (IRLAS). They used a plasma reactor of a rather large power density (discharge power per volume of the discharge chamber) of the order of 100 W per liter. They managed to derive the concentrations of the various stable and unstable plasma species, which were found to be in the range between 10^{17} and 10^{21} m^{-3}. They also studied the influence of the discharge parameters, such as power, pressure and gas mixture, on the molecular concentrations. Based on the construction principle of the reactor, the plasma generation was characterized by a certain degree of inhomogeneity with different temperature zones, i.e., hottest, hot and colder zones. This complexity was characterized by the multiple molecular species, including the HMDSO precursor and products in the ground and excited states existing in the plasma.

Plasma-enhanced chemical vapor deposition (PECVD) technique for the deposition of protective coatings from HMDSO was commercialized decades ago despite the experimentally observed non-homogeneities and instabilities, which may lead to inadequate properties of the deposited films. Recently, Gosar et al. [16] reported that the composition of the deposited films depended on the time-evolution of the plasma parameters, although the discharge parameters (power, pressure, flow rate, pumping speed) remained fairly constant. The time evolution was explained by the drifting plasma parameters, which was detrimental to the quality of the protective films, especially where a rather high power density was used to sustain the gaseous plasma. At low discharge powers, however, the properties of the deposited films were not time dependent. The quality of the films is a crucial parameter in the industrial application of the PECVD technique using HMDSO, so many industrial reactors operate at a very low power density to minimize the risk [23]. On the other hand, the low power density results in a poor deposition rate, as explained by the above-cited authors.

The problem of plasma non-uniformity and the resultant deviations of the film thickness from the desired value in large plasma reactors may be suppressed by rotating samples upon plasma processing [24]. This is a standard solution in commercial reactors for depositing protective coatings in batch mode. The samples are mounted on planetaria and moved through zones with different plasma parameters. The relatively long treatment time (several minutes in commercial plasma reactors) ensures a reasonable coating thickness and uniformity. Still, the problem arising from plasma inhomogeneities is not solved, so there is a need to develop configurations of plasma reactors with deposition rates that are as uniform as possible throughout the entire reactor.

Commercial reactors for the deposition of the protective coatings using the HMDSO as the precursor may be upgraded if the non-uniformities are known and understood. Several groups have already reported the non-uniformity in plasma parameters, but only a few have measured the deposition rates in different parts of the plasma reactor [12,13,20]. The present paper provides measurements of the deposition rate performed with several sensors mounted in selected positions within a large plasma reactor. The deposition rates for an empty and a fully loaded reactor were measured to reveal the influence of the samples on the non-uniformity of the deposition rates.

2. Materials and Methods

2.1. Plasma-Enhanced Chemical Vapor Deposition Reactor

The industrial PECVD reactor useful for the deposition of PDMSO-like coatings was presented in detail in our previous paper [25]. The reactor has a cylindrical shape with a diameter of 1.9 m and a height of 1.8 m. During the deposition, the reactor was pumped with two roots pumps with a total nominal pumping speed 8800 $m^3\ h^{-1}$, backed by two rotary pumps of a total nominal pumping speed 1260 $m^3\ h^{-1}$. Before the deposition, in order to get the base pressure as low as possible (around 0.02 Pa), the reactor was also pumped with two diffusion pumps with a total pumping speed 35,000 L/s. HMDSO was the only gas that was introduced into the plasma reactor. It was introduced through a calibrated flow controller. The pressure was measured with a Pirani gauge. At the HMDSO inlet of 130 sccm (cm^3/min_{STP}), which is the standard flow rate used in mass production, the pressure was about 4 Pa. Plasma was characterized by optical emission spectroscopy (OES) AvaSpec-Mini4096CL (Avantes, Apeldoorn, Netherlands) near one of the powered electrodes as shown in Figure 1.

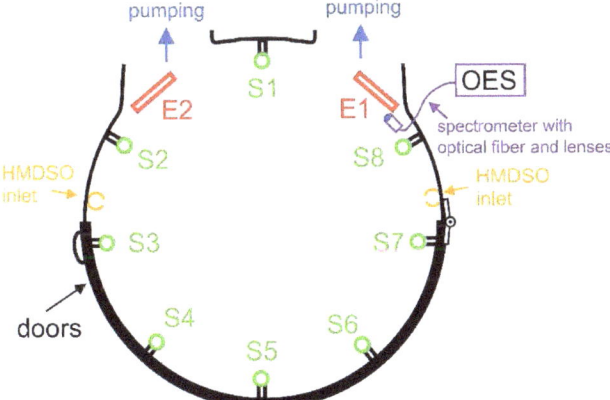

Figure 1. Cross-section of the cylindrical PECVD reactor with the position of the pump ducts, powered electrodes (E1, E2), HMDSO inlet, sensors for deposition rate measurements (S1–S8), OES lens, optical fiber and OES spectrometer.

An asymmetric capacitively coupled RF discharge was used for sustaining gaseous plasma. The discharge was powered by an RF generator (PE II 10K, Advanced Energy, Denver, CO, USA) operating at 40 kHz and adjustable power between 1 and 8 kW. A couple of powered electrodes were mounted close to the pump duct. The area of each electrode was approximately 0.4 m^2. The area of the grounded electrode (housing) was approximately 16 m^2. The ratio between the areas of the powered and grounded electrodes was approximately 40. Therefore, the plasma was sustained by an asymmetrical capacitive coupled RF discharge, and the gradients in the plasma parameters were expected.

The HMDSO inlet was provided through vertically oriented grounded metallic tubes, as shown in Figure 1. The tubes were positioned close to the grounded walls of the plasma reactor. They had small holes separated by 15 cm. The precursor was thus introduced into the reactor unevenly.

2.2. Sensors of the Deposition Rate

Eight sensors were fixed on the sidewalls of the plasma reactor (BDS-MF, Arzuffi, Vallezzo Bellini, Italy) for the real-time monitoring of the deposition rate, as shown in Figure 1 (marked with S1 to S8). The sensor S1 was positioned on the rough grid, which separates the discharge chamber from the polycold pump duct, which was not used in this experiment. A photo of the sensor S1 is shown in Figure 2a. Other sensors were fixed on the chamber walls on the grounded housing.

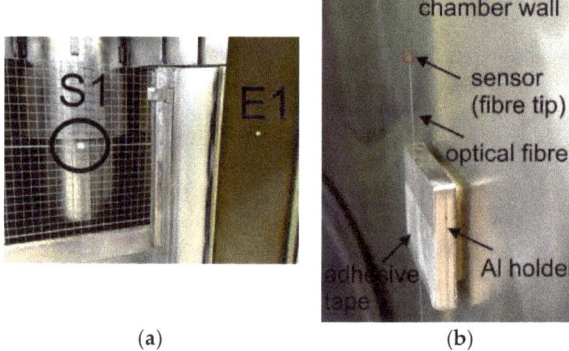

Figure 2. (**a**) Fixation of the sensor S1 and (**b**) the photo of a sensor mounting.

Each sensor essentially consisted of a single-mode optical fiber, which was cleaved and exposed to the processing chamber on one side, while being connected to an appropriate opto-electronics signal integration system on the other side. Opto-electronics signal integration system launched light into the fiber, while acquiring and processing back-reflected optical power from cleaved fiber end. Since the deposited PDMSO-like layer had a different refractive index than vitreous silica, the back-reflectance from the cleaved fiber end changed during the PDMSO deposition. This change was correlated with the change in thickness of the deposited material. The correlation was obtained by an appropriate calibration and processing of acquired signals. One such sensor was already used in our previous work [26], where the deposition rates measured with such sensor in real time were the same as those measured with time-consuming post-deposition surface analysis such as atomic force microscopy (AFM) (Solver PRO, NT-MDT, Moscow, Russia), X-ray photoelectron spectroscopy (XPS) (TFA XPS Physical Electronics, Münich, Germany) and time-of-flight secondary ion mass spectrometry (ToF-SIMS) depth profiles (ToF-SIMS 5 instrument, ION-TOF GmbH, Münster, Germany).

Figure 2b shows a photo of an optical fiber sensor fixed on the aluminum holder, which was fixed on the wall of the plasma reactor.

2.3. Optical Emission Spectroscopy (OES)

An optical lens was mounted in the PECVD reactor (Figure 1) and connected with optical fiber through optical feedthrough to a standard low-resolution optical spectrometer Avantes AvaSpec-Mini4096CL (Avantes, Apeldoorn, Netherlands). The spectrometer measures light emission spectra. The device is based on AvaBench 75 symmetrical Czerny Turner design with a 4096-pixel CCD detector with a focal length of 75 mm. The range of measurable wavelengths is from 200 nm to 1100 nm, and the wavelength resolution is 0.5 nm. The spectrometer has a USB2.0 interface, enabling high sampling rates up to 150 spectra per second. Signal-to-noise ratio is 300:1. Integration time is adjustable from 30 μs to 50 s. At integration times below 6.5 ms, the spectrometer itself performs internal averaging of spectra before transmitting them through the USB interface. The spectrometer was connected to the process computer via USB. The integration time was set to 5 s.

3. Results and Discussion

Plasma in the empty discharge chamber was characterized by OES. Here it should be stressed that an empty chamber means that there are no samples and no planetaria (sample holders) inside the reactor. A typical OES spectrum is shown in Figure 3. The spectrum consists of Balmer series of radiative transitions of H atoms from excited states to the first excited state. The next prominent spectral feature arises from the relaxation of the CH radicals with the bandhead at 431 nm. Other features are marginal. The OES indicates partial dissociation of the precursor molecules, but otherwise, it does not provide any additional significant information. Other radicals are also in the reactor, but their emission is marginal. More interesting is the intensity of the spectral features versus the discharge power. Figure 4 shows quite linear curves. The emission intensity depends on the electron density and temperature as well as the density of radicals in the ground state, and the dependence is not trivial. Still, the behavior of the lines in Figure 4 indicates either more extensive dissociation of the precursor molecules or higher electron density/temperature or both at higher power. This observation is expected, considering that the optical lens for acquiring spectra was mounted just next to the powered electrode.

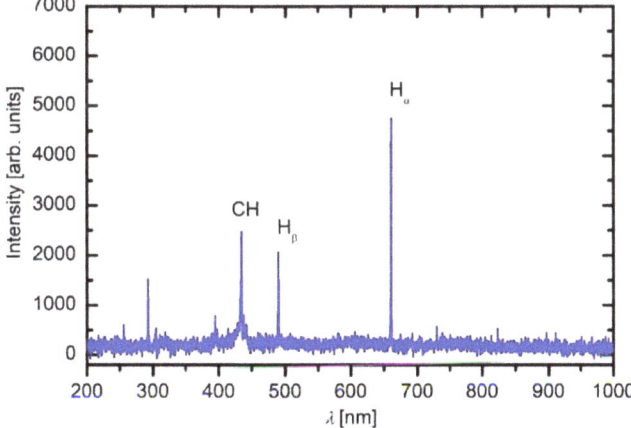

Figure 3. An optical spectrum of the plasma at the discharge power of 5 kW and 130 sccm of HMDSO.

Figure 5 shows the measured deposition rate versus the discharge power. Interestingly enough, the deposition rate is rather constant in the broad range of powers from approximately 2 to 7 kW. This observation is not correlated with data in Figure 4, which shows a gradual increase in the emission intensity. This paradox can be explained by a fact already reported for small experimental systems [16]: only moderate dissociation of the precursor is sufficient for a reasonable deposition rate. Extensive dissociation of the precursor leads to the formation of various radicals that do not stick to the sample surface but are pumped out from the system; therefore, in cases where large power densities are used for sustaining plasma in HMDSO, the deposit does not resemble PDMS but rather silica. Detailed study of the transition from polymer-like films to films rich in silicon oxides was reported in [16]. The power density used in this study was at least 10 times lower than the power density needed for such full transition; however, there are still mild transitions, towards films richer in silicon, that can affect the deposition rates seen in Figure 5.

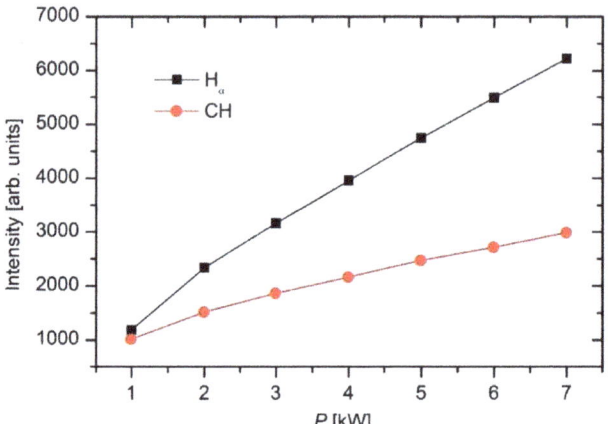

Figure 4. The intensity of the H_α and CH lines at 656 nm and 431 nm as a function of the discharge power at 130 sccm HMDSO.

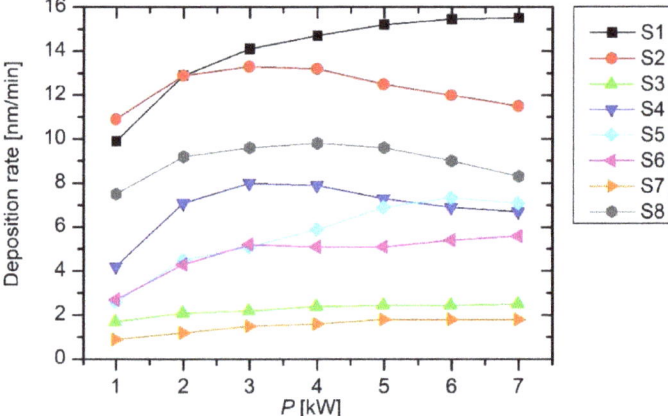

Figure 5. The deposition rate versus the discharge power in the empty reactor.

Both Figures 5 and 6 indicate large differences in the deposition rate at different locations ranging from 1.6 to 14.7 nm min^{-1}. The deposition rate is the largest for sensor S1. This sensor was placed on the grid between the electrodes, as shown in Figures 1 and 2. The highest deposition rate is on the surface, where it is not needed because the radicals at the position of S1 are likely to be pumped away from the system. The high deposition rate indicates a high density of radicals that are capable of forming the protective coating. According to the state-of-the-art, such radicals are partially dissociated HMDSO molecules, including those found at the mass of 148 Da [19]. In the empty chamber, these radicals are denser or more concentrated at the position near the pump ducts than anywhere else in the system, as revealed in Figures 5 and 6.

Figure 6 shows the thickness of the coating obtained from the sensors' signals versus the treatment time for the empty plasma reactor. One can observe almost perfectly linear behavior, which indicates excellent stability of plasma parameters during the deposition of the protective coatings. The stability may be a consequence of the appropriately low pressure in the reactor, which prohibits instabilities that may appear because of the cluster formation [27] and thus the loss of radicals useful for the deposition of the protective coating.

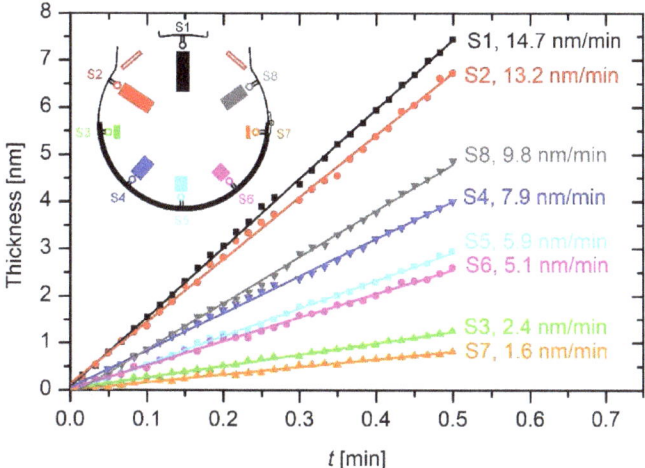

Figure 6. The thickness of the deposited films derived from the sensors' signals (points) with linear fits (lines) versus the plasma treatment time in an empty reactor at a power of 4 kW at 130 sccm HMDSO. In the inset figure, a deposition rate is presented with the height of the column at a sensor position.

Examining Figure 5 and compared to Figure 1, one observes the next largest deposition rate at sensors S2 and S8, which were located a bit farther from the pump ducts. In fact, sensors S2 and S8 were located between the gas inlet and the powered electrodes, as shown in Figure 1. The possible reasons for favored deposition rate at these positions will be discussed later in this report.

The deposition rates at the position of sensors far from the electrodes are lower but still reasonably high. For example, Figure 5 reveals the deposition rates of about 6 nm min^{-1} for the sensors S4, S5, and S6. Conversely, sensors S3 and S7, which were placed close to the gas inlet but away from the powered electrodes, show a poor deposition of approximately 2 nm min^{-1}.

The distribution of the deposition rate in the plasma reactor provides a qualitative model of the gas kinetics that allows the most reasonable degree of fragmentation of the precursor molecules. The injected HMDSO molecules do not interact with the solid materials but should be partially dissociated to radicals with a reasonable sticking coefficient. The plasma density far from the powered electrodes in the reactor used for these experiments is only on the order of 10^{14} m^{-3} [23]. Such a low density of electrons does not enable immediate dissociation to useful fragments. This may explain the poor deposition rates detected by sensors S3 and S7, located close to the gas inlet but away from the powered electrodes. The molecules should be allowed a prolonged residence time in the weakly ionized gaseous plasma to dissociate into useful radicals. The residence time will be estimated later in this paper. The injected precursor molecules enter the plasma reactor with a significant drift velocity but quickly thermalize (assume the random motion after a few elastic collisions). The motion is then governed by diffusion, i.e., it is random. The molecules suffer numerous collisions with plasma electrons while diffusing from the source (gas inlet) to the position of the sensors S4, S5, and S6. The gas at the position of these sensors is thus reasonably well dissociated, which favors the deposition on the surfaces far away from the electrodes. As mentioned above, the residence time of the injected molecules is too short to cause significant deposition at the positions of sensors S3 and S7.

Sensors S2 and S8 are as close to the gas inlet as S3 and S7, but Figure 5 indicates a deposition rate several times higher at S2 and S8 compared to S3 or S7. This paradox may be explained by the larger residence time of molecules striking the surface of the sensors at positions S2 and S8, but the variation of the plasma density versus the distance from the powered electrode may be more important. The asymmetric capacitively coupled RF discharge is characterized by an oscillating sheath next to the powered electrode. Since the frequency of these oscillations is rather low (the RF generator operates at 40 kHz), the electrons oscillate within the sheath and gain energy enough for a rather extensive dissociation and ionization of the gaseous molecules within the oscillating sheath [28]. Therefore, the dissociation of the precursor molecules is more extensive next to the electrodes than in the bulk plasma far away from the powered electrodes. As a result, the deposition rate at the sensors S2 and S8 is favorable despite the proximity of the gas inlet.

The radicals stick to surfaces of any material facing plasma; therefore, the deposition rate as determined by the sensors located in the reactor according to Figure 1 should be lower if the reactor is additionally loaded with samples. To study the influence of samples on the deposition rate, samples were mounted on the planetaria, as shown in Figure 7. About 250 medium-sized, approximately 40-cm-long samples, which represented about 100% of the total chamber capacity, were evenly distributed inside the chamber. The height and the diameter of the planetaria were 160 cm and 55 cm, respectively, and the distance between axles was around 60 cm. The planetaria were spinning at a speed of 6 rpm. The deposition rate measurements were repeated with sensors located at the same positions as in the empty chamber. The results are shown in Figure 8. The highest deposition rate was observed for the sensors S2 and S8. These sensors are located between the gas inlet and the powered electrode (Figure 1). The deposition rate at the positions S2 and S8 are about an order of magnitude greater than at any other position except near the pump ducts. The presence of samples in the plasma reactor, therefore, influences the deposition rate significantly. Not only is it lower than in the empty reactor (compare Figures 6 and 9), but a reasonably large deposition rate is observed only in the region close to the electrodes (S2, S8, and S1). Elsewhere, the deposition rate is below 1 nm min^{-1}.

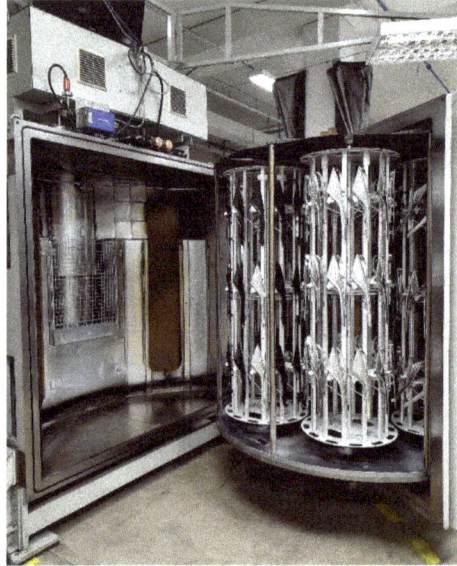

Figure 7. A photo of the fully loaded chamber with samples mounted on planetaria.

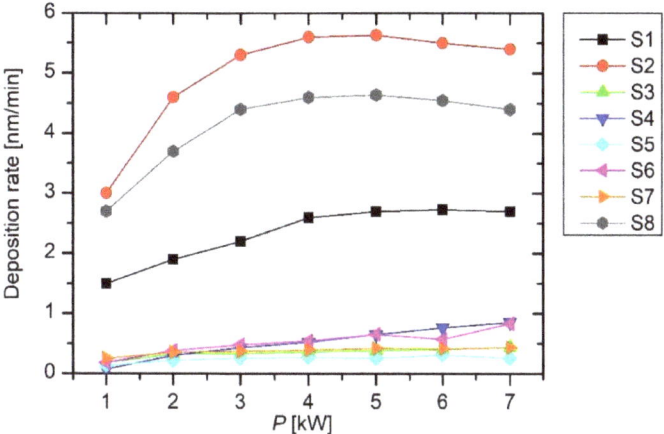

Figure 8. The deposition rate versus the discharge power in reactor loaded with samples.

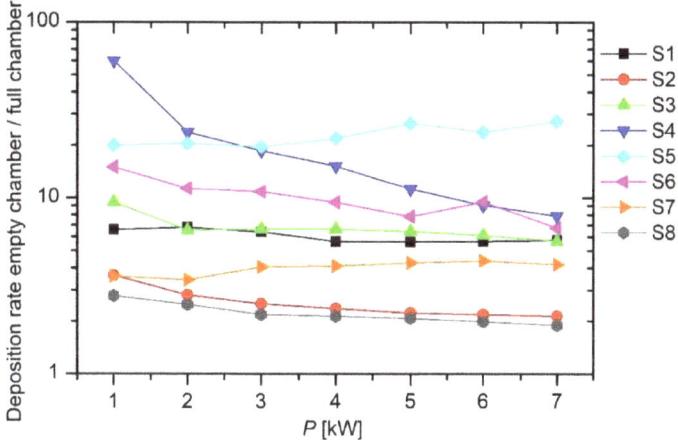

Figure 9. The deposition rate versus the discharge power in reactor loaded with samples.

The very low deposition rate at S4, S5, and S6, as observed in Figure 8, is explained by the loss of radicals on the surfaces of the samples. As discussed above, the plasma density away from the electrodes is low, so the loss of radicals useful for depositing protective coating cannot be balanced by production because of electron-impact dissociation. Conversely, the deposition rate close to the powered electrode (sensors S2 and S8) remains reasonably high because of the higher electron energy in the oscillating sheath.

The ratio between the deposition rate in an empty reactor and a full reactor is shown in Figure 9. The highest ratio of 10–20 is observed for sensors positioned far from the electrodes. This observation was already explained by the loss of radicals on the surface of the samples. However, the ratio is much lower for the sensors positioned close to the powered electrodes. For sensors S2 and S8, the ratio is approximately 3 for the lowest power of 1 kW and only 2 for the highest power of 7 kW. The power-dependence of the ratio is explained by the fact that the electron energy in the vicinity of the powered electrodes is much higher than far from the electrodes, so a significant fraction of injected HMDSO molecules get dissociated and thus contribute to the film growth.

The upper discussion reveals the crucial role of the residence time of molecules in the plasma reactor. Gaseous molecules diffuse in the plasma reactor because the random velocity is much higher than the drifting from the gas inlet to the pump ducts. The drift velocity of gaseous molecules at the entrance to the pump ducts can be calculated if the effective pumping speed at that position is known. The effective pumping speed depends on the nominal pumping speed of the roots pumps and the conductivity of any vacuum elements mounted between the roots pumps and the plasma reactor. The conductivity is difficult to determine, but one can also determine the effective pumping speed from the measured gas flow and pressure inside the reactor by considering the constant mass flow:

$$p_1 \, S_1 = p_2 \, S_2. \tag{1}$$

Here, p_1 is the atmospheric pressure, S_1 is the gas flow as measured by the flow controller, p_2 is the measured pressure in the plasma reactor, and S_2 is the effective pumping speed at the grid which separates the plasma reactor and the pump ducts. Taking into account the measured values, i.e., $p_1 = 10^5$ Pa, $S_1 = 130$ cm^3/min $= 2 \times 10^{-6}$ m^3 s^{-1}, $p_2 = 4$ Pa, one can estimate the effective pumping speed as:

$$S_2 = p_1 S_1 / p_2 = 0.05 \text{ m}^3 \text{ s}^{-1} = 180 \text{ m}^3 \text{ h}^{-1}. \tag{2}$$

As calculated from Equation (1), the effective pumping speed is an order of magnitude lower than the nominal pumping speed of the roots pumps. This observation may be explained by the deviation of the real pumping speed of the roots pumps from the nominal value (the latter is just the maximum pumping speed at optimal conditions) and the limited conductivity of vacuum elements mounted between the plasma reactor and the roots pumps.

There is a negligible pressure gradient throughout the plasma reactor, because the conductivity is orders of magnitude greater than the effective pumping speed. The cross-section of the plasma reactor is a product of the reactor diameter and height, i.e., $A = 3.5$ m^2. The gas drift velocity from the source to the pump ducts is:

$$v = S_2 / A = 0.014 \text{ m s}^{-1}. \tag{3}$$

This value is orders of magnitude lower than the random velocity due to the thermal motion of the molecules, which is:

$$\overline{v} = \sqrt{\frac{8kT}{\pi m}} = 200 \ m \ s^{-1}. \tag{4}$$

In Equation (4), we considered the room temperature ($T = 300$ K) and the HMDSO mass $m = 162$ Da. By considering the distance between the gas inlet and the grid separating the reactor from the pump ducts of $l = 1$ m, one can estimate the average residence time of gaseous molecules as:

$$\tau = l/v = 80 \text{ s}. \tag{5}$$

The residence time as calculated from Equation (5) is an averaged value taking into consideration the simple calculations. Because the random velocity as calculated from Equation (4) is orders of magnitude higher than the drift velocity as determined from Equation (3), the residence time is spread broadly from the value calculated using Equation (5), and thus it should be taken just as an estimation. In any case, the residence time is long enough to assure for numerous collisions with plasma electrons. The large residence time is the reason for the rather large deposition rate at any position far from the gas inlet in the empty reactor. The maximal deposition is observed on the grid near the pump ducts (sensor S1) in the empty reactor. The radicals entering the pump ducts are likely to have been created well before reaching the grid.

Plasma reactors are useful only when the coatings are deposited on various products mounted on the planetaria. Technologically relevant results are presented in Figure 8. The deposition rate at sensor S1 (mounted on the grid near the pump ducts) is moderate at about 2 nm min^{-1}, which is favorable from the technological point of view. Still, a significant fraction of the radicals useful for the thin film deposition is pumped out from the reactor. However, the major deficiency of the plasma reactor is the poor deposition rate at any other position. Despite the long residence time of gaseous radicals, the deposition rate is poor because of the loss of radicals on the samples placed on the planetaria. The only useful part of the reactor, when loaded with samples, is at positions S2 and S8, so close to the powered electrodes. The discharge configuration in this reactor is, therefore, inadequate. The configuration with electrodes placed opposite to the pump duct should be better.

No sensor was placed on a powered electrode because it would heat significantly. Still, according to the measured deposition rates and according to the above discussion, it is reasonable to assume the large deposition rate on the powered electrodes. In fact, the electrodes should occasionally be etched in chemical baths to remove the excessive deposits. The extensive deposition of thin films on the electrodes and thus loss of radicals for coating the samples is a major drawback of the reactor used in this study. The problem could be minimized using symmetric discharge, but it is often not feasible as in our PECVD reactor.

Despite the large dissipation of the deposition rate, the composition of the deposited films remains similar for all films at the positions of different sensors. Figure 10 represents the composition of the films as deduced from XPS survey spectra. The measurements were performed in the reactor loaded with samples. The concentration of carbon is close to 50 at.%, while the concentrations of oxygen and silicon is between 25 and 30 at.% for all samples. The small variations in the composition may be attributed to the accuracy of the XPS technique or to actual variation in the composition, but because the differences are marginal it is possible to conclude that the stoichiometry of the deposited films does not vary significantly between different positions in the plasma reactor.

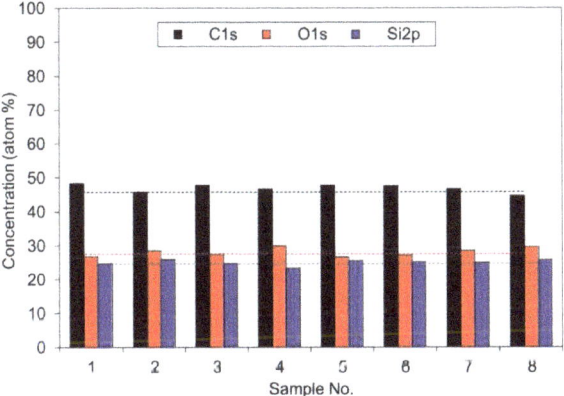

Figure 10. Concentrations of carbon, oxygen and silicon of deposited films at different positions in the PECVD reactor as deduced from XPS survey spectra.

4. Conclusions

Many commercial plasma reactors for the deposition of thin films from organic precursors using the PECVD technique suffer from non-uniform deposition rates. Moving the products to be coated by placing them on planetaria enables reasonable coating uniformity, but the efficiency is poor, because a significant fraction of the precursor radicals used as building blocks of the protective coatings are lost by adsorption on the powered electrodes and/or by pumping out from the reactor. An attempt was made to measure the deposition rates at various locations inside an industrial reactor powered by a capacitively coupled RF discharge. The plasma reactor had a volume of approximately 5 m^3. The maximum deposition rate for an empty reactor was measured on a grid near the pump ducts. The next highest rates were measured close to the powered electrodes, but a reasonable deposition rate was also observed far from the powered electrodes or the pump duct. The observation was interpreted by the formation of radicals useful for the deposition of the thin films throughout the reactor. The average residence time of approximately 80 s ensured a reasonably large production rate, despite the very low electron density in the plasma away from the oscillating sheaths next to the powered electrodes. Loading the reactor with numerous samples caused a significant difference in the deposition rates. Not only were they lower, but the distribution changed significantly. The deposition rates far from the powered electrodes dropped by more than an order of magnitude for a fully loaded chamber. Deposition rates above about 1 nm min^{-1} were only observed close to the powered electrodes. These observations indicate the need for modification of the discharge configuration in the industrial plasma reactor for depositing protective coatings from HMDSO precursor using the PECVD technique.

Author Contributions: Conceptualization, R.Z., D.Ð. and S.P.; methodology, R.Z., Ž.G. and D.Ð.; software, G.P. and D.Ð.; validation, D.Ð., B.G. and A.V.; formal analysis, A.V., Ž.G. and B.G.; investigation, R.Z., G.P. and Ž.G.; resources, Ž.G. and D.Ð.; data curation, M.M., S.P. and B.G.; writing—original draft preparation, M.M.; writing—review and editing, A.V.; supervision, M.M.; project administration, G.P.; funding acquisition, D.Ð. and M.M; All authors have read and agreed to the published version of the manuscript.

Funding: This research was funded by the Slovenian Research Agency, grant number L2-1835 (Innovative sensors for real-time monitoring of deposition rates in plasma-enhanced chemical vapor deposition (PECVD) systems) and research core funding grant number P2-0082 (Thin-film structures and plasma surface engineering).

Institutional Review Board Statement: Not applicable.

Informed Consent Statement: Not applicable.

Data Availability Statement: Not applicable.

Conflicts of Interest: The authors declare no conflict of interest.

References

1. Kotte, L.; Althues, H.; Mäder, G.; Roch, J.; Kaskel, S.; Dani, I.; Mertens, T.; Gammel, F.J. Atmospheric pressure PECVD based on a linearly extended DC arc for adhesion promotion applications. *Surf. Coat. Technol.* **2013**, *234*, 8–13. [CrossRef]
2. Schwarz, J.; Schmidt, M.; Ohl, A. Synthesis of plasma-polymerized hexamethyldisiloxane (HMDSO) films by microwave discharge. *Surf. Coat. Technol.* **1998**, *98*, 859–864. [CrossRef]
3. Morent, R.; de Geyter, N.; van Vlierberghe, S.; Dubruel, P.; Leys, C.; Schacht, E. Organic–inorganic behaviour of HMDSO films plasma-polymerized at atmospheric pressure. *Surf. Coat. Technol.* **2009**, *203*, 1366–1372. [CrossRef]
4. Choudhury, A.J.; Barve, S.A.; Chutia, J.; Pal, A.R.; Kishore, R.; Pande, M.; Patil, D.S. RF-PACVD of water repellent and protective HMDSO coatings on bell metal surfaces: Correlation between discharge parameters and film properties. *Appl. Surf. Sci.* **2011**, *257*, 8469–8477. [CrossRef]
5. Benítez, F.; Martínez, E.; Esteve, J. Improvement of hardness in plasma polymerized hexamethyldisiloxane coatings by silica-like surface modification. *Thin Solid Film.* **2000**, *377–378*, 109–114. [CrossRef]
6. Huang, J.; Cai, Y.; Xue, C.; Ge, J.; Zhao, H.; Yu, S.-H. Highly stretchable, soft and sticky PDMS elastomer by solvothermal polymerization process. *Nano Res.* **2021**. [CrossRef]
7. Mohania, V.; Deshpande, T.D.; Singh, Y.R.G.; Patil, S.; Mangal, R.; Sharma, A. Fabrication and characterization of porous poly(dimethylsiloxane) (PDMS) adhesives. *ACS Appl. Polym. Mater.* **2021**, *3*, 130–140. [CrossRef]
8. Pezzana, L.; Riccucci, G.; Spriano, S.; Battegazzore, D.; Sangermano, M.; Chiappone, A. 3D printing of PDMS-like polymer nanocomposites with enhanced thermal conductivity: Boron nitride based photocuring system. *Nanomaterials* **2021**, *11*, 373. [CrossRef] [PubMed]
9. Schäfer, J.; Foest, R.; Quade, A.; Ohl, A.; Weltmann, K.D. Local deposition of SiO$_x$ plasma polymer films by a miniaturized atmospheric pressure plasma jet (APPJ). *J. Phys. D Appl. Phys.* **2008**, *41*, 194010. [CrossRef]
10. Belmonte, T.; Henrion, G.; Gries, T. Nonequilibrium atmospheric plasma deposition. *J. Therm. Spray Technol.* **2011**, *20*, 744. [CrossRef]
11. Hamedani, Y.; Macha, P.; Bunning, T.J.; Naik, R.R.; Vasudev, M.C. Chemical Vapor Deposition - Recent Advances and Applications in Optical, Solar Cells and Solid State Devices. In *Plasma-Enhanced Chemical Vapor Deposition: Where we are and the Outlook for the Future*; Neralla, S., Ed.; IntechOpen: Rijeka, Croatia, 2016.

12. Hegemann, D.; Bülbül, E.; Hanselmann, B.; Schütz, U.; Amberg, M.; Gaiser, S. Plasma polymerization of hexamethyldisiloxane: Revisited. *Plasma Processes and Polymers.* **2021**, *18*, 2000176. [CrossRef]
13. Top, M.; Schönfeld, S.; Fahlteich, J.; Bunk, S.; Kühnel, T.; Straach, S.; De Hosson, J.T. Hollow-cathode activated PECVD for the high-rate deposition of permeation barrier films. *Surf. Coat. Technol.* **2017**, *314*, 155–159. [CrossRef]
14. Trunec, D.; Navrátil, Z.; Stahel, P.; Zajíčková, L.; Buríková, V.; Cech, J. Deposition of thin organosilicon polymer films in atmospheric pressure glow discharge. *J. Phys. D Appl. Phys.* **2004**, *37*, 2112–2120. [CrossRef]
15. Zajíčková, L.; Buršíková, V.; Kučerová, Z.; Franta, D.; Dvořák, P.; Šmíd, R.; Peřina, V.; Macková, A. Deposition of protective coatings in rf organosilicon discharges. *Plasma Sources Sci. Technol.* **2007**, *16*, S123–S132. [CrossRef]
16. Gosar, Ž.; Kovač, J.; Đonlagić, D.; Pevec, S.; Primc, G.; Junkar, I.; Vesel, A.; Zaplotnik, R. PECVD of hexamethyldisiloxane coatings using extremely asymmetric capacitive RF discharge. *Materials* **2020**, *13*, 2147. [CrossRef]
17. Hnilica, J.; Schäfer, J.; Foest, R.; Zajíčková, L.; Kudrle, V. PECVD of nanostructured SiO_2 in a modulated microwave plasma jet at atmospheric pressure. *J. Phys. D Appl. Phys.* **2013**, *46*, 335202. [CrossRef]
18. Bourreau, C.; Catherine, Y.; Garcia, P. Growth kinetics and step coverage in plasma deposition of silicon dioxide from organosilicon compounds. *Mat. Sci. Eng. A* **1991**, *139*, 376–379. [CrossRef]
19. Theirich, D.; Soll, C.; Leu, F.; Engemann, J. Intermediate gas phase precursors during plasma CVD of HMDSO. *Vacuum* **2003**, *71*, 349–359. [CrossRef]
20. Hegemann, D.; Brunner, H.; Oehr, C. Deposition rate and three-dimensional uniformity of RF plasma deposited SiO_x films. *Surf. Coat. Technol.* **2001**, *142–144*, 849–855. [CrossRef]
21. Hegemann, D.; Brunner, H.; Oehr, C. Evaluation of deposition conditions to design plasma coatings like SiO_x and a-C:H on polymers. *Surf. Coat. Technol.* **2003**, *174–175*, 253–260. [CrossRef]
22. Nave, A.S.C.; Mitschker, F.; Awakowicz, P.; Röpcke, J. Spectroscopic studies of microwave plasmas containing hexamethyldisiloxane. *J. Phys. D Appl. Phys.* **2016**, *49*, 395206. [CrossRef]
23. Gosar, Ž.; Kovač, J.; Mozetič, M.; Primc, G.; Vesel, A.; Zaplotnik, R. Characterization of gaseous plasma sustained in mixtures of HMDSO and O_2 in an industrial-scale reactor. *Plasma Chem. Plasma Process.* **2020**, *40*, 25–42. [CrossRef]
24. Wang, B.; Fu, X.; Song, S.; Chu, H.O.; Gibson, D.; Li, C.; Shi, Y.; Wu, Z. Simulation and optimization of film thickness uniformity in physical vapor deposition. *Coatings* **2018**, *8*, 325. [CrossRef]
25. Gosar, Ž.; Kovač, J.; Mozetič, M.; Primc, G.; Vesel, A.; Zaplotnik, R. Deposition of $SiO_xC_yH_z$ protective coatings on polymer substrates in an industrial-scale PECVD reactor. *Coatings* **2019**, *9*, 234. [CrossRef]
26. Gosar, Ž.; Đonlagić, D.; Pevec, S.; Kovač, J.; Mozetič, M.; Primc, G.; Vesel, A.; Zaplotnik, R. Deposition kinetics of thin silica-like coatings in a large plasma reactor. *Materials* **2019**, *12*, 3238. [CrossRef] [PubMed]
27. Despax, B.; Gaboriau, F.; Caquineau, H.; Makasheva, K. Influence of the temporal variations of plasma composition on the cyclic formation of dust in hexamethyldisiloxane-argon radiofrequency discharges: Analysis by time-resolved mass spectrometry. *Aip Adv.* **2016**, *6*, 105111. [CrossRef]
28. Chabert, P.; Tsankov, T.V.; Czarnetzki, U. Foundations of capacitive and inductive radio-frequency discharges. *Plasma Sources Sci. Technol.* **2021**, *30*, 024001. [CrossRef]

MDPI
St. Alban-Anlage 66
4052 Basel
Switzerland
www.mdpi.com

Coatings Editorial Office
E-mail: coatings@mdpi.com
www.mdpi.com/journal/coatings

Disclaimer/Publisher's Note: The statements, opinions and data contained in all publications are solely those of the individual author(s) and contributor(s) and not of MDPI and/or the editor(s). MDPI and/or the editor(s) disclaim responsibility for any injury to people or property resulting from any ideas, methods, instructions or products referred to in the content.